I0891112

Machine Learning and Computer Vision for Renewable Energy

Pinaki Pratim Acharjya
Haldia Institute of Technology, India

Santanu Koley
Haldia Institute of Technology, India

Subhabrata Barman
Haldia Institute of Technology, India

A volume in the Advances in Environmental Engineering and Green Technologies (AEEGT) Book Series

IGI Global
PUBLISHER of TIMELY KNOWLEDGE

Published in the United States of America by
IGI Global
Engineering Science Reference (an imprint of IGI Global)
701 E. Chocolate Avenue
Hershey PA, USA 17033
Tel: 717-533-8845
Fax: 717-533-8661
E-mail: cust@igi-global.com
Web site: http://www.igi-global.com

Library of Congress Cataloging-in-Publication Data

CIP DATA PROCESSING

Machine Learning and Computer Vision for Renewable Energy
Pinaki Pratim Acharjya, Santanu Koley, Subhabrata Barman
2024 Engineering Science Reference

ISBN: 9798369323557(hc) | ISBN: 9798369347041(sc) | eISBN: 9798369323564

This book is published in the IGI Global book series Advances in Environmental Engineering and Green Technologies (AEEGT) (ISSN: 2326-9162; eISSN: 2326-9170)

British Cataloguing in Publication Data
A Cataloguing in Publication record for this book is available from the British Library.

All work contributed to this book is new, previously-unpublished material. The views expressed in this book are those of the authors, but not necessarily of the publisher.

For electronic access to this publication, please contact: eresources@igi-global.com.

Advances in Environmental Engineering and Green Technologies (AEEGT) Book Series

Sang-Bing (Jason) Tsai
Zhongshan Institute, University of Electronic Science and Technology of China, China & Wuyi University, China
Ming-Lang Tseng
Lunghwa University of Science and Technology, Taiwan
Yuchi Wang
University of Electronic Science and Technology of China Zhongshan Institute, China

ISSN:2326-9162
EISSN:2326-9170

Mission

Growing awareness and an increased focus on environmental issues such as climate change, energy use, and loss of non-renewable resources have brought about a greater need for research that provides potential solutions to these problems. Research in environmental science and engineering continues to play a vital role in uncovering new opportunities for a "green" future.

The **Advances in Environmental Engineering and Green Technologies (AEEGT)** book series is a mouthpiece for research in all aspects of environmental science, earth science, and green initiatives. This series supports the ongoing research in this field through publishing books that discuss topics within environmental engineering or that deal with the interdisciplinary field of green technologies.

Coverage

- Pollution Management
- Electric Vehicles
- Radioactive Waste Treatment
- Water Supply and Treatment
- Contaminated Site Remediation
- Green Transportation
- Air Quality
- Waste Management
- Policies Involving Green Technologies and Environmental Engineering
- Green Technology

IGI Global is currently accepting manuscripts for publication within this series. To submit a proposal for a volume in this series, please contact our Acquisition Editors at Acquisitions@igi-global.com or visit: http://www.igi-global.com/publish/.

The Advances in Environmental Engineering and Green Technologies (AEEGT) Book Series (ISSN 2326-9162) is published by IGI Global, 701 E. Chocolate Avenue, Hershey, PA 17033-1240, USA, www.igi-global.com. This series is composed of titles available for purchase individually; each title is edited to be contextually exclusive from any other title within the series. For pricing and ordering information please visit http://www.igi-global.com/book-series/advances-environmental-engineering-green-technologies/73679. Postmaster: Send all address changes to above address.

Titles in this Series

For a list of additional titles in this series, please visit: http://www.igi-global.com/book-series/advances-environmental-engineering-green-technologies/73679

Convergence Strategies for Green Computing and Sustainable Development
Vishal Jain (Sharda University, India) Murali Raman (Asia Pacific University of Technology, Malaysia) Akshat Agrawal (Amity University, Guragon, India) Meenu Hans (K.R. Mangalam University, India) and Swati Gupta (Center of Excellence, K.R. Mangalam University, India)
Engineering Science Reference • copyright 2024 • 332pp • H/C (ISBN: 9798369303382) • US $270.00 (our price)

Biosorption Processes for Heavy Metal Removal
Pinki Saini (University of Allahabad, India)
Engineering Science Reference • copyright 2024 • 371pp • H/C (ISBN: 9798369316184) • US $275.00 (our price)

Agriculture and Aquaculture Applications of Biosensors and Bioelectronics
Alex Khang (Global Research Institute of Technology and Engineering, USA)
Engineering Science Reference • copyright 2024 • 576pp • H/C (ISBN: 9798369320693) • US $265.00 (our price)

Innovations in Engineering and Food Science
Shilpa Mehta (Auckland University of Technology, New Zealand) Fakhar Islam (NUR International University, Pakistan) and Ali Imran (Government College University, Pakistan)
Engineering Science Reference • copyright 2024 • 424pp • H/C (ISBN: 9798369308196) • US $265.00 (our price)

Modeling and Monitoring Extreme Hydrometeorological Events
Carmen Maftei (Transilvania University of Brasov, Romania) Radu Muntean (Transilvania University of Brașov, Romania) and Ashok Vaseashta (International Clean Water Institute, USA)
Engineering Science Reference • copyright 2024 • 338pp • H/C (ISBN: 9781668487716) • US $250.00 (our price)

Novel AI Applications for Advancing Earth Sciences
Sudesh Yadav (Department of Higher Education, Government PG College, Ateli, India) Satya Prakash Yadav (GL Bajaj Institute of Technology and Management, India) Pethuru Raj (Edge AI Division, Reliance Jio Platforms Ltd., Bangalore, India) Prabhakar Tiwari (Madan Mohan Malaviya University of Technology, India) and Victor Hugo C. de Albuquerque (Department of Teleinformatics Engineering (DETI), Federal University of Ceará, Brazil)
Engineering Science Reference • copyright 2024 • 405pp • H/C (ISBN: 9798369318508) • US $265.00 (our price)

Exploring Ethical Dimensions of Environmental Sustainability and Use of AI

IGI Global
PUBLISHER of TIMELY KNOWLEDGE

701 East Chocolate Avenue, Hershey, PA 17033, USA
Tel: 717-533-8845 x100 • Fax: 717-533-8661
E-Mail: cust@igi-global.com • www.igi-global.com

Table of Contents

Detailed Table of Contents

The chapter compiles the existing landscape of research and uses in energy from undepletable sources, highlighting arising flows and determined challenges. It explores the vital type of renewable energy sciences, stressing the increasing unification of leading fabrics, storage answers, and smart gridiron technologies. Despite notable progress, barriers to a degree intermittency, scalability, and business-related viability persevere. The chapter determines a concise survey of the developing research landscape and the complicatedness guide the extensive adoption of energy from undepletable source beginnings.

This project aims to develop a method for wind turbine blade (WTB) inspection using machine learning and computer vision that would allow early detection and diagnosis of structural faults in WTBs, aiding in condition-based maintenance in the industry. At present, the industry relies on the use of manual inspections of blades for fault detection and diagnosis. The use of drones for inspection has been proven for bridges and dams and is in the process of being implemented in the OSW industry. However, current methods of inspection require huge volumes of data and labour-intensive pre-processing. This project aims to utilise machine learning methods, to reduce human input required in the detection and diagnosis of faults in WTBs. This will consist of developing a set of novel computer vision algorithms that can achieve high accuracies of fault detection and classification from limited datasets to introduce prior knowledge into the learning process.

Claris Shoko, University of Botswana, Botswana
Ntebogang Moroke, North West University, South Africa
Katleho Makatjane, University of Botswana, Botswana

The limitations of traditional deep learning models in processing vast volumes of data and modelling complicated temporal dependencies make it difficult to effectively satisfy these objectives for short-term load forecasting (STLF). This chapter utilises deep learning, which enables the following: k-means clustering to comprehend hourly electricity demand load trend, extraction of complex features with non-linear interactions that impact electricity demand load, handling of long-term dependencies through the modelling of temporal hierarchies in the time series data via long short-term memory, and recurrent neural network to capture dependencies across time steps. The performance of the state-of-the-art hourly demand forecast models with the k-means variant is compared with that of the RNN-LSTM-CNN-copula. Furthermore, it is noted that RNN-LSTM-CNN-copula is a viable deep learning model for energy consumption forecast problems because of its capacity to learn the spatio-temporal dependencies in the hourly electrical demand load data, which results in an accurate energy demand forecast.

Jayeeta Majumder, Haldia Institute of Technology, India
Pinaki Pratim Acharjya, Haldia Institute of Technology, India
Subhabrata Barman, Haldia Institute of Technology, India
Santanu Koley, Haldia Institute of Technology, India

Utilising renewable energy sources is becoming more popular as a way to mitigate the effects of climate change and global warming. In an effort to make renewable energy more predictable, numerous prediction techniques have been developed. The objectives of this study are best illustrated by this chapter, which aims to provide a review and analysis of machine-learning and computer vision techniques in renewable solar energy projections. In addition to machine-learning and computer vision techniques for renewable solar energy projections, this chapter also focuses on the objective to deliver an optimized academic outcome, potentially necessary for the development of new solar energy fields. This could significantly contribute to the amplified usage of solar energy, which is a sustainable and cleaner energy source.

Pampa Saha, Brainware University, India
Anshit Mukherjee, Abacus Institute of Engineering and Management, India
Gunjan Mukherjee, Brainware University, India

The smart energy domain poses substantial challenges in future research, necessitating advanced investigation into optimizing smart customizable networks using artificial intelligence (AI) and machine learning (ML). With renewable energy (RE) being pivotal for global development amid climate change, AI introduces new paradigms to reshape activities, demanding revamped energy infrastructure and RE deployment strategies. The chapter explores the adoption of AI in future smart cities research with considerable economic benefits. Moreover, in the field of environmental science and engineering

(ESE), the ML's potential in revolutionizing ESE by addressing complex problems and outlining crucial components for successful ML implementation in ESE, correct model development, proper interpretation, and applicability analysis has been done. The renewable source of energy like solar and wind energy can be generated in place where they are found in plenty. The ML application of energy source tracing and the prediction of energy type with multiple models of efficiencies has been highlighted.

Renewable energy comes from natural resources that replenish more quickly than they are used up. Compared with burning fossil fuels, producing renewable energy produces a lot fewer emissions. The key to solving the climate catastrophe is shifting away from fossil fuels, which presently provide the majority of emissions, and toward renewable energy. The facility has global impact and recent computer vision based on machine learning creates the business opportunity within all countries and regions globally. So there should be a communication between all countries and regions to exchange information between different global languages. In this study, an analytical survey was done to create the communication between countries and regions. For this purpose, different languages and datasets of the languages are studied where OCR-based character recognition can be done.

Wireless sensor networks (WSN) have emerged as an effective solution for monitoring and recording physical conditions of the environment. Forests, being an important part of the ecosystem, are prone to several challenges including forest fires and tree poaching. Uncontrolled anthropogenic activities, abnormal natural conditions, and tree smuggling are some of the factors that contribute to these problems. To address these issues, the authors propose a WSN-based system for preventing tree poaching and forest fires. The system will consist of spatially distributed sensors that will continuously monitor the physical conditions of the forest, including temperature, humidity, and presence of human activity. The data collected from these sensors will be processed in a central location to detect any abnormal changes in the environment. In case of a potential forest fire or tree poaching activity, the system will send alerts to the relevant authorities, allowing them to take timely action. By leveraging the power of WSN, the proposed system aims to prevent and monitor forest fires and tree poaching.

Green energy deployment in any country has the primary goal of assisting economic development, improving energy security, increasing energy access, and combating global climate change. Sustainable development can be achieved through utilizing renewable energy sources and ensuring inhabitants' access to affordable, reliable, and modern electricity. The term "green energy" refers to energy produced by natural resources such as sunshine, wind, and water. Environmentally hazardous mining and drilling operations are typically avoided by green providers. The advancement of technologies such as automation, additive manufacturing, artificial intelligence (AI), and others can facilitate the shift to a cleaner future by improving environmental quality and controlling and monitoring infrastructures through the replacement of outdated equipment with more advanced options.

The urgent need for a cleaner, more sustainable energy source has been sparked by the fast-dwindling supply of fossil fuels and the catastrophic consequences of climate change. The thriving market of hydrogen as a fruitful alternative is experiencing as an important environmentally friendly industrial feedstock. There are significant techno-financial and environmental obstacles to the commercialization of hydrogen as a fuel. Microalgal biohydrogen generation has drawn increasing attention as a means of overcoming these difficulties. Different metabolic pathways can be used to make microalgal biohydrogen, but a concise explanation of the economics and technical aspects of improving microalgae-driven biohydrogen generation is highly needed. Thus, assured prolonged sustainability can be achieved by improvising economical hydrogen generation technologies through the amendment of vital trial and error factors, possible gene-level, metabolic engineering modification and AIML employment.

Blue wind energy, a pioneering concept, amalgamates offshore wind energy generation with cutting-edge technologies. There is a dire need to address the growing energy demands of urbanization while mitigating the environmental impact of conventional energy sources. Urban areas are substantial energy consumers, and traditional methods are both unsustainable and environmentally damaging. Blue wind energy, alongside AI and IoT integration, presents a compelling solution, but the challenge is in effectively implementing and optimizing these technologies to ensure sustainable, efficient, and eco-friendly energy management in cities. This chapter introduces a set of essential performance indicators to assess the advantages of diverse AI-accelerated workflows in blue-wind energy and the latest developments in applications of computer vision and employing AI in the enhancement of energy harvesting like as photovoltaics, storage including batteries, conversion, and management in smart grids.

Chapter 11
Detecting Aberrations in Renewable Energy With the One-Class Support Vector Machine Model.. 186
Ntebogang Moroke, North West University, South Africa
Katleho Makatjane, University of Botswana, Botswana

The traditional energy time series forecasting methods use accurate input data for prediction and then make single-step or multi-step predictions based on the established regression model. But because of the complexity of financial markets, the traditional forecasting methods are less reliable. Unsupervised anomaly detection techniques operate directly on an unseen dataset, under the assumption that outliers are sparsely present in it. A one-class support vector machine (SVM) is an extension of support vector machines for unlabeled data and can be used for anomaly detection. Even though outliers are accounted for in one-class SVMs, they greatly influence the learned model. Hence, one modification to one-class SVMs is presented here: eta one-class SVMs. For evaluation, eta one support vector machine-based algorithm is compared using four kernels. The results affirmed the proficiency of the RBF kernel, with the eta one-class SVM performing best. In regards to subspace division, the preliminary results showed a decreasing nonlinear trend.

Chapter 12
Electricity Demand Forecasting using Dual Stream TBATS-CNN-LSTM Architecture 206
Katleho Makatjane, University of Botswana, Botswana
Ditebo Xaba, University of South Africa, South Africa
Modisane Bennett Seitshiro, North West University, South Africa

The problem is the model's reliability, accuracy, and meaningfulness to convince decision-makers of the actions to be taken when seasonality is one of the features in the findings of the existing deep learning forecasts. The purpose of this chapter is to come up with a novel dual-stream hybrid architecture that is capable of predicting electricity demand and accessing its accuracy levels by benchmarking it with individual architecture model's forecasting accuracy levels using out-of-sample time series. The approach in this chapter uses time series and convolutional neural network (CNN)-based long short-term memory with various configurations to construct a forecasting model for short- to medium-term aggregate load forecasting. The obtained results show that the TBATS-CNN-LSTM-based model has shown high accuracy as compared to the base learner, and the model is optimised with hyperparameter tuning. Only optimally selected time-lag features captured all the characteristics of complex time series in South Africa.

In this chapter, the forecasting of electricity consumption and production is conducted by analyzing indicators from previous years. The problem is addressed using machine learning within Microsoft Azure Machine Learning Studio. The outcome is an independent service integrated into Excel, enabling consumption forecasting for specified dates. The Excel user interface is developed using Visual Basic for Applications. Python was used to create user blocks for modifying input data pools and forming graphical dependencies, seamlessly integrated into the original modules of Microsoft Azure Machine Learning Studio. An additional aspect of the forecast results involves evaluating the quality of the predicted electricity consumption indicators. The materials used for this chapter were sourced with the support of Ukraine's National Power Company UKRENERGO.

Forecasters must make some educated assumptions about future electricity demand and effectively explain them because real electricity demand and forecasted electricity demand always differ. In order to forecast future wind power load base stations in South Africa, the authors evaluate hourly wind power generation in this chapter. The predicted time series can be used to show the flow of the load demand trend for electricity. Because of the noise in the original time series, they provide the enhanced gradient-boosted decision tree algorithm based on Kalman filter (GBDT-KF). They compare how well the proposed GBDT-KF method performs with varying numbers of decision trees. Re-sampling is cross-validated three times at a 10 fold interval. The employment of MSE, RMSE, MAE, and MAPE allowed for the selection of the best model. As a result, the total findings showed that 90% of the test data and 92% of the training data were successful. The findings of this study will be helpful to the energy sector and decision-makers for planning and future use in the hourly electricity forecasting domain.

Foreword

It gives me pleasure to say few words regarding this book titled *Machine Learning and Computer Vision for Renewable Energy* and its editors. This book focuses on the potentially revolutionary applications of Computer Vision in the field of Renewable Energy, moving from the challenges of energy prediction to the opportunities presented by cutting-edge technology. It focuses Better energy management, the ability to forecast variables influencing renewable energy, and the optimization of sustainability all benefit from this data. Readers will become thoroughly versed in the Artificial Intelligence techniques and computer vision methodologies that are essential to solving the problems that renewable energy systems encounter, including researchers, academics, and students. For scientists, biological engineers, academics, students, and researchers who are interested in learning how natural models have influenced the modern technology-centric world, this book is appropriate. A record-breaking increase in interest in machine learning has occurred in recent years as a result of the accessibility of massive datasets, advancements in algorithms, and exponential development in computing power. This book also highlights advances in Computer Vision and Artificial Intelligence: to propel social growth via innovation, research, and technological achievements in the field of Renewable Energy. For academic researchers, educators, college students, and corporate leaders interested to combine the use of renewable energy and digital transformation, this book will be a priceless tool for navigating the challenges and opportunities that lie ahead.

The editors are widely experienced in this area with long research and teaching experiences.

I feel extremely happy to see the topics of applications of superbly efficient technological systems for the benefit of human civilisations. I hope the editors will be very successful in their great effort and students, teachers and professionals will find this book very useful.

Subhankar Joardar
Computer Science and Engineering, Haldia Institute of Technology, India

Preface

The shortcomings of conventional methods for predicting renewable energy are becoming more and more obvious as the globe struggles to meet the pressing need for sustainable energy alternatives. More than ever, there is a need for more precise forecasts in the areas of net load forecasting, line loss forecasts, and the smooth integration of hybrid solar and battery storage systems. Advanced Artificial Intelligence approaches are developing as a potential solution to address this difficulty and have the potential to completely transform the renewable energy industry. In order to address the critical problems facing renewable energy systems, Machine Learning and Computer Vision for Renewable Energy offers a thorough examination of AI modelling, analysis, performance prediction, and control techniques.

The book focuses on the potentially revolutionary applications of Computer Vision in the field of Renewable Energy, moving from the challenges of energy prediction to the opportunities presented by cutting-edge technology. In the fight to improve sustainability in all sectors of the economy, Computer Vision technology comes out on top, gathering priceless information from digital images and videos. Better energy management, the ability to forecast variables influencing renewable energy, and the optimization of sustainability all benefit from this data. Readers will become thoroughly versed in the Artificial Intelligence techniques and computer vision methodologies that are essential to solving the problems that renewable energy systems encounter, including researchers, academics, and students. The field of Machine Learning and Computer Vision for Renewable Energy presents itself as the driving force behind this transition.

In addition to discussing the energy industry's current issues, the book provides information on how to create a more sustainable future. The goal is made evident by highlighting advances in Computer Vision and Artificial Intelligence: to propel social growth via innovation, research, and technological achievements in the field of Renewable Energy. This book will be an invaluable resource for academic researchers, educators, college students, and business executives who are interested in the convergence of digital transformation and renewable energy to navigate the opportunities and difficulties that lie ahead. This book, which offers a wide range of suggested readings, is a monument to the rapidly changing fields of artificial intelligence and computer vision, helping to shape the future of sustainable energy for future generations.

The following research topics are well covered in this book:

- AI Techniques in Renewable Energy
- Computer Vision in Renewable Energy
- Regression Methods in Computer Vision for Renewable Energy
- Solar Power Prediction with AI and Vision
- Technologies for Renewable Energy Sources
- Trends in Renewable Energy Research and Applications

Organization of the Book

The book is organized into 14 chapters. A brief description of each of the chapters follows:

Chapter 1 compiles the existing landscape of research and uses in energy from undepletable source, highlighting arising flows and determined challenges. It explores the vital type of renewable energy sciences, stressing the increasing unification of leading fabrics, storage answers, and smart gridiron technologies. Despite notable progress, barriers to a degree intermittency, scalability, and business-related viability persevere. The chapter determines a concise survey of the developing research landscape and the complicatedness guide the extensive adoption of energy from undepletable source beginnings.

Chapter 2 aims to develop a method for Wind Turbine Blade (WTB) inspection using Machine Learning and Computer Vision, that would allow early detection and diagnosis of structural faults in WTBs, aiding in condition-based maintenance in the Industry. At present, the industry relies on the use of manual inspections of blades for fault detection and diagnosis. The use of drones for inspection has been proven for bridges and dams and is in the process of being implemented in the OSW industry. However, current methods of inspection require huge volumes of data and labour -intensive pre-processing. This chapter aims to utilise machine learning methods, to reduce human input required in the detection and diagnosis of faults in WTBs.

Chapter 3 focuses on the limitations of traditional deep learning models in processing vast volumes of data and modelling complicated temporal dependencies make it difficult to effectively satisfy these objective for short-term load forecasting (STLF). This chapter utilises deep learning, which enables the following: k-means clustering to comprehend hourly electricity demand load trend; extraction of complex features with non-linear interactions that impact electricity demand load; handling of long-term dependencies through the modelling of temporal hierarchies in the time series data via Long Short Term Memory; and recurrent neural network to capture dependencies across time steps.

Chapter 4 deals with utilising renewable energy sources is becoming more popular as a way to mitigate the effects of climate change and global warming. In an effort to make renewable energy more predictable, numerous prediction techniques have been developed. The objectives of this study are best illustrated by this chapter, which aims to provide a review and analysis of machine-learning and computer vision techniques in renewable solar energy projections. In addition to machine-learning and computer vision techniques for renewable solar energy projections, this chapter also focuses on the objective to deliver an optimized academic outcome, potentially necessary for the development of new solar energy fields.

Chapter 5 deals with the Smart Energy domain which poses substantial challenges in future Research, necessitating advanced investigation into optimizing smart customizable networks using artificial intelligence (AI) and machine learning (ML). With renewable energy (RE) being pivotal for global development amid climate change, AI introduces new paradigms to reshape activities, demanding revamped energy infrastructure and RE deployment strategies. This chapter explores the adoption of AI in Future Smart Cities Research with considerable economic benefits.

Chapter 6 deals with renewable energy which comes from natural resources that replenish more quickly than they are used up. Compared with burning fossil fuels, producing renewable energy produces a lot fewer emissions. The key to solving the climate catastrophe is shifting away from fossil fuels, which presently provide the majority of emissions, and toward renewable energy.

Chapter 7 expedites the use of Wireless Sensor Networks (WSN) which have emerged as an effective solution for monitoring and recording physical conditions of the environment. Forests, being an important part of the ecosystem, are prone to several challenges including forest fires and tree poaching. Uncontrolled anthropogenic activities, abnormal natural conditions, and tree smuggling are some of the factors that contribute to these problems. To address these issues, a WSN-based system for preventing tree poaching and forest fires have been proposed.

Chapter 8 deals with Green energy deployment in any country has the primary goal of assisting economic development, improving energy security, increasing energy access, and combating global climate change. Sustainable development can be achieved through utilizing renewable energy sources and ensuring inhabitants' access to affordable, reliable, and modern electricity. The term "green energy" refers to energy produced by natural resources such as sunshine, wind, and water. Environmentally hazardous mining and drilling operations are typically avoided by green providers.

Chapter 9 aims to provide the urgent need for a cleaner, more sustainable energy source has been sparked by the fast dwindling supply of fossil fuels and the catastrophic consequences of climate change. The thriving market of hydrogen as a fruitful alternative is experiencing as an important environmental friendly industrial feedstock. There are significant techno-financial and environmental obstacles to the commercialization of hydrogen as a fuel. Microalgal bio hydrogen generation has drawn increasing attention as a means of overcoming these difficulties.

Chapter 10 deals with Blue Wind Energy, a pioneering concept, amalgamates offshore wind energy generation with cutting-edge technologies. There is a dire need to address the growing energy demands of urbanization while mitigating the environmental impact of conventional energy sources. Urban areas are substantial energy consumers, and traditional methods are both unsustainable and environmentally damaging. Blue Wind Energy, alongside AI and IoT integration, presents a compelling solution, but the challenge is in effectively implementing and optimizing these technologies to ensure sustainable, efficient, and eco-friendly energy management in cities.

Chapter 11 deals with the traditional energy time series forecasting methods use accurate input data for prediction and then make single-step or multi-step predictions based on the established regression model. But because of the complexity of financial markets, the traditional forecasting methods are less reliable. Unsupervised anomaly detection techniques operate directly on an unseen dataset, under the assumption that outliers are sparsely present in it.

Chapter 12 aims at providing the model's reliability, accuracy, and meaningfulness to convince decision-makers of the actions to be taken when seasonality is one of the features in the findings of the existing deep learning forecasts. The purpose of this chapter is to come up with a novel dual-stream hybrid architecture that is capable of predicting electricity demand and accessing its accuracy levels by benchmarking it with individual architecture model's forecasting accuracy levels using out-of-sample time series.

Chapter 13 deals with the forecasting of electricity consumption and production is conducted by analyzing indicators from previous years. The problem is addressed using machine learning within Microsoft Azure Machine Learning Studio. The outcome is an independent service integrated into Excel, enabling consumption forecasting for specified dates.

Chapter 14 deals with Forecasters who must make some educated assumptions about future electricity demand and effectively explain them because real electricity demand and forecasted electricity demand always differ. In order to forecast future wind power load base stations in South Africa, we evaluate hourly wind power generation in this chapter. The predicted time series can be used to show the flow of the load demand trend for electricity.

Pinaki Pratim Acharjya
Haldia Institute of Technology, India

Santanu Koley
Haldia Institute of Technology, India

Subhabrata Barman
Haldia Institute of Technology, India

Chapter 1
Trends and Difficulties in Research and Applications of Renewable Energy

Himadri Sekhar Das
https://orcid.org/0000-0002-3509-3388
Haldia Institute of Technology, India

ABSTRACT

The chapter compiles the existing landscape of research and uses in energy from undepletable sources, highlighting arising flows and determined challenges. It explores the vital type of renewable energy sciences, stressing the increasing unification of leading fabrics, storage answers, and smart gridiron technologies. Despite notable progress, barriers to a degree intermittency, scalability, and business-related viability persevere. The chapter determines a concise survey of the developing research landscape and the complicatedness guide the extensive adoption of energy from undepletable source beginnings.

1. INTRODUCTION

The worldwide push towards tenable and detergent energy beginnings has surpassed to a surge in research and uses of energy from undepletable source. As the globe grapples with the challenges of humidity change and the exhaustion of established hydrocarbon deposits, renewable energy electronics have win meaningful consideration. This episode explores the flows and troubles in the continuous research and uses of energy from undepletable source, shedding come to rest on the progresses, challenges, and potential resolutions in this place rapidly progressing field.

1.1. Trends in Renewable Energy Research

Solar Energy: Solar energy remnants a key focus of research, accompanying progresses in photovoltaic (PV) electronics, solar warm plans, and arising matters. The current towards growing efficiency, lowering costs, and surveying creative battery powered by the sun designs is forceful the progress of this science

DOI: 10.4018/979-8-3693-2355-7.ch001

Figure 1. Solar energy system

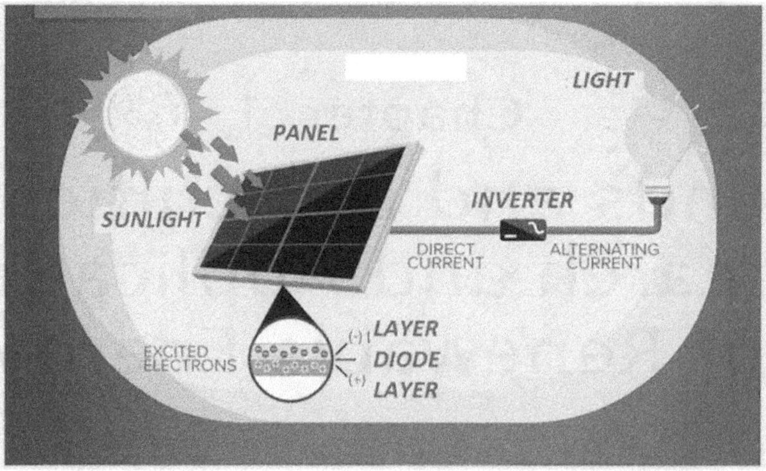

(Ali O M Maka,et al.,2022) and (Naveen Kumar Sharma,et al.,2012). The unification of cosmic panels into construction fabrics, smart grids, and strength storage methods is likewise acquire distinction (Figure 1. shows the solar energy System).

Historical Evolution of Solar Energy

Ancient Solar Technologies: Overview of early cosmic sciences, containing the use of magnifying lorgnette to concentrate light part of every 24 hours for explosion.

Examples of old civilizations, to a degree the Greeks and Romans, utilizing inactive solar designs in design.

The Birth of Photovoltaics: Exploration of the finding of the photovoltaic effect and the creation of the first battery powered by the sun by Edmond Becquerel in the 19th of one hundred years (Ashwani Kumar, et al.,2010). The gifts of scientists like Albert Einstein and Bell Labs in numbering photovoltaic electronics (Kanna, et al.,2016).

Solar Thermal Technologies: Overview of cosmic warm sciences, from early uses in the 19th centennial to the growth of concentrated energy (chow, et al.,2018) from undepletable source (CSP) plants (Huide Fu, et al., 2017).

Modern Solar Technologies: The contemporary age has visualized extraordinary progresses in cosmic sciences, making them more effective, inexpensive, and approachable. This division investigate the key elements of modern solar power wholes.

Photovoltaic Solar Cells: In-depth study of the various types of photovoltaic containers, containing monocrystalline, polycrystalline, and thin-film technologies. Discussion on the adeptness betterings and cost reductions (Fang, et al., 2017) in cosmic panels over the age (Mills, et al., 2004).

Solar Thermal Systems: Overview of up-to-date cosmic thermal sciences, in the way that symbolic depressions, cosmic towers, and platter collectors (Baljit, et al., 2016). Examining the requests of cosmic warm structures in power creation, warming, and desalination (Mahmud, et al., 2018).

Energy Storage Solutions: Exploration of energy depository sciences, in the way that batteries and melted seasoning depository, to address the intermittent type of solar power (Tyagi, et al., 2012).

Discussion on the significance of depository in reinforcing the dependability and scalability of solar power (Platzer, 2014).

Environmental Benefits of Solar Energy: One of ultimate irresistible facets of solar power is allure helpful impact on the atmosphere. This portion delves into the material benefits of controlling energy from the sun as a clean and tenable energy beginning.

Reduced Greenhouse Gas Emissions: Examination of by means of what solar power provides to checking surroundings change by reducing colorless odorless gas diffusions. Comparative reasoning of the material footmark of energy from the sun versus established nonrenewable fuel sources.

Land Use and Biodiversity: Discussion on the slightest land footmark of cosmic establishments and their rapport with land actions. Addressing concerns had connection with the impact of cosmic projects on local environments and biodiversity.

Water Conservation: Exploration of by what method solar power helps in water preservation, specifically in the circumstances of cosmic warm electronics (Anand, et al.,2021). Case studies highlighting the water-conditional benefits of solar power in dry domains (Kazem, et al., 2016).

Challenges and Opportunities: While the development of solar power has been influential, it is not outside challenges. This portion resolves the barriers that solar radiation faces and the opportunities for further progresses.

Intermittency and Grid Integration: Discussion on the irregular type of energy from the sun and designs for merging it into existing strength grids. Examination of smart gridiron sciences and strength administration orders to enhance gridiron dependability.

Technological Innovations: Overview of continuous test in cosmic electronics, containing perovskite solar containers, cosmic paint, and cosmic casements. Exploration of in what way or manner these changes could transform the solar power countryside.

Policy and Economic Considerations: Analysis of the part of management procedures and incentives in advancing solar radiation enactment (Ogunmodimu, et al., 2018). Discussion on the financial being of energy from undepletable source and the decreasing costs guide cosmic establishments (Ramos, et al., 2017).

The Future of Solar Energy: Solar Energy in Developing Countries: Examination of the part of solar radiation in providing approach to power in remote and underserved domains. Case studies reveal profitable cosmic projects in underdeveloped countries. Integration accompanying Other Renewable Sources: Exploration of the synergies 'tween solar radiation and different sustainable beginnings, in the way that wind and hydropower (Kumar, et al., 2011). Discussion on the idea of hybrid energy from undepletable source plans for improved dependability (Eicker, et al.,2015).

1.2. Wind Energy

Wind strength has knowledgeable significant tumor, accompanying best and more effective wind turbines becoming commonplace (Figure 2. Shows the wind energy system) (Blanco, et al., 2009) (Herbert, et al.,2007) (Manwell, et al.,2010).

Wind Turbine Technology: Modern wind turbines are sophisticated machines created to capture and convert the moving power of the wind into energetic power. These divisions investigate the differing types of wind turbines, containing horizontal-shaft and upright-pole designs, as well as the detracting parts to degree rotor blades, generators, and control arrangements. Additionally, progresses in fabrics and design that improve efficiency and dependability are argued.

Figure 2. Wind energy

1.3 Hydropower and Ocean Energy

Hydropower persists expected a bigger beginning of energy from undepletable source, accompanying ongoing research into reconstructing adeptness and underrating material impacts. Ocean strength, containing gushing and wave strength, is an arising field accompanying the potential to supply regular and predictable capacity. Research is concentrated on evolving trustworthy sciences and discussing referring to practices or policies that do not negatively affect the environment concerns (Tong, et al.,2019) (Thorburn, et al.,2006) (Sipahutar, et al.,2013).Hydropower, individual of the most aged forms of strength change, traces allure ancestries back to old water mills secondhand for abrasive piece (Figure 3. shows the Hydropower system) (Ibrahim, et al., 2019).

Overview of Ocean Energy: Ocean strength surrounds differing sciences that harness the capacity of the sea's unaffected drives, containing waves, tides, currents, and hotness differentials. Unlike hydropower, sea strength is an arising field accompanying different electronics at various stages of happening (Zhang, et al., 2022).

Wave Energy: *Point Absorbers*: These ploys use buoyant forms to consume energy from the upright motion of waves (Figure 4. shows the wave energy). The campaign is convinced into power through hydraulic structures and engine converting energy (Falco, et al., 2010). *Attenuators*: Attenuators loiter, multi-slice makeups that move accompanying the motion of waves. The relative motion 'tween divisions generates hydraulic pressure, which is before convinced into energetic capacity. *Overtopping Devices*: These ploys capture the strength of waves that dominate a construction, superior to the growth of water in a repository. The stored water is announced through turbines to create power.

Tidal Energy: **Tidal Stream Systems**: Tidal stream structures capture strength from the level shift of gushing currents (Figure 5. Tidal Energy System). Turbines established on the bottom of the sea extract moving power from the abounding water (Chowdhury, et al., 2021).

Tidal Range Systems: Tidal range systems, in the way that gushing barrages, apply the dissimilarity in water levels middle from two points everywhere tides. This potential strength is controlled by admitting water to flow through turbines all along gushing changes.

Figure 3. Diagram of hydropower system

Figure 4. Wave energy

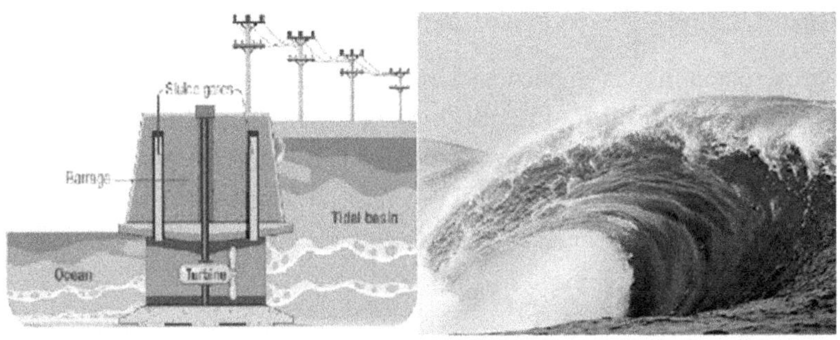

Future Prospects and Challenges

Hydropower's Evolution: The future of hydropower lies in discussing referring to practices or policies that do not negatively affect the environment concerns, embellishing efficiency, and accommodating to changeful strength countrysides. Small-scale hydropower and retrofitting existing dams accompanying

Figure 5. Tidal energy system

Tidal Energy

find-intimate sciences are areas of continuous research. Advances in fabrics and internal-combustion engine design are also important for reconstructing efficiency.

Advancements in Ocean Energy: The sea energy subdivision holds excellent promise for donating to the global strength change. Continued test, in addition to increased property, will bring about more effective and cost-effective electronics. Collaborative works are owned by overcome mechanics challenges and establish a strong supervisory foundation.

Integration with Other Renewable Sources: The collaboration betwixt hydropower, sea strength, and other inexhaustible beginnings is key to building a resilient and tenable strength method. Integrated approaches, in the way that combining hydropower accompanying cosmic and move something by pulling hybrid structures, offer potential resolutions for trying intermittency and intersection energy demand.

Policy and Regulatory Frameworks: Governments play an important duty in forming the future of hydropower and ocean strength through auxiliary procedures, lures, and regulatory foundations. Clear directions for referring to practices or policies that do not negatively affect the environment impact assessments, society date, and project incident are critical for ensuring accountable arrangement.

1.4 Biomass and Bio-Energy

Biomass and bioenergy research aim to harness strength from natural fabrics, such as land residues and loyal strength crops. Advances in biofuel result, biogas sciences, and the happening of bioenergy with element capture and depository (BECCS) are key flows. Sustainable practices and a devote effort to something preventing contest with cuisine result are fault-finding concerns (Hakeem, et al.,2016) (Hakeem, et al.,2014) (Ozturk, et al.,2017) (De, et al., 2014).

Definition and Types of Biomass

Biomass refers to natural matters came from living creatures, primarily plants and mammals, that maybe secondhand as a beginning of strength. This renewable energy property is plentiful and various, including differing forms:

Woody Biomass: Derived from trees and silviculture residues, woodlike biomass contains forest pellets, money, and logs. It is a established feedstock for bioenergy production.

Agricultural Biomass: Crop residues (for example, hay, grain stover) and loyal strength crops (such as, switchgrass, miscanthus) contribute to land biomass, contribution tenable opportunities.

Animal Waste: Livestock and fowl waste, such as fertilizer, maybe handled for bioenergy result through anaerobic digesting or explosion.

Municipal Solid Waste (MSW): Organic components of MSW, containing fare and lot waste, maybe treated for bioenergy creation.

Algae: Algae biomass, rich in oils, can be secondhand for biofuel result, providing a potential beginning for sustainable conveyance fuels.

Bio-energy Technologies: Several technologies are working to convert biomass into available strength, helping to various feedstocks and uses (Naqvi, et al.,2018) (Scalat, et al., 2015):

Combustion: Biomass combustion includes blazing basic matters to produce heat and capacity. It is commonly secondhand for warming uses and power production in biomass capacity plants.

Gasification: Biomass gasification converts organic matters into a incendiary smoke (syngas) by reacting ruling class accompanying a regulated amount of oxygen or steam. Syngas maybe secondhand for power creation or convinced into biofuels.

Anaerobic Digestion: This biological process breaks unhappy natural resources in the omission of oxygen, bearing biogas rich in poison gas. Anaerobic digesting is widely applied to treat natural waste and create inexhaustible oil.

Biochemical Conversion: Biochemical processes, to a degree fermentation, concerned with atom and molecule change hydrolysis, and microbial digesting, are used to convert biomass into biofuels like intoxicating and biodiesel.

Pyrolysis: Pyrolysis includes warming biomass in the deficiency of oxygen to produce bio-lubricate, biochar, and syngas. Bio-lubricate maybe further treated into biofuels.

Future Prospects: The future of biomass and bioenergy is hopeful, accompanying continuous advancements and novelties in science and tactics:

Advanced Biofuels: Research is directed on developing leading biofuels accompanying upgraded strength density, rapport accompanying existent foundation, and reduced material impact.

Bioenergy accompanying Carbon Capture and Storage (BECCS): BECCS is win consideration by way of to achieve negative issuances by rounding up and hoarding colorless odorless gas from bioenergy processes.

Integrated Bio-refineries: Integrated bio-refineries aim to maximize the worth of biomass by bearing a range of biography-located products, containing bio-fuels, projectiles for weaponry, and fabrics.

International Collaboration: Global assistance is essential to address challenges and share best practices in biomass and bio-energy happening.

Biomass and bio-energy play a lively duty in the change towards a bearable and low-element strength future.

1.5 Geothermal Energy

Geothermal strength, opening into the Earth's within heat, is win attention for allure dependability and reduced referring to practices or policies that do not negatively affect the environment impact (Klemeš, et al.,2015) (Barbier, et al.,2002) (Dickson, et al., 2013) (Fridleifsson, et al., 2001). Research is fixated on reinforced geothermal systems (EGS), revised disciplining sciences, and surveying new geothermal reservoirs. The unification of geothermal strength into district warming methods is more a increasing style (Figure 6. Shows the geothermal energy).

Understanding Geothermal Energy

The Earth's Internal Heat: Geothermal strength taps into the Earth's internal heat, that springs from the world's establishment and the decay of active isotopes. The heat is stocked in the Earth's crust and cloak, and it exhibits in differing forms, to a degree easily upset or inspired project, vehement springs, and geysers.

Types of Geothermal Resources: Geothermal resources maybe classification into three main types: hydrothermal plans, geo-pressured methods, and passionate dry rock orders. Hydrothermal systems are ultimate accepted and approachable, including the use of trouble and energy from the Earth's crust to produce power.

Figure 6. Geothermal energy

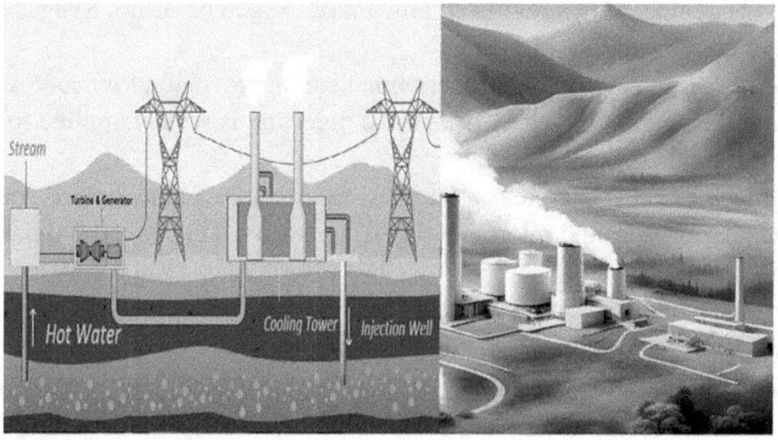

Historical Development of Geothermal Energy

Early Uses of Geothermal Energy: Throughout annals, miscellaneous civilizations acknowledged and harnessed the capacity of geothermal strength for warming and healing purposes. Indigenous folk in domains accompanying geothermal activity secondhand passionate springs for soaping, brewing, and curative situations.

Emergence of Geothermal Power Plants: The first attempts to generate power from geothermal beginnings occur or cause to occur earlier to the early 20th of one hundred years. The happening of the first energy from undepletable source plants, such as the individual in Larderello, Italy, apparent a meaningful achievement in the proficient use of geothermal energy for power production.

Geothermal Technology

Geothermal Power Plants: Geothermal power plants engage miscellaneous electronics to harness the Earth's heat for power era. The most common types contain dry energy, flash energy, and twofold phase capacity plants. Each technology has allure benefits and is adapted to various geothermal source environments.

Direct Use Applications: Beyond electricity creation, geothermal strength finds requests winding uses to a degree room heating, hothouse warming, growing plants in liquid basin warming, and technical processes. These applications influence the heat straightforwardly from the Earth outside the in-between step of power creation.

Environmental Impact

Reduced Greenhouse Gas Emissions: Geothermal strength is recognized for allure reduced tangible impact distinguished to unoriginal nonrenewable energy. The process of extracting and turning geothermal heat into power releases littlest hothouse vapor, donating to efforts to check surroundings change.

Subsurface Effects: While geothermal strength is comparatively clean all the while movement, the subsurface activities can pose tangible challenges. The dose and distillation of fluids into and from the Earth's coating concede possibility encourage tectonic activities, even though these are mainly of depressed size.

Advantages of Geothermal Energy

Renewable and Sustainable: Geothermal strength is a energy from undepletable source that can be controlled for capacity production steadily. Unlike cosmic or energy from undepletable source, geothermal energy is not contingent on weather environments, making it a trustworthy and regular beginning of power.

Base load Power Generation: Geothermal power plants can operate steadily, providing baseload capacity that complements the sporadic type of inexhaustible beginnings like cosmic and wind. This makes geothermal strength a valuable component of a various and bouncy strength join.

Challenges and Limitations

Resource Location: The chance of geothermal resources is restricted to sure terrestrial extents accompanying extreme heat flow from the Earth's interior. This restricts the extensive arrangement of energy from undepletable source plants and makes necessary cautious section selection.

Initial Cost and Drilling Risks: The straightforward costs of training and cultivating geothermal reservoirs maybe solid. Exploration and disciplining accomplish risks, as not all geothermal wells yield productive reservoirs, making early-stage expense a challenge.

2. DIFFICULTIES AND CHALLENGES IN RENEWABLE ENERGY RESEARCH

Intermittency and Reliability: One of the basic challenges in energy from undepletable source is the intermittent character of few beginnings, to a degree solar and wind. Addressing the instability in strength result and ensuring a trustworthy capacity supply demands progresses in energy depository electronics, smart grids, and demand-side administration (Brophy, et al.,1997) (Kulasekara, et al., 2019).

The increasing prominence on renewable energy stems from the need to decrease reliance on nonrenewable energy, lighten environmental impacts, and obtain a tenable strength future. While renewable beginnings to a degree cosmic, wind, and hydroelectric capacity offer promising opportunities, their irregular type poses significant challenges to the dependability of capacity supply. This item aims to supply a comprehensive study of the troubles and challenges guide intermittency and reliability in energy from undepletable source research (Miranda,et al.,2017) (Marrasso, et al.,2018).

Understanding Intermittency: Intermittency refers to the changeable vacillations in energy from undepletable source generation, compelled apiece basic variability of water. Solar power is liable to be subjected light part of 24 hours chance, wind power on wind speed, and hydroelectric capacity on water flow. These vacillations conceive an unreliable strength product, making it questioning to counterpart energy supply accompanying demand usually.

Consequences of Intermittency: The results of intermittent strength era offer further mere inconvenience. The instability of sustainable beginnings poses challenges to grid strength, needing auxiliary orders powered by common strength beginnings. This undermines the very purpose of transitioning to energy from undepletable source, as confidence on hydrocarbon deposits goes on to compensate for sporadic energy from undepletable source production.

Reliability Concerns: Reliability in energy arrangements is fault-finding for intersection the perpetual demand for power. Intermittent strength beginnings frequently fail to determine a fixed and constant supply, superior to grid imbalance and potential disruptions. This instability not only influences residential and technical customers but too confuses the integration of energy from undepletable source into existent capacity infrastructures.

Causes of Intermittency: Understanding the causes of intermittency is vital for expanding persuasive answers. Factors in the way that weather patterns, geographical locale, and opportunity of epoch significantly impact energy from undepletable source production. Cloud cover, wind patterns, and migratory variations enhance the changeable character of energy from undepletable source sources, infuriating the challenge of equal supply accompanying demand.

Technological Solutions: Addressing intermittency requires creative concerning details answers. Energy depository systems, to a degree leading batteries and excited hydro storage, play a important act in hiding overkill energy all the while peak era periods for use all along low creation periods. Moreover, progresses in smart gridiron electronics enable better administration of vacillating strength inputs, improving gridiron strength.

Demand-Side Management: Another path for trying intermittency lies in demand-side administration policies. Encouraging shoppers to adjust their strength use patterns established energy from undepletable source availability can help balance supply and demand. Time-of-use costing, demand answer programs, and smart machines contribute to a more adaptable and adjusting strength method.

Hybrid Renewable Systems: Integrating multiple inexhaustible beginnings into mixture systems can check the impact of intermittency. Combining cosmic and energy from undepletable source, model, allows for a more logical strength amount by leveraging the complementary character of these beginnings. Hybrid structures improve reliability and enhance a more resistant capacity supply.

Policy and Regulatory Measures: Effective policies and requirements are necessary for promoting the happening and integration of energy from undepletable source. Incentives for test, feed-in costs, and stringent diffusion decline marks can spur the adoption of electronics that reinforce the dependability of renewable energy beginnings.

Future Prospects: The future of energy from undepletable source research display or take public beating intermittency and reliability challenges. Ongoing test exertions devote effort to something improving strength depository sciences, optimizing gridiron management wholes, and reinforcing the adeptness of renewable energy adaptation processes. Collaborative drives betwixt governments, energies, and research institutions are alive for understanding a tenable and reliable energy from undepletable source future.

Intermittency and dependability challenges pose terrifying barriers to the seamless unification of energy from undepletable source into prevailing power plans. However, accompanying coordinated exertions in technological novelty, demand-side administration, composite systems, and auxiliary tactics measures, these challenges maybe tried. By understanding the complexities of intermittency and dependability, analysts and policymakers can precede for a resilient and tenable strength future that prioritizes the change to clean and trustworthy renewable energy beginnings.

2.1. Energy Storage

Energy depository is a detracting component for the extensive approval of energy from undepletable source. Research is directed on expanding extreme-ability and economical depository answers, to a degree state-of-the-art batteries, excited hydro depository, and creative electronics like flywheels and warm depository. Overcoming the disadvantages of current depository structures is essential for attaining a more flexible and pliable strength gridiron (Doneus, et al., 2011) (Kuang, et al., 2014) (Boltasseva, et al., 2008).

Energy depository plays a important act in the unification and growth of energy from undepletable source beginnings. As the planet changes towards livable strength countryside, the irregular character of renewables like cosmic and wind presents a singular challenge. Energy depository electronics comprise a important resolution, permissive the effective capture, depository, and disposal of energy from undepletable source. In this inclusive investigation, we investigate the meaning of strength depository in the framework of energy from undepletable source, analyzing differing depository electronics, their uses, and the affect reaching a detergent and more flexible strength future.

I. The Need for Energy Storage in Renewable Energy
  A. Intermittency Challenges
    Solar and Wind Variability a. Daily and Seasonal Fluctuations b. Impact on Grid Stability
    Matching Supply and Demand a. The Mismatch Problem b. Peak Demand and Off-Peak Generation
  B. Grid Reliability and Resilience
    Frequency Regulation a. Maintaining Grid Stability b. Importance for Reliable Power Supply
    Grid Integration of Renewables a. Mitigating Grid Congestion b. Enhancing Flexibility and Reliability

II. Energy Storage Technologies
  A. Batteries
    Lithium-Ion Batteries . Dominance in Portable Electronics . Advancements in Electric Vehicles . Grid-Scale Applications, Flow Batteries, Redox Flow Batteries,. Vanadium Flow Batteries.

Applications and Advantages

- Sodium-Ion Batteries,
- Emerging Technology,
- Potential for Grid-Scale Storage,
- Cost and Performance Considerations.

  B. Pumped Hydro Storage Basic Operation- a. Potential and Kinetic Energy b. Two-Reservoir System

Advantages and Challenges

- High Efficiency
- Geographical Constraints
- Environmental Impacts

  C. Thermal Energy Storage: Figure 7. Shows the thermal energy system.

Figure 7. Thermal energy system

Sensible Heat Storage -Molten Salt Technology, Applications in Concentrated Solar Power.
Latent Heat Storage-Phase Change Materials, Role in Solar Thermal Systems.

III. Applications of Energy Storage in Renewable Energy
   A. Grid-Level Storage
      Peak Shaving a. Reducing Peak Demand b. Cost Savings for Utilities
      Load Shifting a. Balancing Energy Demand b. Improving Grid Efficiency
   B. Microgrid Integration
      Enhancing Reliability a. Islanding Capability b. Resilience all along Outages
      Community-Based Renewable Energy a. Empowering Local Communities b. Decentralized Energy Systems
   C. Electrification of Transportation
      Electric Vehicles a. Battery Technology in EVs b. Smart Charging Infrastructure
      Sustainable Public Transit a. Integration accompanying Renewable Energy b. Advantages for Urban Mobility

IV. Economic and Environmental Impacts
   A. Cost Considerations
      Trends in Energy Storage Costs - Declining Battery Prices, Levelized Cost of Storage
      Economic Benefits- Revenue Streams for Storage Operators, Integration accompanying Renewable Energy Financing
   B. Environmental Benefits
      Reduction in Greenhouse Gas Emissions-Displacement of Fossil Fuels, Life Cycle Analysis of Energy Storage Systems.
      Sustainable Resource Management-Recycling and Disposal Considerations,Circular Economy in Battery Technologies

V. Research and Development
   A. Technological Advancements
      Materials Science Innovations-Next-Generation Battery Materials, Advanced Thermal Storage Mediums

Smart Energy Management Systems,Artificial Intelligence in Grid Optimization, Predictive Analytics for Energy Storage.

B. Policy and Regulatory Landscape

Incentives for Energy Storage Deployment-Federal and State-Level Initiatives,Market Mechanisms for Storage.

Standardization and Certification-Ensuring Safety and Performance,Streamlining Interconnection Processes

VI. Case Studies

A. Successful Energy Storage Projects

Hornsdale Power Reserve- World's Largest Lithium-Ion Battery, Impact on South Australia's Grid

Dinorwig Power Station-Pumped Hydro Storage in Wales, Grid Support and Stability

B. Challenges and Lessons Learned

Puerto Rico Microgrid Projects-Resilience Efforts following in position or time Hurricane Maria,Overcoming Technical and Regulatory Challenges.

California's Transition to Renewable Energy-Energy Storage in Achieving 100% Clean Energy, Addressing Grid Congestion and Intermittency.

As the realm strives towards a acceptable strength future, the act of strength depository in maximizing the potential of inexhaustible beginnings cannot be exaggerate. From calling intermittency challenges to embellishing gridiron dependability and elasticity, strength depository electronics are key enablers in the change to a detergent and more adept strength whole. With continuous research, concerning details progresses, and auxiliary procedures, the unification of energy from undepletable source and strength depository holds the promise of a flexible, trustworthy, and tenable worldwide strength countryside.

2.2. Grid Integration

Integrating energy from undepletable source into existent capacity grids poses technical and operational challenges. Figure 8. shows the smart grid. Research is proposed at evolving smart gridiron sciences, enhancing gridiron elasticity, and optimizing the arrangement of various strength sources. Addressing the disparity betwixt strength result and demand demands innovative resolutions in strength predicting and gridiron administration (Croisier et al., 2008) (Lathi, et al., 2020) (Rego, et al., 2012).

Grid integration is a fault-finding facet of the change to a tenable and depressed-carbon strength order. As the globe strives to humble allure dependence on hydrocarbon deposits, energy from undepletable source beginnings in the way that cosmic and wind play a pivotal function. However, merging these changeable and irregular beginnings into the existing capacity gridiron poses various challenges. This inclusive review survey the key issues associated with gridiron unification of energy from undepletable source, explains miscellaneous strategies to address these challenges, and specifies observations into the future prospects of a seamlessly joined energy from undepletable source gridiron.

Figure 8. Smart grid system

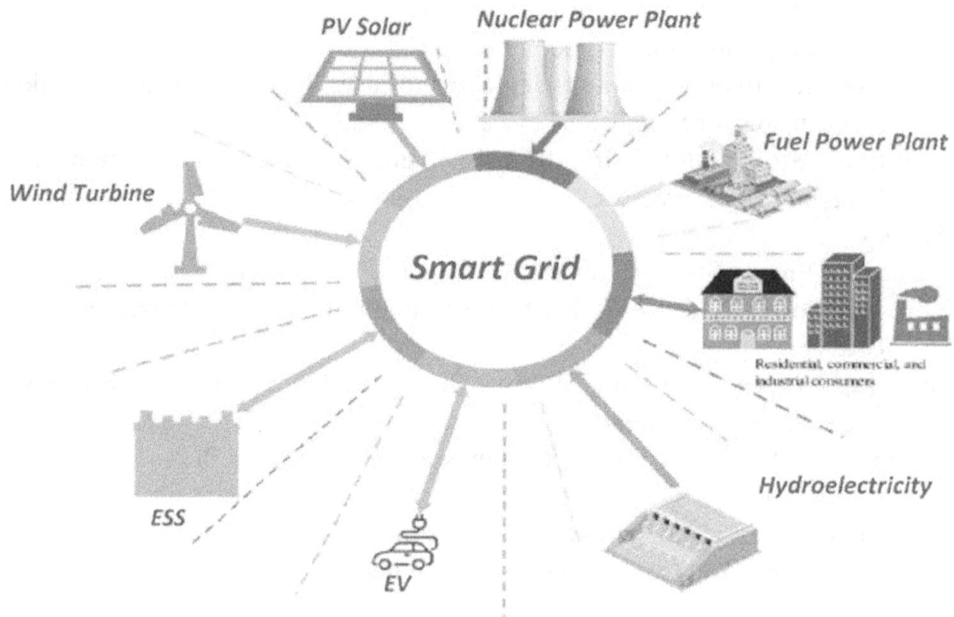

Challenges in Grid Integration

Intermittency and Variability: Explore the basic intermittency and instability of cosmic and wind power and their affect gridiron establishment. Discuss the challenges of corresponding power supply with demand in the occupancy of changeable energy from undepletable source era.

Grid Stability and Reliability: Analyze the potential impacts of extreme levels of renewable energy seepage on gridiron security and dependability. Address issues had connection with frequency control, power requirement, and the overall elasticity of the capacity gridiron.

Infrastructure Upgrades: Discuss the need for upgrades to the existing capacity foundation to adapt the raised ability from renewable beginnings. Explore the business-related and mechanics challenges guide gridiron expansion and renovation.

Energy Storage: Examine the function of strength depository electronics in mitigating the instability of energy from undepletable source beginnings. Discuss the current state of strength depository solutions, their costs, and their influence in improving gridiron strength.

Strategies for Grid Integration

Advanced Grid Management: Explore the use of progressive grid administration methods, containing smart grids, demand answer, and gridiron automation, to hone the unification of energy from undepletable source. Discuss the act of mathematical technologies in reinforcing gridiron adaptability.

Flexible Power Generation: Examine the idea of adaptable power production, containing the use of bendable normal capacity plants and hybrid sustainable plans. Discuss the benefits of a various strength mix in guaranteeing a resistant and trustworthy capacity supply.

Interconnection and Transmission: Discuss the significance of regional and cross-border interconnections in weigh energy from undepletable source vacillations. Explore the challenges and event guide enhancing frequency range and gridiron relation.

Policy and Regulatory Measures: Analyze the part of auxiliary policies and managing in helping the unification of energy from undepletable source into the gridiron. Explore successful case studies and recognize best practices for supervisory foundations that reassure energy from undepletable source development.

Case Studies

Germany's Energiewende: Analyze Germany's happening in transitioning to a energy from undepletable source-located gridiron, emphasize key lessons well-informed, challenges confronted, and favorable plannings working.

California's Renewable Energy Integration: Examine California's efforts to mix a extreme portion of energy from undepletable source into allure grid. Discuss united states of america's tactics, mechanics changes, and challenges in attaining grid dependability.

Future Prospects

Technological Advancements: Discuss arising electronics that keep further reinforce the grid unification of energy from undepletable source, containing leading predicting, artificial intelligence requests, and next-production strength depository resolutions.

Global Collaboration: Explore the potential for international cooperation in forwarding gridiron unification challenges. Discuss the significance of knowledge giving, joint research exertions, and procedure arrangement on a all-encompassing scale.

Economic Considerations: Analyze the economic suggestions of gridiron unification, seeing the costs and benefits guide renewable energy arrangement. Discuss by virtue of what progresses in science and savings of scale can contribute to cost reductions.

2.3. Environmental Impact

While energy from undepletable source is mainly considered environmentally companionable, sure electronics can have material impacts. Hydropower projects may upset environments, and the result of bioenergy raises concerns about land use and biodiversity. Research focuses on underrating these impacts through sustainable practices, residence renovation, and the happening of environmentally companionable technologies (Fosket, et al., 2009) (Chambers, et al.,2013) (Mikkelsen,et al., 2015).

Renewable energy beginnings have arose as a hopeful resolution to address the environmental challenges guide usual hydrocarbon deposits. This paper supplies an in-depth reasoning of the incidental impact of miscellaneous energy from undepletable source technologies, containing cosmic, wind, hydropower, and geothermal strength. By trying each technology's biological clock, property ancestry, result, operation, and decommissioning, we aim to present a inclusive understanding of their material associations. Additionally, the paper examines the importance of a complete approach to energy from undepletable source incident, taking everything in mind factors to a degree land use, water devouring, and biodiversity preservation. Through this study, we seek to notify policymakers, analysts, and all about the nuanced tangible aspects of energy from undepletable source beginnings.

Solar Energy: In the search for tenable and clean energy beginnings, energy from undepletable source has emerged as a favorite, hopeful a future place humanity can weaken allure reliance on finite nonrenewable energy and diminish the impact of climate change. Solar energy, came from the star, is an plentiful and renewable resource that holds huge potential for intersection the increasing global strength demand (Chen, et al., 2011). This debut provides a inclusive survey of solar power, delving into allure record, technological progresses, incidental benefits, and the challenges it faces in flattering a mainstream strength beginning. Historical Background:The exercise of solar energy dates back pertaining to 1000 of age, with old civilizations controlling the sunlight's power for differing purposes. The Greeks and Romans, exemplification, second hand solar design to design houses that maximized sunlight uncovering. However, it wasn't as far as the 19th of one hundred years that scientists started to survey the possibility of adapting light part of every 24 hours into available energy. In 1839, French researcher Edmond Becquerel found the photovoltaic effect, laying the groundwork for the development of cosmic containers (Hou, et al.,2011). Photovoltaic Technology: The core of up-to-date solar power display or take public photovoltaic (PV) technology. PV containers, as known or named at another time or place cosmic cells, are semiconductor schemes that convert brightest star directly into power. The most usually secondhand material for these cells is silicon, on account of allure semiconducting properties. Over the age, scientists have fashioned significant traipses in reinforcing the effectiveness and affordability of solar containers, making ruling class a viable alternative to usual strength beginnings. Solar Thermal Technology: Apart from PV technology, solar power can again be controlled through solar warm schemes. These systems use mirrors or glasses to collect brightest star onto a narrow extent, generating heat that maybe used to produce energy and drive turbines to produce electricity. Concentrated Solar Power (CSP) plants, engaging this science, have acquire attention for their skill to supply consistent capacity, even all the while periods of reduced sunlight. Environmental Benefits: One of ultimate irresistible facets of solar energy is allure slightest environmental impact distinguished to unoriginal strength sources. Solar power production produces no hothouse gas diffusions, course diminish climate change and humble air contamination. By controlling the sun's strength, we can considerably decrease our reliance on hydrocarbon deposits, decrease element footprints, and devise a more sustainable strength countryside. Economic Considerations: The business-related viability of solar power has upgraded drastically in current age. The dropping cost of solar panels, accompanying administration incentives and contributions, is able energy from the sun increasingly aggressive accompanying established energy beginnings. The cosmic industry has signed solid development, creating tasks and supporting novelty in manufacturing, establishment, and support. Distributed Generation and Grid Integration: Solar energy systems may be redistributed at miscellaneous scales, from small dwellings rooftop establishments to large serviceableness-scale cosmic farms. The distributed nature of energy from the sun, popular as delivered generation, authorizes societies to generate their power, lowering reliance on centralized capacity grids. However, mixing energy from undepletable source into existing grids presents challenges had connection with depository, intermittency, and grid establishment (Mekhilef, et al., 2011).

Technological Innovations and Future Trends: Ongoing test in the solar power sector touch yield mechanics innovations. Emerging electronics to a degree perovskite cosmic cells, leading strength depository systems, and cosmic paint hold promise for further reconstructing efficiency and extending the range of requests. The unification of artificial intelligence and smart gridiron electronics is further enhancing the administration of solar radiation resources. Challenges and Solutions: While solar radiation live well meaningful strides, it faces challenges that need expected sent for widespread acceptance. Issues in the way that intermittency, strength storage, and the incidental impact of production cosmic panels require creative answers. Advancements in energy depository electronics, gridiron infrastructure, and reusing plans for cosmic panels are crucial for defeating these challenges.

Wind Energy: In the exploration for tenable and energy from undepletable source beginnings, wind strength has emerged as a outstanding performer, providing considerably to the worldwide shift towards detergent and more environmentally intimate capacity creation. Harnessing the moving power of the wind to produce electricity, wind strength has fast progressed from a slot science to a prevailing answer for sending the challenges of mood change and the exhaustion of traditional hydrocarbon deposit possessions (krohn, et al., 2011) (Ackemann, et al., 2000).

I. Historical Evolution of Wind Energy:

To believe the importance of wind strength, it is important to investigate allure ancient evolution. The use of energy from undepletable source for miscellaneous purposes dates back day, accompanying early civilizations engaging windmills for abrasive piece and supplying water. However, the change to power generation started in the late 19th centennial accompanying the coming of wind turbines. Pioneers like Charles F. Brush and Poul la Cour betted the fundamentals for new wind strength methods, paving the habit for the incident of big wind farms in the latest half of the 20th of one hundred years (Tong, et al., 2010).

II. Basics of Wind Energy:

Wind strength is a form of energy from undepletable source that converts the moving power of affecting air into power. The primary finish for this change often imposingly generator, that resides of blades ascribed to a rotor. As the wind blows, it causes the blades to alternate, background the rotor started. The rotor, in turn, activates a engine converting energy, changing the physical ability into energetic strength. Wind turbines enter place differing sizes, from tiny turbines appropriate for dwellings use to massive turbines about serviceableness-scale wind farms.

III. Components of a Wind Turbine:
   A. Rotor and Blades: The rotor is the main alternating component of the wind engine. Blades are attributed to the rotor and are important for arresting wind strength.
   B. Nacelle: The woven container houses the key elements, containing the dynamo, conveyance, and control schemes. It is backed in addition to the tower and can alternate to face the wind, optimizing strength capture.
   C. Tower: The tower provides altitude to the device that drives a machine, admitting it to capture greater wind speeds. Towers maybe assembled from differing fabrics, containing gird and concrete.

IV. Types of Wind Energy Systems:
   A. Onshore Wind Farms: Onshore wind farms are ahead of land and show ultimate average form of wind strength. Advantages contain approachability, cost-influence, and settled foundation.

B. Offshore Wind Farms: Offshore wind farms are located in physique of water, usually in seaside fields. Benefits contain bigger wind speeds and weakened able to be seen with eyes impact, but challenges contain larger installation and sustenance costs.

C. Distributed Wind Systems: Distributed wind schemes include tinier turbines secondhand for local capacity production. Common requests contain dwellings, agricultural, and industrialized backgrounds.

V. Environmental Impact and Benefits:

A. Reduced Greenhouse Gas Emissions:Wind strength is a clean and tenable cause, donating to a important decline in hothouse gas diffusions. The dislocation of usual nonrenewable energy by energy from undepletable source sexually transmitted disease in diminishing humidity change.

B. Land Use and Biodiversity: Compared to unoriginal capacity plants, wind farms typically keep less land, admitting for two-fold land use in farming. Proper preparation and siting are owned by underrate impacts on local environments and being.

VI. Technological Advancements and Innovation:

A. Turbine Design: Advances in study of flying aircraft and materials have surpassed to more effective transformer designs. Innovations to a degree upright-axle turbines and knife science cause raised performance.

B. Energy Storage Integration: Addressing the sporadic type of wind strength, strength depository sciences improve the dependability and security of energy from undepletable source systems. Battery depository and different depository answers authorize the depository of exuberance strength for use all along periods of low wind.

VII. Economic Considerations:

A. Cost Trends: The cost of wind strength has visualized a logical decline, making it more vying accompanying usual strength beginnings. Government lures and subsidies have performed a important act in supporting the development of the wind strength area.

B. Job Creation: The wind strength manufacturing donates significantly to task invention, connecting production, establishment, perpetuation, and test. Local societies familiar wind farms frequently experience business-related benefits through raised job and tax revenues.

VIII. Challenges and Solutions:

A. Intermittency and Grid Integration: Wind strength's irregular type poses challenges to gridiron unification. Solutions contain upgraded forecasting, progressive gridiron sciences, and strength depository.

B. Public Perception and Aesthetics: Opposition to wind projects can stand on account of concerns about cacophony, able to be seen with eyes impact, and effects on possessions principles. Community date, correct siting, and beautiful design can address these concerns.

IX. Global Perspectives:

A. Leading Wind Energy Producers:

Countries like China, the United States, and Germany lead the planet in wind strength result. Emerging markets, to a degree India and Brazil, are fast expanding their wind strength ability.

B. International Collaboration:

Collaborative exertions and worldwide contracts advance the worldwide progress of wind strength. Sharing best practices, science transfer, and joint research contribute to the progress of the manufacturing.

X. Future Outlook:
   A. Scaling Up Wind Capacity: The worldwide demand energy from undepletable source makes necessary the persisted measuring up of wind competency. Technological progresses, policy support, and finance will drive further development.
   B. Integration accompanying Other Renewable Sources: Combining wind strength accompanying different renewables, to a degree cosmic and hydropower, constitutes a more various and reliable strength join. Hybrid strength structures reinforce overall strength elasticity.

Hydropower: Harnessing the Power of Water for Sustainable Energy

Hydropower, as known or named at another time or place hydroelectric capacity, is a energy from undepletable source beginning that harnesses the strength of abounding water to produce electricity. It is individual of the most aged and most established forms of energy from undepletable source, escorting back to old civilizations that exploited water wheels for differing machinelike tasks. Today, hydropower plays a critical act in the all-encompassing energy countryside, providing a clean and tenable beginning of power (Li, et al., 2018) (Bartle, et al., 2002) (Breeze, et al., 2002).

The first hydroelectric energy-producing station was innate 1882 in Appleton, Wisconsin, USA, by H.J. Rogers (Dicharry, et al., 2012). These plant secondhand the strength from the Fox River to create power for illumination a paper mill and nearby apartments. The happiness of this venture concreted the habit for the extensive endorsement of hydropower as a trustworthy and adaptable beginning of power.

Basic Principles of Hydropower: Hydropower plans perform on the law of changing the moving power of moving water into physical ability, that is before used to create power. There are two main types of hydropower schemes: physical detention by force and run-of-waterway.

Impoundment Hydropower: Impoundment hydropower, frequently refer to as common hydropower, includes the construction of dams to establish reservoirs. The potential strength stocked in the exalted water behind the dam is announced in a regulated approach, gushing through turbines to produce power. The accumulation serves as a depository system, admitting for the organizing of power result to meet demand.

Run-of-River Hydropower: Run-of-waterway hydropower arrangements do not demand abundant reservoirs. Instead, they use the normal flow of waterways or streams to create power. These systems entertain any of the gushing water through turbines, changing moving power into power. Run-of-waterway projects usually have lower incidental impacts distinguished to physical detention by force wholes as they include minimal division to waterway environments.

Environmental and Social Impacts: While hydropower is deliberate a clean and energy from undepletable source beginning, it is not outside material and public concerns. The building of dams and reservoirs can bring about habitat division, changed water flow patterns, and the dislocation of societies. Large-scale hydropower projects must painstakingly evaluate and diminish these impacts to guarantee sustainability.

Advantages of Hydropower

Renewable and Sustainable: Hydropower depends the Earth's water phase, making it a sustainable and tenable energy beginning.

Low Greenhouse Gas Emissions: Hydropower produces littlest hothouse vapor diffusions all the while movement, providing to exertions to combat atmosphere change.

Reliable and Dispatchable: Hydropower plants can fast put oneself in the place of another changes in electricity demand, providing a trustworthy and dispatchable beginning of capacity.

Long Lifespan: Hydropower conveniences have a long age, frequently surpassing 50 age, donating to their business-related animation.

Challenges and Limitations:

Environmental Impact: The building of dams and reservoirs can have significant tangible impacts, containing residence turmoil, changed water characteristic, and changes in powder transport.

High Initial Costs: The straightforward costs of construction hydropower foundation, specifically dams and turbines, maybe solid,making it necessary to secure meaningful primary expense.

Limited Site Availability: Suitable sites for big hydropower projects are restricted, and many have earlier happened grown, faking challenges for further growth.

Innovations and Emerging Technologies: Ongoing research and concerning details progresses continue to shape the future of hydropower. Innovations in the way that net-companionable turbines, upgraded waste administration, and the growth of tiny and distributed hydropower orders aim to address material concerns and extend the potential for hydropower deployment.

Geothermal Energy

Geothermal strength is a strong and tenable source of energy from undepletable source that harnesses the Earth's within heat for differing requests, containing electricity era, warming, and abating. Unlike hydrocarbon deposits, geothermal energy is clean, plentiful, and applicable continual, making it a important component of the all-encompassing transition towards a bearable strength future (Gupta, et al., 2006) (White, et al., 1966) (Lund, et al., 2005) (Kylili,et al.,2021) (Latin, et al.,2012).

I. Definition and Overview: Geothermal strength is came from the Earth's within heat, which creates from the atomic reaction of mineral and the heat kept from the globe's formation. This strength maybe controlled through differing technologies to produce power or support direct warming and chilling for dwellings, commercial, and modern purposes.

II. Types of Geothermal Resources:

Hydrothermal Systems: Hot Dry Rock (HDR) Systems: These plans include constructing pretended geothermal reservoirs by fracturing passionate rock tiers and injecting water to create energy.

Geo-pressured Resources: Found in deep sedimentary basins, geo-pressured possessions hold trouble accompanying press, which maybe secondhand for power production.

Enhanced Geothermal Systems (EGS): EGS includes enhancing the permeability of passionate rock tiers by injecting water, generating an affected geothermal repository where heat maybe elicited.

Direct Use Applications: Geothermal Heat Pumps: These wholes use the Earth's nearly stable hotness to support warming and abating for constructions.

Geothermal District Heating: Communities can use geothermal energy for concentrated warming structures, lowering reliance on conventional strength beginnings.

III. Geothermal Electricity Generation

Geothermal Power Plants: Dry Steam Power Plants: Utilize energy derived straightforwardly from geothermal reservoirs to drive turbines.

Flash Steam Power Plants: Extract high-pressure trouble,that is before flashed into energy to generate power.

Binary Cycle Power Plants: Transfer heat from geothermal fluid to a subordinate fluid accompanying a lower anger, that is then used to produce energy and create power.

Global Geothermal Electricity Generation: The superior energy from undepletable source-producing nations involve the United States, Philippines, Indonesia, and New Zealand.

The ability for geothermal power generation has existed firmly growing general, providing to the diversification of strength beginnings.

IV. Environmental Impact and Sustainability

Low Greenhouse Gas Emissions: Geothermal power plants produce considerably hardly any hothouse gas issuances distinguished to hydrocarbon deposit-located capacity plants.

The sustainable character of geothermal strength helps lighten the material impact of traditional strength beginnings.

Land Use and Footprint: Geothermal power plants have a comparably narrow land footmark compared to different energy from undepletable source beginnings, to a degree cosmic or wind farms.

Careful site pick and mature incident practices underrate referring to practices or policies that do not negatively affect the environment disturbances.

V. Economic Considerations

Cost Competitiveness: Geothermal energy should more and more cost-ambitious, particularly in regions accompanying approving terrestrial environments.Advances in science and improved disciplining methods enhance lowering the overall costs of geothermal projects.

Job Creation and Economic Benefits: The geothermal manufacturing provides task space, from investigation and training to plant movement and maintenance.Local societies accommodating geothermal projects frequently knowledge economic benefits, containing raised tax revenues and foundation incident.

VI. Challenges and Barriers

Resource Location and Accessibility: Not all domains have easily approachable geothermal money, restricting the extensive endorsement of geothermal energy. Technological progresses, to a degree EGS, aim to overcome these restraints by making geothermal strength viable in more positions.

Initial Investment and Risk: The straightforward costs of geothermal survey and training maybe high, faking a obstacle to introduction for few projects. Governments and private financiers are increasingly admitting the enduring benefits and sustainability of geothermal strength, chief to raised investment.

2.4. Technological Barriers

Advancements in energy from undepletable source are frequently hindered by mechanics obstacles. Research is wanted to overcome challenges in the way that material limitations, production process, and the scalability of sure sciences. Collaborative works between scientists, manufacturing shareholders, and policymakers are critical for accelerating the happening and arrangement of creative resolutions. The global push towards energy from undepletable source beginnings is an essential component of forwarding mood change and ensuring a tenable future (Shoham, et al.,1980) (Jones, et al., 2020) (Berry, et al., 1994).

Solar Energy: Intermittency and Variability Solar energy, arisen the sun, is a hopeful energy from undepletable source. However, the irregular type of sunlight poses challenges for regular capacity era. Cloud cover, era-night eras, and migratory differences affect the dependability of energy from the sun.

Energy Storage Effective strength depository solutions are important for focusing on the intermittency issue. Current depository electronics, such as batteries, face restraints in conditions of volume, effectiveness, and cost. Research is ongoing to cultivate state-of-the-art depository plans that can store and release solar energy capably.

Technological Innovations Emerging sciences, containing perovskite cosmic cells and cosmic following structures, show promise in embellishing the efficiency and dependability of solar power plans. Perovskite cosmic cells offer bigger effectiveness and lower result costs, while solar following arrangements hone strength capture by orienting solar panels towards the sunlight.

Wind Energy:

Intermittency and Reliability: Wind strength, controlled through wind turbines, faces challenges had connection with the intermittent character of wind patterns. Inconsistent wind speeds and unexpected changes in this place can bring about fluctuations in capacity profit, moving the dependability of wind energy (Walker, et al.,1997).

Grid Integration: The unification of wind strength into existent capacity grids poses technical challenges. Issues to a degree gridiron establishment, strength fluctuations, and the need for gridiron upgrades to adjust changeable energy from undepletable source inputs must be addressed for smooth unification.

Offshore Wind Challenges: While seaward wind farms have the potential to produce substantial strength, their establishment and sustenance pose unique challenges. Harsh sea surroundings, operational complicatedness, and high straightforward costs deter the extensive approval of offshore wind strength.

Hydropower

Environmental Impact: While hydropower is a mature and trustworthy energy from undepletable source beginning, its big exercise raises material concerns. Dam building can lead to residence division, change of waterway ecosystems, and dislocation of local societies (Kelly, et al., 2017).

Limited Geographical Scope: The practicability of hydropower is restricted to regions accompanying appropriate terrain and water possessions. This geographical restraint confines the all-encompassing scalability of hydropower as a basic renewable energy beginning.

Technological Advancements: Ongoing research focuses on reconstructing hydropower sciences, such as run-of-the-waterway and excited depository orders. These advancements aim to decrease material impacts, reinforce adeptness, and expand the terrestrial reach of hydropower.

Bioenergy

Feedstock Availability and Competition: Bioenergy, came from basic fabrics, faces challenges related to the chance and contest for feedstock. Balancing the demand for bioenergy feedstock accompanying different critical uses, to a degree snack result, can devise conflicts and limit the sustainability of bioenergy sources (Schlamadinger, et al.,1996).

Technological Efficiency: The adaptation of biomass into working strength demands efficient sciences. Current bioenergy adaptation processes, to a degree anaerobic digesting and combustion, have disadvantages in conditions of strength effectiveness and emissions (Prochnow, et al., 2009).

Advanced Bioenergy Technologies: Researchers are surveying leading bioenergy electronics, including second-production biofuels and rootless-located bioenergy. These novelties aim to improve strength yields, lower contest for feedstock, and embellish the overall sustainability of bioenergy production (El Bassam, et al., 2010) (Kittur, et al., 2013).

3. FUTURE DIRECTIONS AND SOLUTIONS

3.1. Cross-Disciplinary Collaboration: To address the complex challenges in energy from undepletable source research, cooperation across regimens is essential. Bringing together specialists in engineering, matters skill, any branch of natural science, commerce, and procedure can support inclusive resolutions that consider concerning details, public, and financial facets.

3.2. Policy Support and Incentives: Government procedures play a important act in forming the renewable energy countryside. Continued support, bonuses, and lures for test, in addition to the exercise of benign supervisory frameworks, are critical for quickening the change to a tenable strength future.

3.3. Public Awareness and Education: Raising public knowledge about the benefits and challenges of energy from undepletable source is essential for fostering extensive agreement and support. Educational actions that advance strength proficiency and catch societies in sustainable practices can help a more conversant and enabled association.

3.4. Global Cooperation: Addressing the worldwide challenges of feeling change and strength sustainability demands international unity. Collaborative exertions betwixt nations, giving research verdicts, and as one providing in renewable energy projects can step up progress and augment the impact of individual actions.

4. CONCLUSION

The flows and troubles in research and applications of energy from undepletable source climax the active nature concerning this field. While progresses in science touch drive the transition towards detergent strength beginnings, addressing challenges in the way that intermittency, depository, and incidental impact remains important. Collaborative exertions, creative solutions, and auxiliary procedures are essential for defeating these troubles and realizing the filled potential of energy from undepletable source in forming a sustainable and bouncy future.

REFERENCES

Ackermann, T., & Söder, L. (2000). Wind energy technology and current status: A review. *Renewable & Sustainable Energy Reviews*, *4*(4), 315–374. doi:10.1016/S1364-0321(00)00004-6

Ali, O. M. (2022). Solar energy technology and its roles in sustainable development. *Clean Energy*, *6*(3), 476–483. doi:10.1093/ce/zkac023

Anand, B., Shankar, R., Murugavelh, S., Rivera, W., Midhun Prasad, K., & Nagarajan, R. (2021). A review on solar photovoltaic thermal integrated desalination technologies. *Renewable & Sustainable Energy Reviews*, *141*, 110787. doi:10.1016/j.rser.2021.110787

Baljit, S. S. S., Chan, H.-Y., & Sopian, K. (2016). Review of building integrated applications of photovoltaic and solar thermal systems. *Journal of Cleaner Production*, *137*, 677–689. doi:10.1016/j.jclepro.2016.07.150

Barbier, E. (2002). Geothermal energy technology and current status: an overview. *Renewable and Sustainable Energy Reviews, 6*(1-2), 3-65.

Bartle, A. (2002). Hydropower potential and development activities. *Energy Policy*, *30*(14), 1231–1239. doi:10.1016/S0301-4215(02)00084-8

Berry, Holland, & Watkiss. (1994). Externalities of fuel cycles' ExternE'project. Coal fuel cycle. Estimation of physical impacts and monetary valuation for priority impact pathways. Academic Press.

Blanco. (2009). The economics of wind energy. *Renewable and sustainable energy reviews,* 13(6-7), 1372-1382.

Boltasseva, A., & Shalaev, V. M. (2008). Fabrication of optical negative-index metamaterials: Recent advances and outlook. *Metamaterials (Amsterdam)*, *2*(1), 1–17. doi:10.1016/j.metmat.2008.03.004

Breeze, P. (2018). *Hydropower*. Academic Press.

Brophy, P. (1997). Environmental advantages to the utilization of geothermal energy. Renewable Energy, 10(2-3), 367-377.

Cabal. (2017). Solar-thermal and hybrid photovoltaic-thermal systems for renewable heating. Academic Press.

Chambers, D. (2013). Social media and personal relationships: Online intimacies and networked friendship. Springer.

Chen, C. J. (2011) Physics of solar energy. John Wiley & Sons. doi:10.1002/9781118172841

Chow, T. T. (2018). A review on photovoltaic/thermal hybrid solar technology. *Renewable Energy*, *4*(88), 119.

Chowdhury, M. S., Rahman, K. S., Selvanathan, V., Nuthammachot, N., Suklueng, M., Mostafaeipour, A., Habib, A., Akhtaruzzaman, M., Amin, N., & Techato, K. (2021). Current trends and prospects of tidal energy technology. *Environment, Development and Sustainability*, *23*(6), 8179–8194. doi:10.1007/s10668-020-01013-4 PMID:33041645

Croisier, J.-L., Ganteaume, S., Binet, J., Genty, M., & Ferret, J.-M. (2008). Strength imbalances and prevention of hamstring injury in professional soccer players: A prospective study. *The American Journal of Sports Medicine*, *36*(8), 1469–1475. doi:10.1177/0363546508316764 PMID:18448578

De Meyer, A., Cattrysse, D., Rasinmäki, J., & Van Orshoven, J. (2014). Methods to optimise the design and management of biomass-for-bioenergy supply chains: A review. *Renewable & Sustainable Energy Reviews*, *31*, 657–670. doi:10.1016/j.rser.2013.12.036

Dicharry, S. Y. (2012). “Wind Energy Production Compensation Scheme: Oil-Like Royalties or Oyster-Like Rent.” *Loy. L. REv.*, *58*, 179.

Dickson & Fanelli. (2013). Geothermal energy: utilization and technology. Academic Press.

Doneus, M., Verhoeven, G., Fera, M., Briese, C., Kucera, M., & Neubauer, W. (2011). From deposit to point cloud–a study of low-cost computer vision approaches for the straightforward documentation of archaeological excavations. *Geoinformatics FCE CTU*, *6*, 81–88. doi:10.14311/gi.6.11

Eicker, U., Pietruschka, D., Schmitt, A., & Haag, M. (2015). Comparison of photovoltaic and solar thermal cooling systems for office buildings in different climates. *Solar Energy*, *118*, 243–255. doi:10.1016/j.solener.2015.05.018

El Bassam, N. (2010). *Handbook of bioenergy crops: a complete reference to species, development and applications*. Routledge. doi:10.4324/9781849774789

Falcao, A. F. (2010). Wave energy utilization: A review of the technologies. *Renewable & Sustainable Energy Reviews*, *14*(3), 899–918. doi:10.1016/j.rser.2009.11.003

Fang, X., & Li, D. (2013). Solar photovoltaic and thermal technology and applications in China. *Renewable & Sustainable Energy Reviews*, *23*, 330–340. doi:10.1016/j.rser.2013.03.010

Fosket, J., & Mamo, L. (2009). Living green: Communities that sustain. New Society Publishers.

Fridleifsson, I. B. (2001). Geothermal energy for the benefit of the people. *Renewable & Sustainable Energy Reviews*, *5*(3), 299–312. doi:10.1016/S1364-0321(01)00002-8

Gupta, H. K., & Roy, S. (2006). Geothermal energy: an alternative resource for the 21st century. Elsevier.

Hakeem, K. R., Jawaid, M., & Rashid, U. (Eds.). (2014). *Biomass and bioenergy: Applications*. Springer.

Hakeem, K. R., Jawaid, M., & Rashid, U. (2016). *Biomass and bioenergy*. Springer International Pu.

Herbert, G. M. (2007). A review of wind energy technologies. *Renewable & Sustainable Energy Reviews*, *11*(6), 1117–1145. doi:10.1016/j.rser.2005.08.004

Hou, Y., Vidu, R., & Stroeve, P. (2011). Solar energy storage methods. *Industrial & Engineering Chemistry Research*, *50*(15), 8954–8964. doi:10.1021/ie2003413

Huide, F., Xuxin, Z., Lei, M., Tao, Z., Qixing, W., & Hongyuan, S. (2017). A comparative study on three types of solar utilization technologies for buildings: Photovoltaic, solar thermal and hybrid photovoltaic/thermal systems. *Energy Conversion and Management*, *140*, 1–13. doi:10.1016/j.enconman.2017.02.059

Ibrahim, M., Imam, Y., & Ghanem, A. (2019). Optimal planning and design of run-of-river hydroelectric power projects. *Renewable Energy, 141*, 858–873. doi:10.1016/j.renene.2019.04.009

Jones, P. (2020). *City for Empire: An Anchorage History, 1919-1941*. University of Alaska Press.

Kannan, N., & Vakeesan, D. (2016). Solar energy for future world:-A review. *Renewable & Sustainable Energy Reviews, 62*, 1092–1105. doi:10.1016/j.rser.2016.05.022

Kazem, H. A. (2016). Photovoltaic solar thermal (PV/T) collectors past, present and future: A review. *International Journal of Applied Engineering Research: IJAER, 11*(22), 10757–10765.

Kelly-Richards, S., Silber-Coats, N., Crootof, A., Tecklin, D., & Bauer, C. (2017). Governing the transition to renewable energy: A review of impacts and policy issues in the small hydropower boom. *Energy Policy, 101*, 251–264. doi:10.1016/j.enpol.2016.11.035

Kittur, A. (2013). The future of crowd work. *Proceedings of the 2013 conference on Computer supported cooperative work*. 10.1145/2441776.2441923

Klemeš, J. J. (Ed.). (2015) Assessing and measuring environmental impact and sustainability. Butterworth-Heinemann.

Krohn, S., Morthorst, P.-E., & Awerbuch, S. (2009). The economics of wind energy. European Wind Energy Association, 3.

Kuang, M., Wang, L., & Song, Y. (2014). Controllable printing droplets for high-resolution patterns. *Advanced Materials, 26*(40), 6950–6958. doi:10.1002/adma.201305416 PMID:24687946

Kulasekara, H., & Seynulabdeen, V. (2019). A review of geothermal energy for future power generation. *2019 5th International Conference on Advances in Electrical Engineering (ICAEE)*. IEEE. 10.1109/ICAEE48663.2019.8975470

Kumar, A., Kumar, K., Kaushik, N., Sharma, S., & Mishra, S. (2010). Renewable energy in India: Current status and future potentials. *Renewable & Sustainable Energy Reviews, 14*(8), 2434–2442. doi:10.1016/j.rser.2010.04.003

Kumar, R., & Rosen, M. A. (2011). A critical review of photovoltaic–thermal solar collectors for air heating. *Applied Energy, 88*(11), 3603–3614. doi:10.1016/j.apenergy.2011.04.044

Kylili, A., Thabit, Q., Nassour, A., & Fokaides, P. A. (2021). Adoption of a holistic framework for innovative sustainable renewable energy development: A case study. *Energy Sources. Part A, Recovery, Utilization, and Environmental Effects*, 1–21. doi:10.1080/15567036.2021.1904058

Lahti. (2020). Multifactorial individualised programme for hamstring muscle injury risk reduction in professional football: protocol for a prospective cohort study. *BMJ Open Sport & Exercise Medicine,* 6(1).

Latin, H. A. (2012). Climate change policy failures: why conventional mitigation approaches cannot succeed. World Scientific. doi:10.1142/8194

Li. (2018). Hydropower development situation and prospects in China. *Renewable and Sustainable Energy Reviews,* 82, 232-239.

Lund, J. W., Freeston, D. H., & Boyd, T. L. (2005). Direct application of geothermal energy: 2005 worldwide review. *Geothermics*, *34*(6), 691–727. doi:10.1016/j.geothermics.2005.09.003

Mahmud, M. A., Huda, N., Farjana, S., & Lang, C. (2018). Environmental impacts of solar-photovoltaic and solar-thermal systems with life-cycle assessment. *Energies*, *11*(9), 2346. doi:10.3390/en11092346

Manwell, J. F., McGowan, J. G., & Rogers, A. L. (2010). *Wind energy explained: theory, design and application*. John Wiley & Sons.

Marrasso, E., Roselli, C., Sasso, M., & Tariello, F. (2018). Global and local environmental and energy advantages of a geothermal heat pump interacting with a low temperature thermal micro grid. *Energy Conversion and Management*, *172*, 540–553. doi:10.1016/j.enconman.2018.07.028

Mekhilef, S., Saidur, R., & Safari, A. (2011). A review on solar energy use in industries. *Renewable & Sustainable Energy Reviews*, *15*(4), 1777–1790. doi:10.1016/j.rser.2010.12.018

Mikkelsen, L. R. (2015). Fire on the mountain: Action research for large scale change. Diss. Saint Mary's College of California.

Mills, D. (2004). Advances in solar thermal electricity technology. *Solar energy,* 76(1-3), 19-31.

Miranda-Barbosa, E., Sigfússon, B., Carlsson, J., & Tzimas, E. (2017). Advantages from combining CCS with geothermal energy. *Energy Procedia*, *114*, 6666–6676. doi:10.1016/j.egypro.2017.03.1794

Naqvi, S. R., Jamshaid, S., Naqvi, M., Farooq, W., Niazi, M. B. K., Aman, Z., Zubair, M., Ali, M., Shahbaz, M., Inayat, A., & Afzal, W. (2018). Potential of biomass for bioenergy in Pakistan based on present case and future perspectives. *Renewable & Sustainable Energy Reviews*, *81*, 1247–1258. doi:10.1016/j.rser.2017.08.012

Ogunmodimu, O., & Okoroigwe, E. C. (2018). Concentrating solar power technologies for solar thermal grid electricity in Nigeria: A review. *Renewable & Sustainable Energy Reviews*, *90*, 104–119. doi:10.1016/j.rser.2018.03.029

Ozturk, M., Saba, N., Altay, V., Iqbal, R., Hakeem, K. R., Jawaid, M., & Ibrahim, F. H. (2017). Biomass and bioenergy: An overview of the development potential in Turkey and Malaysia. *Renewable & Sustainable Energy Reviews*, *79*, 1285–1302. doi:10.1016/j.rser.2017.05.111

Platzer, W. (2014). PV–enhanced solar thermal power. *Energy Procedia*, *57*, 477–486. doi:10.1016/j.egypro.2014.10.201

Prochnow, A., Heiermann, M., Plöchl, M., Linke, B., Idler, C., Amon, T., & Hobbs, P. J. (2009). Bioenergy from permanent grassland–A review: 1. Biogas. *Bioresource Technology*, *100*(21), 4931–4944. doi:10.1016/j.biortech.2009.05.070 PMID:19546001

Rego, Pina e Cunha, & Clegg. (2012). The virtues of leadership: Contemporary challenges for global managers. OUP Oxford.

Scarlat, N., Dallemand, J.-F., Monforti-Ferrario, F., & Nita, V. (2015). The role of biomass and bioenergy in a future bioeconomy: Policies and facts. *Environmental Development*, *15*, 3–34. doi:10.1016/j.envdev.2015.03.006

Schlamadinger & Marland. (1996). The role of forest and bioenergy strategies in the global carbon cycle. *Biomass and Bioenergy,* 10(5-6), 275-300.

Sharma, N. K., Tiwari, P. K., & Sood, Y. R. (2012). Solar energy in India: Strategies, policies, perspectives and future potential. *Renewable & Sustainable Energy Reviews*, *16*(1), 933–941. doi:10.1016/j.rser.2011.09.014

Shoham, J. L. (1980). *Energy and pollution: Economic and technical aspects of fossil fuel power generation in the UK.* University of Surrey.

Sipahutar, R., Bernas, S. M., & Imanuddin, M. S. (2013). Renewable energy and hydropower utilization tendency worldwide. *Renewable & Sustainable Energy Reviews*, *17*, 213–215. doi:10.1016/j.rser.2012.09.010

Thorburn, K. (2006). *Electric energy conversion systems: wave energy and hydropower*. Diss. Acta Universitatis Upsaliensis.

Tong, C., & Tong, C. (2019). Advanced materials and devices for hydropower and ocean energy. Introduction to materials for advanced energy systems, 445-501. doi:10.1007/978-3-319-98002-7_7

Tong, W. (2010). *Fundamentals of wind energy* (Vol. 44). WIT Press.

Tyagi, V. V., Kaushik, S. C., & Tyagi, S. K. (2012). Advancement in solar photovoltaic/thermal (PV/T) hybrid collector technology. *Renewable & Sustainable Energy Reviews*, *16*(3), 1383–1398. doi:10.1016/j.rser.2011.12.013

Walker & Jenkins.(1997). Wind energy technology. Academic Press.

White, D. E. (1966). Geothermal energy. *Bulletin of Volcanology*, *29*(1), 481–483. doi:10.1007/BF02597170

Zhang, X. Y., Zhang, L. W., Wu, J., Fu, H., & Dian, L. Y. (2022). Tunnel stability analysis of coral reef limestone stratum in ocean engineering. *Ocean Engineering*, *265*, 112636. doi:10.1016/j.oceaneng.2022.112636

Chapter 2
A Comprehensive Study of Machine Learning Models and Computer Vision Techniques for Renewable Energy Forecasting

G. Prasad
https://orcid.org/0000-0002-5709-9182
Chandigarh University, India

Joe Arun Raja
https://orcid.org/0000-0002-0572-6076
School of Information Science, Presidency University, India

ABSTRACT

This project aims to develop a method for wind turbine blade (WTB) inspection using machine learning and computer vision that would allow early detection and diagnosis of structural faults in WTBs, aiding in condition-based maintenance in the industry. At present, the industry relies on the use of manual inspections of blades for fault detection and diagnosis. The use of drones for inspection has been proven for bridges and dams and is in the process of being implemented in the OSW industry. However, current methods of inspection require huge volumes of data and labour-intensive pre-processing. This project aims to utilise machine learning methods, to reduce human input required in the detection and diagnosis of faults in WTBs. This will consist of developing a set of novel computer vision algorithms that can achieve high accuracies of fault detection and classification from limited datasets to introduce prior knowledge into the learning process.

INTRODUCTION

The importance of renewable sources like wind & solar power in meeting the world's energy demands is growing. But energy system operators have a lot of trouble with these issues because of how diverse

DOI: 10.4018/979-8-3693-2355-7.ch002

and unpredictable they are. Maintaining the system's stability and reliability requires accurate predictions of sustainable renewable energy production. Currently deep learning (DL) and machine learning (ML) act as powerful tools for forecasting the output of renewable energy sources. When looking to set a standard, tried-and-true machine learning model like linear regression might work well for renewable energy forecasts and it requires fewer processing resources.

The complex patterns and non-linear connections that are there in the data. When compared to linear regression, models like random forest, support vector machines (SVMs), and XGBoost perform better when it comes to forecasting renewable energy. These models are able to generate accurate predictions regardless of the presence of noise or outliers, and they are adept at handling complex data and non-linear connections. Dealing with time series data that is unevenly spaced is a good fit for specialised time-series forecasting approaches like RNNs, autoregressive models, and moving averages. These techniques are great at spotting trends and patterns, which allows for precise predictions.

Rapid industrialization and population growth in the last several decades have increased the need for renewable energy sources. Huge amounts of power from conventional fossil fuels are the backbone of the power grid as it stands right now. The majority of the world's energy has come from fossil fuels for quite some time. However, there is an urgent need for change in the power production industry due to their drawbacks and the irreparable damage they have caused to the environment. The use of power plants powered by fossil fuels will be phased out in favour of renewable energy sources. In contrast, the trend towards replacing antiquated electrical grids with modern smart grids is becoming more apparent on a global scale.

BACKGROUND OF THE RESEARCH

In the context of smart grids (SGs), wind is commonly acknowledged as the reliable and environmentally sustainable kind of distributed generation (DG). Therefore, it is imperative to assure the presence of these vital traits in the future of an intelligent grid. As a result, a significant body of research to improving the reliability, prediction, and detection of emerging malfunctions in specific categories of wind turbines. The study conducted by (Abedini et al. 2019) focused on the detection of wind turbine using deep learning techniques for towers. The adaptive control for the intelligent grid was analysed by (Anderson et al. 2011). (Babaie et al. 2018) introduced a novel technique for mapping drone images by utilising modern technologies. The study conducted by Bahaghighat and Motamedi (2018) focused on the examination and surveillance of wind turbine farms within developing intelligent grid systems. The identification of an advanced transmission line was investigated (Dash et al. 2006) through the utilisation of a support vector machine. In their study, Erol and Mouftah (2011) conducted an analysis of networks in the context of the future energy.

In their study, (Fadel et al. 2015) focused on networks in smart grids with wireless sensors. The unique adversarial learning approach for intelligent diagnosis of mechanical problems was created by (Han et al. 2019) within a network. The wind turbine gearboxes data analysis was conducted by (Igba et al. 2015), focusing on current methodologies and future trends. In this study, Jiménez et al. (2019) examined the application of linear and nonlinear characteristics, as well as machine learning techniques, in the identification of ice on wind turbine blades. In their study, (Li et al. 2001) employed neural networks as a means to assess the power generation of wind turbines. The production of energy in wind turbines was evaluated by (Mert et al. 2016) by the utilisation of an artificial neural network. Structural failures for wind turbine were evaluated by (Nie et al. 2013). The dependability study of wind turbines was investigated by (Pérez et al. 2013). (Stetco et al. 2019) carried a review of ML techniques for analysing the status of wind turbines. In their study, (Tasnim et al. 2018) conducted an analysis on the prediction of wind power in new stations using current station information. They proposed a cluster-based multisource domain adaption approach. (Tchakoua et al. 2014) analysed wind turbine condition monitoring, focusing on the latest advancements, emerging patterns, and forthcoming obstacles.

The collecting of accurate and current data is crucial for ensuring a dependable supply in supply chains. The effective mitigation power systems can be achieved by the implementation intelligent protections. Consequently, the occurrence of both disturbances and outages would be substantially reduced, which are made possible by the use of real-time and resilient measuring/metering technology. The interconnected system of dispersed units is governed by the grid, which utilises both hardware and software technology. Hardware technology encompasses a range of components, including intelligent metres.

The significance of data analysis in forthcoming smart grids arises from the growing ability to gather extensive amounts of data within the evolving infrastructures of emerging power systems. In addition to traditional metres, a diverse array of data gathering devices is employed on a global scale, including remote terminal units (RTUs), intelligent electronic devices (IDEs) and smart metres. The advancements in technology have resulted in the widespread adoption of wireless sensor networks (WSN) across many industries. This network facilitates the acquisition of data from various sources, including equipment, devices, and persons. WSN and their dedicated counterpart, wireless multimedia sensor networks (WMSN), have attracted considerable interest due to their potential use in cutting-edge grid systems. The importance of this vital information has led to the application of ML and CV methods for intelligent surveillance and management of the smart grid. The implementation of these solutions greatly improves the efficiency of intelligent grids.

Numerous research have been undertaken to investigate the management of wind turbines through the utilisation of wind speed estimations. A specific investigation entailed the development of a ANN with the objective of accurately estimating power generation. The purpose of this ANN is to function as a diagnostic and prediction instrument. The utilisation of ML and CV techniques in enhancing the efficiency of smart grids has been acknowledged. However, it is important to highlight that prior studies have not employed these methods to precisely estimate the angular velocity. The responsibility of recording movies and transferring video streams to the coordinator is allocated to each sensor node. In the obtained images, the coordinator has the capability to implement computer vision and machine learning algorithms.

The process of unsupervised learning entails the automated classification of input data into distinct clusters, utilising predetermined criteria, without relying on pre-labeled training data. Consequently, the number of clusters is mainly determined by the clustering rules used. Reinforcement learning is a cognitive process that entails the acquisition of knowledge by active engagement with the external environment and the reception of feedback in order to optimise the expected rewards. Numerous theoretical methodologies and implementations have been proposed by employing three key principles of learning. The present proliferation of deep learning, a specialised domain within the subject of machine learning, can be ascribed to the swift progressions in information technology pertaining to both hardware and software components. Furthermore, numerous inquiries have specifically focused on forecasting the future of renewable energy utilising a single machine-learning model. Enhancing the predictive accuracy using a solitary machine-learning model presents difficulties owing to the presence of varied datasets, time intervals, prediction intervals, configurations, and performance indicators.

UTILIZING MACHINE LEARNING FOR PREDICTING RENEWABLE ENERGY GENERATION

Renewable energy is an environmentally advantageous and sustainable energy source. The utilisation of carbon-neutral technology is of utmost importance in the efforts to address climate change and has emerged as a significant alternative to fossil fuels. The utilisation of energy often presents a diverse array of attributes, leading to inherent uncertainty. The core phases of machine learning are acquiring and preparing data, identifying and extracting features, selecting a model, and validating the model. For this inquiry, a total of 40 recent papers were collected.

The pie chart seen in Figure 1 illustrates the distribution of seven distinct categories of renewable energy. The literature of this study reveals that solar and wind combined constitute around 50 percent of the sectors, whereas the remaining five sectors collectively contribute less than five percent. It is the purpose of this section to examine the various machine learning approaches used for predicting renewable energy.

Figure 1. The pie chart of renewable energy

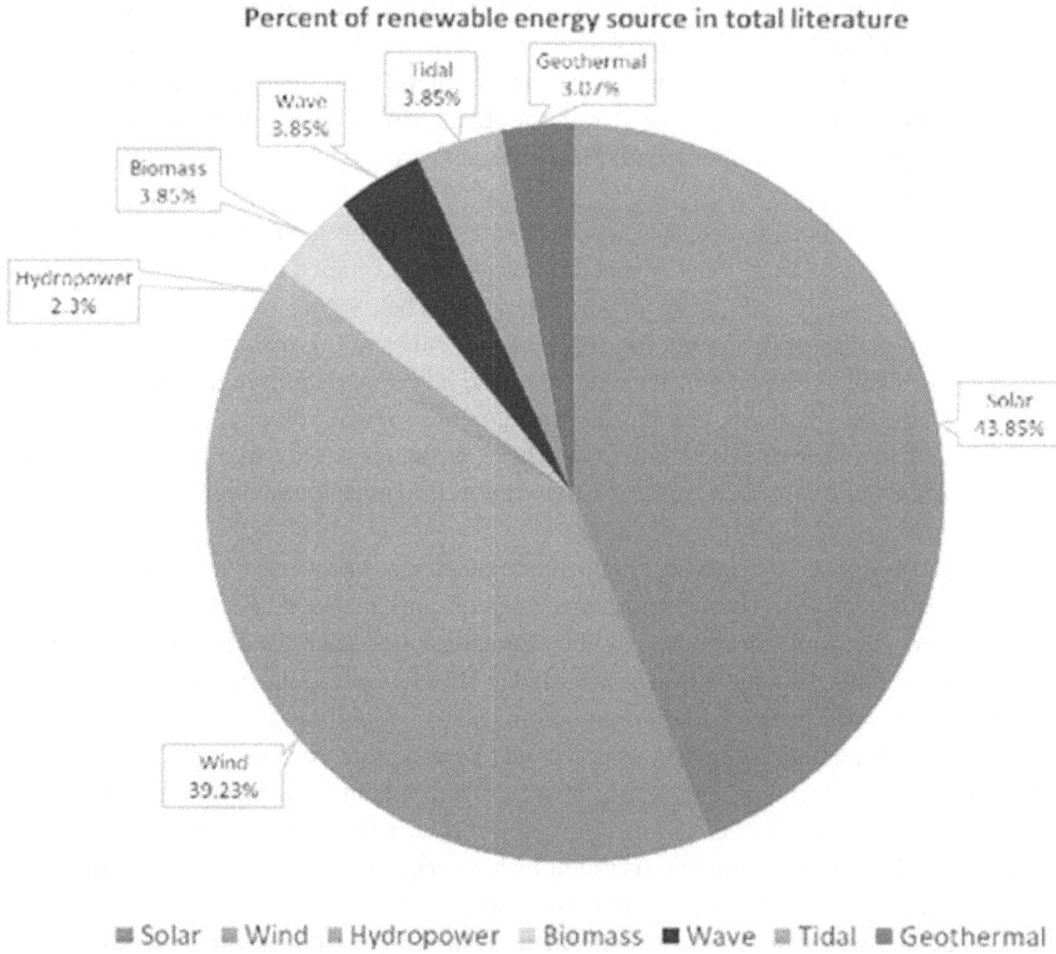

SUPERVISED LEARNING

In the realm of artificial intelligence, the branch known as machine learning is devoted to the development of techniques and models. These are able to learn information from the data and improve performance in the jobs you are performing. The process entails the creation of algorithms and statistical models that make it possible for computers to recognise trends in data and make use of such patterns for the purpose of making choices or forecasting. A computer is taught to recognise and respond to particular kinds of input through the process of machine learning. This is accomplished by supplying the computer with a large number of examples, which are referred to as "training data." Through this process of training, the computer is able to recognise patterns and create predictions or choices based on sources of information that are novel to it. The application of machine learning (ML) is widespread and can be found in a variety of fields, including health care, banking, online shopping, and a great number of other industries.

In addition, the utilisation of machine learning techniques can be utilised to forecast the generation of renewable energy, which ultimately leads to enhanced administration and optimisation of systems that produce electricity from renewable sources. There is a wide variety of machine learning algorithms available, each of which has its own set of advantages and disadvantages. There are three basic classifications that may be applied to algorithms, and these are unsupervised learning, supervised learning, and learning through reinforcement.

The process of training a model through the use of supervised training is a type of artificial intelligence methodology that requires the input of labelled data. The dataset is made up of intake-output pairs, where the input represents the data that was used for training the algorithm, and the output represents the conclusion that was anticipated. The model acquires the ability to construct a correspondence between the inputs and the results by lowering the disparity in the anticipated and observed outputs during the training phase. This is accomplished by reducing the amount of difference between the two. It is possible to use the model for making predictions on new data that has not been labelled once it has been previously trained. In the realm of supervised learning algorithms, the two basic sub-types that are typically encountered are regression and classification.

The categorization of automated learning is illustrated in Figure 2, which can be found here. In the context of supervised learning, regression is a type of learning that makes use of a number of input variables in order to establish predictions for an ongoing output variable. Regression analysis seeks to determine a function of mathematics that may establish an association between the variables that are input and an output that is a constant variable. This constant output variable may represent either a single value or a range of possibilities. The goal of regression is to determine any such function. There are three primary categories of supervised learning algorithms that are utilised in regression analysis.

A straightforward method that is frequently used for forecasting a dependent variable that is continuous by making use of a number of independent variables is the linear regression technique. This technique is widely used since it is straightforward. The study makes use of a model of linear regression in order to highlight the relationship that exists between the variables that are independent and the variables that are dependent on them. Polynomial regression is a type of linear regression that makes use of polynomial functions of the nth degree to illustrate the relationship that exists between the variable in question and the variable that is being studied (the dependent variable). The ability of a model to capture more complicated linkages among the input data and the variable of interest is one way in which the precision of predictions can be improved. For the purpose of predicting the amount of electricity that

Figure 2. Classification of machine learning

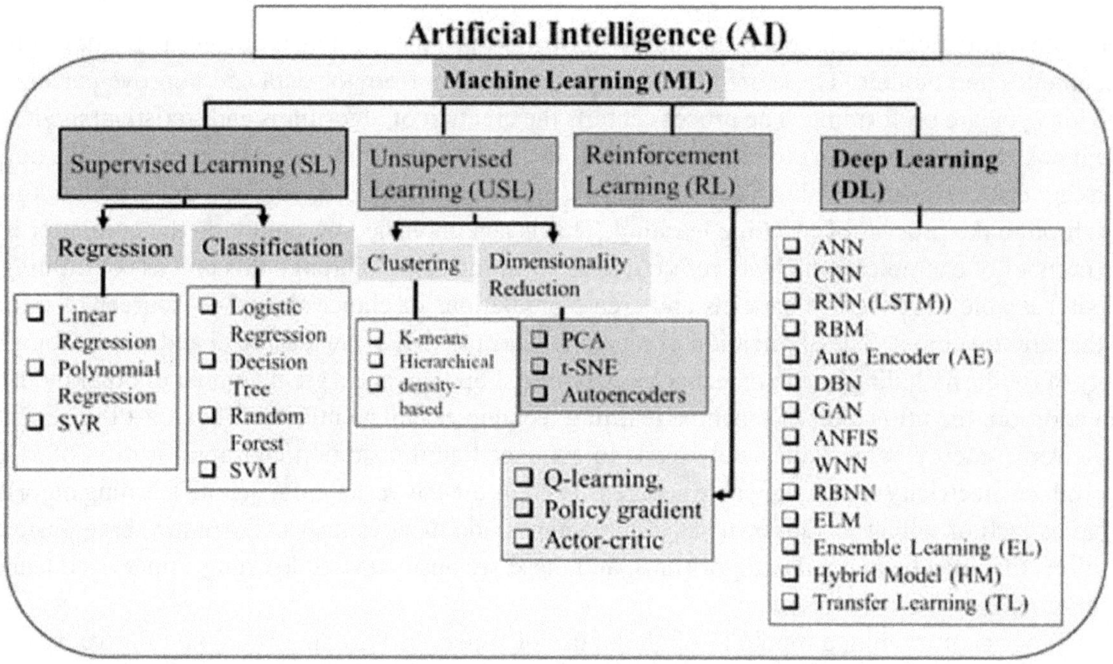

will be generated by sources of clean energy like solar and wind power, linear and quadratic regression techniques are frequently utilised in the field of renewable energy forecasting. There is a common practice of including weather data in the input characteristics. This data may include humidity, wind speed, temperature, and historic power output statistics. The electrical output of the source of clean energy is regarded as the variable that is dependent, and it is possible to make predictions about it by making use of the features of the independent inputs.

When it comes to supervised learning, classification is a type of learning in which one or more input variables are utilised in order to produce predictions regarding a category outcome variable. The purpose of categorization is to identify an equation that is capable of successfully assigning variables that are input to distinct groups that are output. Regression-based random forests, decision trees, and support vector machines are the classification algorithms that are utilised the most commonly for the purpose of predicting renewable energy sources (RES).

The statistical technique known as logistic regression is applied for the aim of classification. In this procedure, one or more input factors are utilised in order predict an outcome that is a binary variable. Within the framework of the sigmoid function, an estimation of the probability that the output variable is either true or false is conducted. According to the findings of the study, the accuracy of forecasting is dependent on the careful selection of appropriate solar factors through careful consideration. In addition to this, it assessed the accuracy of the algorithm as well as the probability that a power plant will function and produce electricity on a specific day in the future.

The research that was carried out by Hassan and colleagues (2024) utilised several methods of data analysis in order to investigate the influence that renewable energy sources have had on the evolution of the worldwide energy system. 2024 on Yi We investigated the impact that the growth of the internet-based economy has had on the growth of renewable energy sources. Moosavian et al. (2024) state that in order to develop an environmentally friendly energy system on an island that is completely self-sufficient, it is necessary to engage in strategic planning and make efficient use of green energy resources. The research was carried out by Ren et al. (2024). Through the application of machine learning, monitoring the health of clean energy installations is being done. Machine learning, on the other hand, is having a big impact on the current era of automated manufacturing for renewable energy materials, according to Hemavathi et al. (2024).

It was Yu et al. (2024). There is a clear indication of the application for machine learning as well as artificial intelligence in the technology, views, and issues that are associated with dispersed renewable energy sources. Specifically, the year 2023, as recorded by Ajibade and others. The purpose of this study is to conduct bibliometric analysis and investigate the patterns of publishing that have been associated with research concerning the use of artificial intelligence in green energy from the years 2012 to 2021. (Klaiber et al., 2023) is the citation. A comprehensive analysis of deep learning strategies is presented in this article, with a focus on their application to different types of renewable energy. A guide to artificial intelligence methodologies is provided by the work that was carried out by Tripathi and colleagues (2021). This is also the case with Rangel-Martinez et al. (2024). The purpose of this article is to provide an in-depth examination and viewpoint on renewable energy sources, catalysts, smart electricity networks, and energy storage. The article focuses on computer learning and environmentally friendly electricity generation. A study that was carried out by Singh and colleagues (2021) The purpose of this essay is to investigate the many issues and challenges that are associated with the implementation of artificial intelligence within the framework of the expanding applications that are seen in the renewable energy sector. The intention of the writers of the research paper titled “An Introduction to AI Approaches for Renewable Energies Networks” (Boza et al., 2021) is to investigate the application of machine learning as a means to facilitate the incorporation of renewable energy sources that are variable into the existing transmission system.

A decision tree is an extra categorization approach that separates the input area into smaller sections based on the values of the variables that are being entered. After that, an identifier or value is assigned to each of these areas specifically. The goal of the decision tree models that have been constructed for the reason of predicting the power produced by different forms of renewable energy has been the subject of multiple research articles. A prediction model has been developed in order to estimate the amount of electricity that can be generated by a solar power plant (PV system). The objective of the researchers is to identify the model that performs the best among artificial neural networks (ANN), randomly generated forests (RF), decision forests (DT), extreme gradient boost (XGB), as well as long short-term memory (LSTM) systems. This will help them fill a gap in the existing body of empirical research. The outcomes of the research reveal that, despite the fact that all approaches are effective, the ANN (Artificial Neural Network) emerges as the most accurate way for estimating the generation of photovoltaic (PV) solar electricity.

Through the creation of a large number of decision tree structures and the subsequent combination of those trees, the algorithm known as random forest, which is a machine learning supervised technique, is able to generate predictions that are accurate and trustworthy. The random forests algorithm uses the bagging approach in order to reduce the unpredictability of the algorithms that are used as the basis for the algorithm. Significant value can be derived from this strategy when it comes to the forecasting of time-series data. employing randomization in two unique ways—sampling from the initial data set and choosing from a selected group of characteristics—the random forest methodology reduces the reliance that exists between trees. This is accomplished by employing randomization. By generating a unique and thorough binary tree for each of the N trees, the Random Forest approach makes it possible to perform simultaneous processing.

UNSUPERVISED LEARNING

The primary objective is to identify and analyse patterns, structures, or correlations present in the dataset. There are two main categories of unsupervised learning methods: clustering and dimensionality reduction. This particular unsupervised learning methodology entails the categorization of similar data points according to their proximity or resemblance. Unsupervised learning methodologies are highly advantageous in scenarios involving vast amounts of disorganised data necessitating investigation, particularly in cases when the precise target variable is ambiguous. The utilisation of unsupervised learning has been extensively employed in the field of renewable energy forecasting, particularly in the clustering of meteorological data.

Within the field of renewable energy forecasting, unsupervised learning has been employed to choose relevant characteristics from a more extensive collection of input variables. The usage of choosing features in the process of predicting renewable energy acts as a strategy to minimise the computational difficulty of models based on machine learning and to increase the precision of predictions pertaining to the output of renewable energy. Within the area of renewable energy forecasting, unsupervised learning has been shown to be significantly effective in the evaluation of large amounts of disorganised data to a significant degree. This topic covers a wide range of applications of autonomous learning, including grouping, identifying anomalies, and feature selection, among others. An overwhelming majority of the efforts that are now being made are focused on the development of innovative algorithms that are intended to handle the specific issues that are inherent in the forecast of renewable energy.

ALGORITHMS FOR REINFORCEMENT LEARNING

Reinforcement learning (RL) is a domain within the science of ML that focuses on the agent's capacity to make optimal decisions within a given environment. The Q-values serve as indicators of the expected future rewards associated with each activity across all states. The algorithm known as the policy gradient is a reinforcement learning methodology that facilitates the direct acquisition of policies without the requirement of calculating Q-values. A reinforcement learning strategy known as the actor-critic algorithm combines elements from value-based and policy-based approaches. This is achieved by employing an actor-network for action generation and a critic network for assessing the value attributed to these acts.

Reinforcement learning (RL) has been utilised in several applications, such as the prediction of renewable energy. A subfield of Machine Learning (ML), Deep Learning (DL) is a subfield that leverages ANN with several layers to successfully capture sophisticated data representations by way of incremental degrees of abstraction. Deep Learning (DL) is a subfield of ML. When referring to ANN, the term "deep" refers to the considerably large number of layers that are included in these networks. These layers can range from just a few levels to a few thousands of layers total. For deep learning algorithms, the act of "training" is what enables them to develop the ability to detect correlations and patterns within data. The synaptic weights that link neurons inside an ANN are changed during the training phase in order to reduce the amount of difference that exists between the output that was anticipated and the output that was actually observed. As a result of the use of deeper learning strategies, significant advancements have been made in a variety of fields, such as power systems, artificial intelligence, natural language processing, recognition of speech, and autonomous systems, among others. The application of this technique has made it possible to make significant progress in a variety of domains, including the processing of natural languages, gaming, recognition of speech, and image recognition, among others.

OBSTACLES AND POTENTIAL OUTLOOK

Forecasting the generation of energy from renewable sources (RES), such as wind and solar energy, can be accomplished through the use of a wide variety of machine learning, or ML, and DL techniques. Every methodology has a unique set of advantages and disadvantages that are specific to it. It is dependent on the particular application as well as the data that is available to determine which method of ML (machine learning) or deep learning (DL) is the most appropriate for predicting renewable energy sources (RES). When it comes to establishing a baseline model, linear regression is a reasonable choice; however, random forests, xgboost, and support vector machines are more suited for dealing with non-linear correlations and complex datasets. When it comes to the field of time-series forecasting, recurrent neural network (RNN) models, which are classified as deep learning (DL) models, provide a number of significant advantages. However, in comparison to traditional machine learning (ML) models, the training process for these models may need a significant amount of computer resources. It is a big problem to ensure a dependable and constant supply of energy that can fulfil the necessary demand, and the fluctuation of wind and solar energy presents a considerable obstacle in this regard. Intermittency is a term that refers to the fluctuation of energy generation that is caused by environmental factors such as weather patterns and diurnal changes. It may be necessary for traditional forecasting methods, which are based on historic data and assumptions, to take into account the complex and ever-changing nature of sources of clean energy in an adequate manner. Using sophisticated approaches that acquire information from real-time data and adapt to shifting situations can provide a promising response to this challenge. These techniques can be used to address this issue. Especially in circumstances that are marked by uncertainty and unpredictability, the models that were stated above are able to efficiently depict the intricate correlations that exist between environmental variables and the generation of energy. This makes it possible to make correct forecasts. It is possible to use typical machine learning methods for the goal of forecasting; however, the effectiveness of these approaches may differ when compared to the effectiveness of specific time-series prediction methodologies such as autoregressive models, averages that move, and recurrent neural networks (RNNs). This problem is mostly caused by the prevalent hypothesis in traditional machine learning models, which is that data points are distinct and evenly distributed. However, this assumption frequently turns out to be not applicable to time series data, which constitutes the primary contributor to this phenomena. There is a strong indication that the values that are observed at a particular time point are affected by the values that were observed at earlier time points, as indicated by the temporal dependencies that are detected in time series data. As a result of the assumption that input variables are independent of one another, traditional algorithms for machine learning have challenges when attempting to comprehend the trends and patterns that are present within the data. Moreover, it is of the utmost importance that traditional machine learning approaches be particularly adjusted in order to effectively address the obstacles that are provided by time-varying features or irregularly spaced data, both of which are frequently encountered in the field of forecasting time series.

For the purpose of effectively managing time series data with irregular intervals and accurately detecting patterns and trends in order to generate precise predictions, it is recommended that specialised time-series forecasting approaches be utilised. The use of hybrid models is becoming more and more accepted in the process of forecasting renewable energy sources (RES). The combination of traditional analysis of time-series with ML approaches in these models greatly enhances accuracy and solves the issues associated with over fitting or under fitting. Recurrent neural networks (RNNs) are primarily intended for the processing of time-series data. They are able to accurately capture temporal correlations, which allows them to learn from over time trends and patterns. As a consequence of this, these models are typically chosen throughout the forecasting process for renewable energy sources. The models that have been discussed above have the potential to improve the operational effectiveness and growth of renewable energy sources, in addition to facilitating their smooth integration into the electrical grid. On the other hand, it is still challenging to accurately forecast RES due to the fact that these sources are unpredictable and it is impossible to forecast them with precision. This demonstrates the importance of conducting continual study and developing more advanced methods of forecasting.

The results of this study provide evidence that ML have a significant promise for accurately anticipating renewable energy sources. Through the analysis of vast volumes of data and the identification of intricate patterns that humans could miss, these algorithms have the potential to enhance the precision of predictions. In addition, these models have real-time forecasting capabilities, which allow them to adjust to different weather conditions. As a result, they enhance the reliability of the power system and make it easier to make decisions that are based on accurate information. Accurate forecasts have the potential to contribute to improved resource allocation, which in turn can improve operational efficiency and save costs. In addition, improved projections can be of assistance in the process of incorporating sources of clean energy (RES) into the electrical grid, which can help to reduce instability and enhance the grid's overall efficiency. Using ML and DL systems has the potential to dramatically increase the efficiency of renewable energy activities and the stability of electrical grids. This might be accomplished through the thorough evaluation of data quality, the complexity of the model, and the validation procedure.

CONCLUSION

In the realm of renewable energy forecasting, hybrid models have been utilised. These models employ a combination of traditional analysis of time series with latest techniques. These models have the capacity to comprehend and illustrate the two types of linear and non-linear correlations that are present in the data. As a result, they are able to generate accurate predictions even when there is noise and outliers present. A thorough knowledge of time-series analysis as well as machine learning and deep learning techniques is required in order to successfully complete the process of constructing hybrid models, which is characterised by its many facets. In spite of the fact that techniques have achieved significant advancements in the field of green energy forecasting, there are still a few problems that require additional attention and solutions. Due to the fact that high-quality data is required for validation as well as training purposes, this presents one of the most significant problems. The availability of data for predicting renewable energy is frequently limited, characterised by substantial levels of random fluctuation, and requires more extensive information. As a result, the creation of reliable models is made more difficult by these factors. There is a barrier presented by the restricted openness and interpretability of machine learning and deep learning models. There is a significant lack of transparency in a number of models, which makes it difficult to completely comprehend the fundamental concepts that are responsible for the predictive capabilities of these models. When it comes to the prediction of renewable energy utilising machine learning (ML) and deep learning (DL) techniques, there are a lot of prospects for future research. It is necessary to conduct additional study in order to improve the creation of models that are capable of properly managing multiple sources of renewable energy simultaneously. The creation of models that are capable of efficiently managing the inherent uncertainty and variance that comes with sources of clean energy is another area of concentration that ought to be given priority. It would be helpful to incorporate weather information, network data, and other outside sources of information in order to increase the accuracy of forecasts for renewable energy.

REFERENCES

Abedini, F., Bahaghighat, M., & S'hoyan, M. (2019). Wind turbine tower detection using feature descriptors and deep learning. Facta Universitatis, Series. *Electronics and Energetics*, *33*(1), 133–153.

Ajibade, S. S. M., Bekun, F. V., Adedoyin, F. F., Gyamfi, B. A., & Adediran, A. O. (2023). Machine learning applications in renewable energy (MLARE) research: A publication trend and bibliometric analysis study (2012–2021). *Cleanroom Technology*, *5*(2), 497–517. doi:10.3390/cleantechnol5020026

Anderson, R. N., Boulanger, A., Powell, W. B., & Scott, W. (2011). Adaptive stochastic control for the smart grid. *Proceedings of the IEEE*, *99*(6), 1098–1115. doi:10.1109/JPROC.2011.2109671

Babaie, M., Shiri, M. E., & Bahaghighat, M. (2018, April). A new descriptor for UAV images mapping by applying discrete local radon. In 2018 8th conference of AI & robotics and 10th RoboCup Iranopen international symposium (IRANOPEN) (pp. 52-56). IEEE. 10.1109/RIOS.2018.8406631

Bahaghighat, M., & Motamedi, S. A. (2018). Vision inspection and monitoring of wind turbine farms in emerging smart grids. Facta Universitatis, Series. *Electronics and Energetics*, *31*(2), 287–301.

Boza, P., & Evgeniou, T. (2021). Artificial intelligence to support the integration of variable renewable energy sources to the power system. *Applied Energy*, *290*, 116754. doi:10.1016/j.apenergy.2021.116754

Dash, P. K., Samantaray, S. R., & Panda, G. (2006). Fault classification and section identification of an advanced series-compensated transmission line using support vector machine. *IEEE Transactions on Power Delivery*, *22*(1), 67–73. doi:10.1109/TPWRD.2006.876695

Erol-Kantarci, M., & Mouftah, H. T. (2011). Wireless multimedia sensor and actor networks for the next generation power grid. *Ad Hoc Networks*, *9*(4), 542–551. doi:10.1016/j.adhoc.2010.08.005

Fadel, E., Gungor, V. C., Nassef, L., Akkari, N., Malik, M. A., Almasri, S., & Akyildiz, I. F. (2015). A survey on wireless sensor networks for smart grid. *Computer Communications*, *71*, 22–33. doi:10.1016/j.comcom.2015.09.006

Han, T., Liu, C., Yang, W., & Jiang, D. (2019). A novel adversarial learning framework in deep convolutional neural network for intelligent diagnosis of mechanical faults. *Knowledge-Based Systems*, *165*, 474–487. doi:10.1016/j.knosys.2018.12.019

Hassan, Q., Viktor, P., Al-Musawi, T. J., Ali, B. M., Algburi, S., Alzoubi, H. M., & Jaszczur, M. (2024). The renewable energy role in the global energy Transformations. *Renewable Energy Focus*, *48*, 100545. doi:10.1016/j.ref.2024.100545

Hemavathi, B., Vidya, G., Anantharaju, K. S., & Pai, R. K. (2024). Machine Learning in the Era of Smart Automation for Renewable Energy Materials. e-Prime-Advances in Electrical Engineering, Electronics and Energy, 100458.

Igba, J., Alemzadeh, K., Durugbo, C., & Henningsen, K. (2015). Performance assessment of wind turbine gearboxes using in-service data: Current approaches and future trends. *Renewable & Sustainable Energy Reviews*, *50*, 144–159. doi:10.1016/j.rser.2015.04.139

Jiménez, A. A., Márquez, F. P. G., Moraleda, V. B., & Muñoz, C. Q. G. (2019). Linear and nonlinear features and machine learning for wind turbine blade ice detection and diagnosis. *Renewable Energy*, *132*, 1034–1048. doi:10.1016/j.renene.2018.08.050

Klaiber, J., & Van Dinther, C. (2023). Deep learning for variable renewable energy: A systematic review. *ACM Computing Surveys*, *56*(1), 1–37. doi:10.1145/3586006

Li, S., Wunsch, D. C., O'Hair, E. A., & Giesselmann, M. G. (2001). Using neural networks to estimate wind turbine power generation. *IEEE Transactions on Energy Conversion*, *16*(3), 276–282. doi:10.1109/60.937208

Mert, I., Karakuş, C., & Üneş, F. (2016). Estimating the energy production of the wind turbine using artificial neural network. *Neural Computing & Applications*, *27*(5), 1231–1244. doi:10.1007/s00521-015-1921-0

Moosavian, S. F., Noorollahi, Y., & Shoaei, M. (2024). Renewable energy resources utilization planning for sustainable energy system development on a stand-alone island. *Journal of Cleaner Production*, *439*, 140892. doi:10.1016/j.jclepro.2024.140892

Nie, M., & Wang, L. (2013). Review of condition monitoring and fault diagnosis technologies for wind turbine gearbox. *Procedia CIRP*, *11*, 287–290. doi:10.1016/j.procir.2013.07.018

Pérez, J. M. P., Márquez, F. P. G., Tobias, A., & Papaelias, M. (2013). Wind turbine reliability analysis. *Renewable & Sustainable Energy Reviews*, *23*, 463–472. doi:10.1016/j.rser.2013.03.018

Rangel-Martinez, D., Nigam, K. D. P., & Ricardez-Sandoval, L. A. (2021). Machine learning on sustainable energy: A review and outlook on renewable energy systems, catalysis, smart grid and energy storage. *Chemical Engineering Research & Design*, *174*, 414–441. doi:10.1016/j.cherd.2021.08.013

Ren, B., Chi, Y., Zhou, N., Wang, Q., Wang, T., Luo, Y., Ye, J., & Zhu, X. (2024). Machine learning applications in health monitoring of renewable energy systems. *Renewable & Sustainable Energy Reviews*, *189*, 114039. doi:10.1016/j.rser.2023.114039

Singh, O., Yadav, M., Yadav, P., & Rishiwal, V. (2021). AI-Based Renewable Energy with Emerging Applications: Issues and Challenges. In Introduction to AI Techniques for Renewable Energy System (pp. 29-42). CRC Press.

Stetco, A., Dinmohammadi, F., Zhao, X., Robu, V., Flynn, D., Barnes, M., Keane, J., & Nenadic, G. (2019). Machine learning methods for wind turbine condition monitoring: A review. *Renewable Energy*, *133*, 620–635. doi:10.1016/j.renene.2018.10.047

Tasnim, S., Rahman, A., Oo, A. M. T., & Haque, M. E. (2018). Wind power prediction in new stations based on knowledge of existing Stations: A cluster based multi source domain adaptation approach. *Knowledge-Based Systems*, *145*, 15–24. doi:10.1016/j.knosys.2017.12.036

Tchakoua, P., Wamkeue, R., Ouhrouche, M., Slaoui-Hasnaoui, F., Tameghe, T. A., & Ekemb, G. (2014). Wind turbine condition monitoring: State-of-the-art review, new trends, and future challenges. *Energies*, *7*(4), 2595–2630. doi:10.3390/en7042595

Tripathi, S. L., Dubey, M. K., Rishiwal, V., & Padmanaban, S. (Eds.). (2021). *Introduction to AI Techniques for Renewable Energy System*. CRC Press. doi:10.1201/9781003104445

Yi, J., Dai, S., Li, L., & Cheng, J. (2024). How does digital economy development affect renewable energy innovation? *Renewable & Sustainable Energy Reviews*, *192*, 114221. doi:10.1016/j.rser.2023.114221

Yu, X., & Zhou, Y. (2024). Machine learning and artificial intelligence-distributed renewable energy sources: technologies, perspectives, and challenges. Advances in Digitalization and Machine Learning for Integrated Building-Transportation Energy Systems, 17-30.

Chapter 3
A Deep Learning Framework for Modeling Temporal Dependencies and Hierarchies in Hourly Electricity Demand Load

Claris Shoko
https://orcid.org/0000-0002-0896-4387
University of Botswana, Botswana

Ntebogang Moroke
https://orcid.org/0000-0001-8545-1860
North West University, South Africa

Katleho Makatjane
https://orcid.org/0000-0002-3687-4098
University of Botswana, Botswana

ABSTRACT

The limitations of traditional deep learning models in processing vast volumes of data and modelling complicated temporal dependencies make it difficult to effectively satisfy these objectives for short-term load forecasting (STLF). This chapter utilises deep learning, which enables the following: k-means clustering to comprehend hourly electricity demand load trend, extraction of complex features with non-linear interactions that impact electricity demand load, handling of long-term dependencies through the modelling of temporal hierarchies in the time series data via long short-term memory, and recurrent neural network to capture dependencies across time steps. The performance of the state-of-the-art hourly demand forecast models with the k-means variant is compared with that of the RNN-LSTM-CNN-copula. Furthermore, it is noted that RNN-LSTM-CNN-copula is a viable deep learning model for energy consumption forecast problems because of its capacity to learn the spatio-temporal dependencies in the hourly electrical demand load data, which results in an accurate energy demand forecast.

DOI: 10.4018/979-8-3693-2355-7.ch003

INTRODUCTION

Buildings, industry, and transportation are the three primary economic sectors that together account for the majority of a country's energy use. Buildings, being the greatest energy consumer, contribute for 36% of all carbon emissions and 40% of global yearly energy consumption, primarily in urban areas, according to data from the World Watch Institute (Aldhshan et al., 2021). Building energy consumption is expected to grow at a pace of 1.5% annual growth in OECD countries and 2.1% annual growth in non-OECD countries between 2012 and 2020. In particular, compared to residential structures, the overall energy consumption of commercial and institutional buildings in higher education is 30% and 45% greater, respectively.

Lower operating expenses are achieved by reducing energy use and avoiding utility penalties through the installation of suitable energy management systems and end-user energy efficiency control schemes. According to Kamal-Chaoui and Robert (2009), numerous countries have increased the enforcement of energy rules and construction norms. This has resulted in the development of innovative, energy-efficient building designs that guarantee lower energy consumption, energy expenses, and environmental effect (carbon emissions). Larger investments in smart infrastructures that support energy efficiency are required since, according to the World Energy Council (2016), "electricity demand would double by 2060, while overall per capita energy demand would begin to fall." The combination of electric vehicles, decentralized generating and storage systems, energy management systems, and cutting-edge information and communication technology has made buildings more intelligent from a smart grid standpoint. Thus, it is now crucial to design and deploy smart technologies for buildings and power grids in order to meet the world's energy needs in a way that is both economically viable and efficient while reducing carbon emissions.

Buildings equipped with smart meters to track energy usage at precise intervals have recently made research on energy consumption forecasting in buildings more important. In order to understand energy consumption patterns for use in energy demand forecasting, demand response, heating, ventilation, and air conditioning (HVAC) optimization, fault diagnosis, and detection, large-scale, high-dimensional energy consumption data from smart meters collected at various granularities is helpful (Razavi et al., 2019). The utilities can plan resources and balance supply and demand with the help of accurate and trustworthy energy demand forecasts, which guarantees the security and stability of the electrical grid as well as the dependability of service delivery. However, Hamed et al. (2022), have declared that overestimation and underestimation of the energy demand lead to a severe impact on the economic and industrial developments. Precise energy demand modeling and prediction facilitates context-aware control methods, precise demand response tactics, electrical supply management, and effective energy management in smart buildings. In order to improve the energy efficiency of buildings for a sustainable economy through a conservation-minded society, reasonable use of available energy resources, and an effective national energy strategy, energy consumption forecast models have thus become an essential component of the Building Energy Management System (BEMS) (Mariano-Hernández et al., 2020). Nevertheless, Somu et al. (2021) avowed that the non-linear, dynamic, and complex nature of energy consumption data, along with the presence of trend, seasonal and irregular patterns, and dependence on various exogenous factors like climatic conditions, nature of the day, socio-economic factors, etc. presents accurate and reliable energy consumption forecast as an interesting research problem.

Three types of energy consumption forecast models may be found in the literature: (1) Engineering methods: these are challenging and time-consuming; Energy Plus and Ecotect are two examples. They rely on complicated building and environmental data and use physical and thermodynamic rules. (2) Statistical methods: Inaccurate and rigid in its relationship between energy use and pertinent variables like as occupancy and climatic data; Examples include regression models (linear regression) and time series (autoregressive models), and (3) artificial intelligence techniques: Learns the consumption patterns from the historical energy consumption data, i.e., discover the non-linear relationship between the input (historical data) and output (target consumption); Examples-artificial neural network, support vector regression, etc. Among these, artificial intelligence approaches have become 'active research hotspot' due to their efficiency and flexibility over engineering and statistical methods (Liu et al., 2020).

However, the recent COVID-19 outbreak has affected the patterns of the electric power system; as a result, the expected demand for a typical weekday now matches that of a typical weekend day, as most workers remain at home and work from there (Jogunola et al., 2022). As a result, the measured difference between the weekend's peak and lowest energy demand was 24 percent. This unforeseen incident has made demand forecasting more challenging, which has exacerbated the electrical grid's instability.

Bunn, and Farmer (1985) show that a 1% increase in load forecasting error can result in a GBP 10 M increase in annual operating costs Chitalia et al (2022). Furthermore, the surge of distributed energy resources, coupled with several constraints such as weather, occupant behaviors, electricity prices, and desired comfort levels, made it more difficult to accurately predict energy consumption to balance supply and demand (Jogunola et al., 2020). Furthermore, precise energy consumption forecast is a crucial prerequisite for energy service providers to supply residential and commercial facilities with a range of flexible and supplementary energy management services. As a result, there would be financial savings, more options for customers, and effective grid management for service providers. Additional advantages encompass a consistent power source, effective energy distribution, effective energy management, and control methodologies Jogunola et al (2021).

LITERATURE REVIEW

Statistical techniques such as linear regression, exponential smoothing, autoregressive integrated moving averages (ARIMA), and seasonal autoregressive integrated moving averages (SARIMA) are frequently utilized to predict past consumption data (Ahmadi et al., 2023). These models are based on linear connections between past data and predicted data. Yet, machine learning and deep learning models have surpassed statistical methods due to the intricate non-linear patterns present in electricity consumption data. Deep learning approaches are straightforward and quickly process calculations, demonstrating their effectiveness as models according to Singaravel et al. (2018). Deep learning models can handle long-term dependencies, in addition to non-linear relationships, making them well-suited for energy forecasting (Ensafi et al., 2022; Kohzadi et al., 1996).

Multilayer perceptions and deep feedforward networks are the cornerstones of deep learning. Consequently, the fundamental components of a deep learning model are layers of algorithms known as artificial neural networks (Wongrin et al., 2022). The CNN and RNN layers are two of the most crucial layers. CNNs have been widely employed recently for a variety of applications, including as image recognition (Sharma, Jain, and Mishra, 2018) and time series forecasting (Le et al., 2019, Nawar et al., 2023) among others. Deep learning has successfully modeled RNNs for sequence models, including the LSTM and the grated recurrent units (GRU) (Shiri et al., 2023). Hochreiter and Schmidhuter (1997) introduced the LSTM architecture as a solution to the declining gradient issue. In order to estimate electricity demand,

Cordeiro-Costas et al. (2023) evaluate the effectiveness of temporal convolution networks, random forests, support vector machines, extreme gradient boosting, multilayer perception, and LSTM. In comparison to the other artificial intelligence models, the study demonstrates that the LSTM has the best forecasting accuracy. Sequence modeling is best left to LSTMs, while declining frequency variation problems are best handled by CNNs. As a result, having two architectures helps the models work well together. According to Slater et al. (2022), hybrid models that incorporate many forecasting methodologies have demonstrated potential in enhancing the efficacy of distinct algorithms for precise forecasting.

The possibility of the empirical mode decomposition embedded with CNNs and LSTM (EMD-CNN-LSTM) for short-term residential load forecasting is investigated by Lotfipoor et al. (2024). Over four distinct forecasting horizons, the hybrid EMD-CNN-LSTM model's performance was evaluated. CNN and LSTM, which are single forecast models, perform worse than the hybrid model. The most popular hybrid deep learning models are CNN-RNN.

A study of machine learning techniques, particularly the CNN, for prediction was given by Rahman et al. in 2021. They investigate the accuracy with which various data-driven artificial neural network models can forecast renewable energy. They made use of the RNN, CNN, and LSTM among other neural network architectures. Their study's findings demonstrate the suitability of the RNN, CNN, and LSTM models for short-term forecasting of renewable energy

This chapter presents RNN-LSTM-CNN-Copulas (herein referenced recurrent neural network-long short-term memory-convolutional neural network) coupled with copulas. This is believed to be a deep learning framework that operates on the electricity consumption data recorded at predefined intervals to provide accurate electricity consumption forecasts. The RNN-LSTM-CNN -Copulas employs (i) k-means clustering to perform cluster analysis to understand hourly electricity demand load pattern/trend; (ii) Convolutional Neural Networks to extract complex features with non-linear interactions that affect hourly electricity demand load; and (iii) long short term memory neural networks to handle long-term dependencies through modeling temporal hierarchies in the time series data. Finally, the copula handles dependencies on hierarchies in time series data.

The efficiency and applicability of RNN-LSTM-CNN-Copula are demonstrated using real-time hourly electricity demand load data acquired from ESKOM, South Africa. The performance of RNN -LSTM-CNN-Copulas is compared with the k-means variant of the state-of-the-art hourly electricity demand load forecast models in terms of well-known quality metrics. Additionally, it is anticipated that CNN-LSTM-Copula will be a good deep learning model for hourly electricity demand load forecast problems because of its capacity to learn the spatiotemporal dependencies in the energy consumption data, which results in an accurate energy demand forecast.

The ESKOM company will benefit from accurate forecasting by being able to recognize future crises and take appropriate action to avoid significant consequences. As a result, the suggested RNN-LSTM-CNN-Copula model can help with future planning, decision-making, and adopting preventative measures to deal with unfavorable events. A sustainable grid within the ESKOM smart grid will be attained by integrating the suggested state-of-the-art hybrid model with their forecasting toolkit.

The limitation of this study is that the proposed framework in this study was evaluated using data from ESKOM, South Africa. As a result, the findings from other countries may be different.

The rest of the chapter is presented as follows; section 2 presents the proposed deep learning models together with the evaluation metrics. Section 3 presents the results and discusses how the results compare with findings from previous studies. The last section concludes and gives recommendations for further studies.

METHODOLOGY

This section of the study presents the methods and procedures followed.

The proposed system is used to develop a combined forecasting model for hourly electricity demand load in South Africa on a real-time basis by analyzing incoming load demand and this designed system consists of two components for hourly load forecasting. These are described or discussed in the next two subsections.

Designing a Framework for Data Pre-Processing

This constituent is legally responsible for the usage of big data effectively and bids it to the analytical server for predictive modeling. The configuration of the system mainly consists of high-performance computing (HPC) under Saturn Cloud/Saturn-python. We utilize TensorFlow because it can be used in a wide variety of programming languages, including Python, JavaScript, C++, and Java. This flexibility lends itself to a range of applications in many different sectors. The fields in a raw data file are separated by a comma delimiter. The raw data file is given to the analytical model for the building of a data model. This adds to the system's remarkable versatility and helps it develop a powerful, real-time self-learning analytical model. It also makes the system considerably scalable.

This element is legally responsible for the effective utilization of massive data and offers the analytical server a bid for predictive modeling. The main platform for high-performance computing (HPC) in the system configuration is Saturn Cloud/Saturn-python. Because TensorFlow can be used with so many different programming languages, like Python, JavaScript, C++, and Java, it is used. Because of its versatility, it can be applied to a variety of tasks in a multitude of sectors. In a raw data file, fields are separated with commas. The raw data file is sent to the analytical model so that it can build data model.

Proposed RNN-LSTM-CNN

This subsection of the chapter presents our proposed deep-learning algorithm.

Recurrent Neural Networks

An artificial neural network type called a recurrent neural network (RNN) is made to handle sequential data, such as time series or natural language. They can detect temporal dependencies because they have feedback links that help them remember data from earlier time steps. We use RNN to help us find what we cannot observe as the correlation between time series data. Our work implements the RNN stacked architecture along with long short-term memory with Peephole connection units. The stacked architecture used in this study closely relates to the architecture mentioned in the work of Bandara et al. (2020). Figure 1 shows the folded version of the RNN. The idea is that the same RNN unit repeats for every time step, sharing the same weights and biases between each of them. The feedback loop of the cell helps the network to propagate at the state h_t to the future time steps. For generalization, only the state h_t is shown in the diagram. However, for an LSTM cell, h_t should be accompanied by the cell state C_t. The notion of stacking means that multiple LSTM layers can be stacked on top of one another. In the most basic setup, the model has only one LSTM layer.

The recursive formulas of RNN are given as follows

Figure 1. Proposed recurrent neural network
Source: Authors

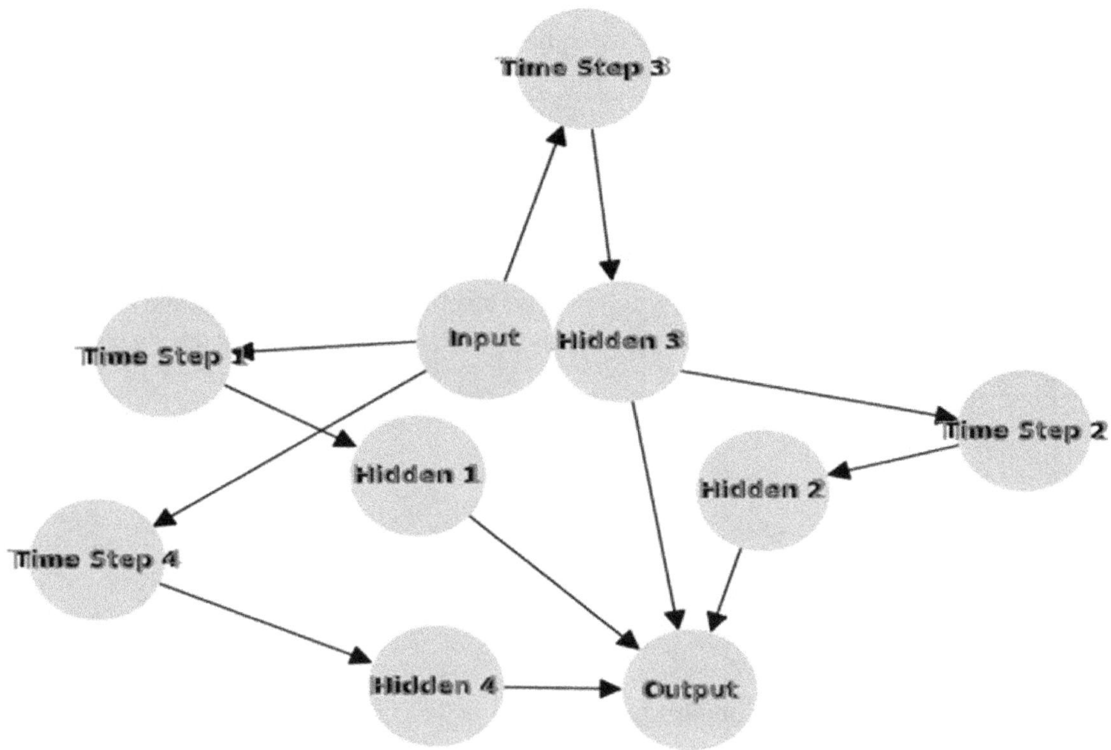

$$h_t = \tanh\left(W_h h_{t-1} + W_x x_t\right) \tag{1}$$

and

$$y_t = W_y h_t \tag{2}$$

where y_t is output vector, h_t is hidden layer vector, x_t is input vector, and W_h is weighting matrix. RNNs can theoretically solve these long dependency problems, but (the weight matrix) keeps getting multiplied by the previous results as the time interval gets longer (reference data is at longer time points). This causes problems with vanishing and exploding gradients. To solve this problem, we use long short-term memory (LSTM) networks.

Long Short-Term Memory

As seen in Figure 2, long short-term memory (LSTM) units are building blocks for layers of an RNN, as defined by Wikipedia. An LSTM network is the term used to describe an RNN made up of LSTM units. Since every neuron in an LSTM network is a memory cell, LSTM neural networks vary from conventional RNN neural networks. The LSTM connects the current neurons to the information from the prior data. The input gate, forget gate and output gate are the three gates found in every neuron. The LSTM can resolve the issue of long-term data reliance by using the input gate. Next, we outline the LSTM architecture's internal gates and explain how it addresses long-term dependency issues. There are three gates in this figure and they are discussed below

1) ***The Forget Gate***: In Figure 2, the forget gate which is defined in equation (3) determines which information will be discarded from the cell. By entering the output h_{t-1} at the previous unit (t–1) and adding the current time t and input X_t into a *Sigmoid* function $S(t)$ that is defined in equation (4) as shown in Figure 3. A value between 0 and 1 is generated and this value will be multiplied with cell state C_{t-1} to determine how much information will be forgotten or remembered. The 0 means completely forgotten, while 1 is completely remembered. Therefore, W and b are the weight matrices and bias vector parameters respectively, which need to be learned during training.

$$f_t = \sigma\left(W_f \cdot [h_{t-1}, x_t] + b_f\right) \quad (3)$$

$$S(t) = \frac{1}{1 + \exp - t} \quad (4)$$

Figure 2. Long short-term memory architecture
Source: Authors

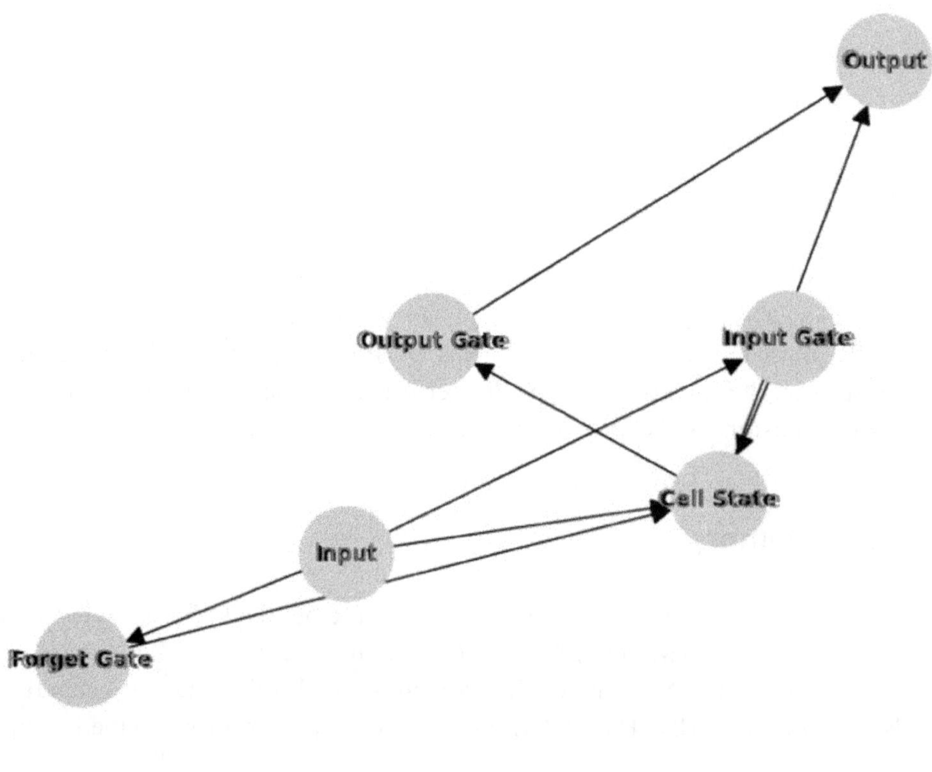

Figure 3. Tanh and sigmoid function

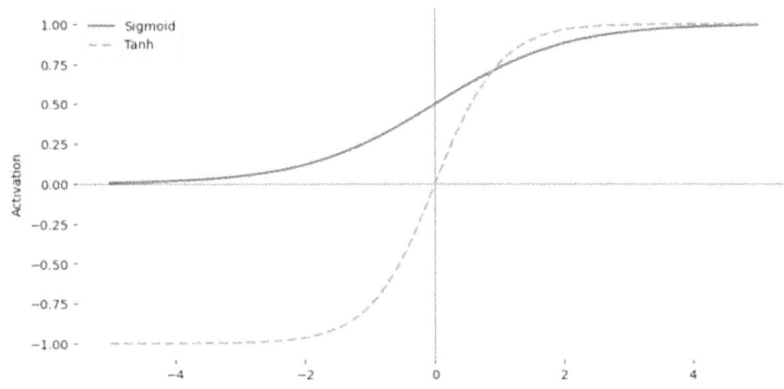

2) ***Input gate***: In Figure 2, the input gate determines which new information to remember in cell state. By entering the output h_{t-1} at the previous time t–1 and adding the current time t and an input X_t into a *Sigmois* function as defined in equation (4) a value i_t in equation (5) that is between 0 and 1 is generated to decide how much new information cell state need to be remembered. At the same time, a tanh in Figure 3 layer obtains an election message $\tilde{C}_t$ defined in equation (6) to be added to the cell state by inputting the output h_{t-1} at the previous time t–1 and adding the current time t and an input information X_t. Multiply the values i_t and $\tilde{C}_t$ just obtained by the *Sigmoid* function to get the updated information that we really want to add to the cell state C_t in equation (7).

$$i_t = \sigma\left(W_t \cdot [h_{t-1}, x_t] + b_i\right) \tag{5}$$

$$\tilde{C}_t = \tanh\left(W_c \cdot [h_{t-1}, x_t] + b_c\right) \tag{6}$$

$$C_t = f_t * C_{t-1} + i_t * \tilde{C}_t \tag{7}$$

3) ***Output gate:*** In Figure 2, the output gate determines which information will be output in the cell state. By entering the output h_{t-1} at the previous point in time t–1 and adding the current time t and input X_t into a *Sigmoid* function, a value i_t as shown in equation (8) which is between 0 and 1 is generated to determine how many cells state information that leads to output. The cell state is first activated in the tanh layer before being multiplied by o_t between 1 and -1. The result of multiplication is the output information h_t in equation (9) of the LSTM block at time t.

$$o_t = \sigma\left(W_0 \cdot [h_{t-1}, x_t] + b_0\right) \tag{8}$$

$$h_t = o_t * \tanh(C_t) \tag{9}$$

Convolutional Neural Networks

In the machine learning field, not only deep learning becomes more and more popular, but also convolutional neural network (CNN) has attracted attention recently. To feed *N*-dimensional data as input, and process with multi-layer of convolution. Each layer of convolution will receive the output of the previous layer of convolution as the input. CNN keeps pooling and convolution to extract useful features for each layer and finally obtain high-quality features, which can get a favorable ability of image classification.

A few research using CNNs for time series forecasting tasks have been conducted recently (Kranjc et al., 2015). These authors suggested a novel method that combines CNN with an autoregressive model and demonstrated that it performs better than both CNN and LSTM NNs. Four artificially generated datasets in addition to two real datasets (household electricity and money use) were used in the evaluation. The single CNN fared better than the LSTM on the electrical data. We observe that the electrical data utilized in Kranjc et al. (2015) includes the values of ten variables at each time step, but our data is univariate. A CNN known as Wave Net was used to analyze univariate and bivariate financial and artificial time series in the work of Hagenau et al. (2013). It is demonstrated to be a successful approach, yielding encouraging outcomes and surpassing LSTM and the other comparative methodologies.

Convolutional and max pooling layers make up the two primary types of layers found in convolutional neural networks. The two primary techniques used by convolutional layers are weight distribution and local connections. Local connections refer to the fact that, as opposed to all inputs as in a typical MLP NN, each convolutional node is linked to a small subset of inputs (corresponding to a local region of the inputs and called the node's receptive field). Furthermore, a convolution is used in place of the weighted sum of a conventional NN. In particular, each convolutional layer convolves the input with a weight matrix (also known as a filter or sliding window) of a preset size in order to compute a feature map. To accomplish this, move the weight matrix over the input and compute the input-weight point derivative.

Convolutional neurons in the same layer detect the same pattern in various regions of the input because they share the same weights. Training happens more quickly when weight sharing and local connectivity are used to minimize the amount of weights that must be learnt and stored. The number of nodes that share weights is determined by the filter's kernel size, which is used to implement the convolution and weight sharing. A max pooling layer, which calculates the maximum value of the chosen pool of nearby neurons from the convolutional layer, usually comes after a convolutional layer (Wang and Oates, 2015). The combination of a convolutional and max-pooling layer ensures that the output of the max-pooling layer is invariant to shifts in the input data, which is a very useful property for processing natural data. A fully connected output layer computes the final prediction.

In our experiment we used CNNs with Conv1D layer with a kernel size of 3, followed by a Flatten layer to transition from convolutional to dense layers. Then, two dense layers with ReLU activation and a final dense layer with linear activation. The training is done with the Adam optimizer, minimizing the mean squared error. To avoid overfitting, a dropout (Huang and Pan, 2015) is employed during the weight optimization at all layers. Dropout is a technique that includes randomly selecting a fraction of neurons at each training epoch and not including them in the weight optimization. We also applied batch normalization which normalizes the layer inputs for each training mini-batch. This technique has been shown to speed up the training and reduce the dependence of gradients on the scale of the parameters and also the sensitivity of CNN to the initial parameter values Huang and Pan (2014).

Parameters: To select an appropriate CNN architecture and parameters for each time series, we evaluated the performance of different architectures and parameter combinations on the validation set and selected the best one. For example, for the solar data, we considered CNN architectures with two convolutional layers with (20, 8) and (64, 32) nodes in the first and second layer respectively; with and without max pooling layers; with a batch size of 1, 4, 8 and 16, and a number of training epochs 10, 50 and 250.

Stacking Ensemble Learning Method

In recent years, overall models have become more important due to improved performance. It is derived from various tasks such as classification or regression problems. It consists of techniques that combine different learning models to improve the results of each model. Two steps are typically used: In the first step, a set of base learners is obtained from the training data, and in the second step, the learners from the first step are combined to create a single prediction model. Therefore, multiple predictions based on different base students are built and combined into a composite model that is more improved than the individual base models. Combining the best individual models into a single improved composite model usually results in higher accuracy. Figure 4 shows the schema of the Stacking-Ensemble algorithm used for predicting hourly electricity demand load in South Africa.

The most used and well-known basic ensemble methods are bagging, boosting and stacking. This paper uses a stacking approach since it is the most suitable for the base learners' problem considered in this work. In the following section, we will specify which learning algorithms have been used in the proposed scheme, as shown in Figure 3. Therefore, the stacking-ensemble approach to time series prediction is when heterogeneous base models are trained and tested on the dataset in parallel or stacking. The assumption is that prediction from the base learners is weak. In stacking, the algorithm takes the outputs of the base learners as input and attempts to learn the best way of combining the heterogeneous

Figure 4. Schema for the stacking-ensemble algorithm

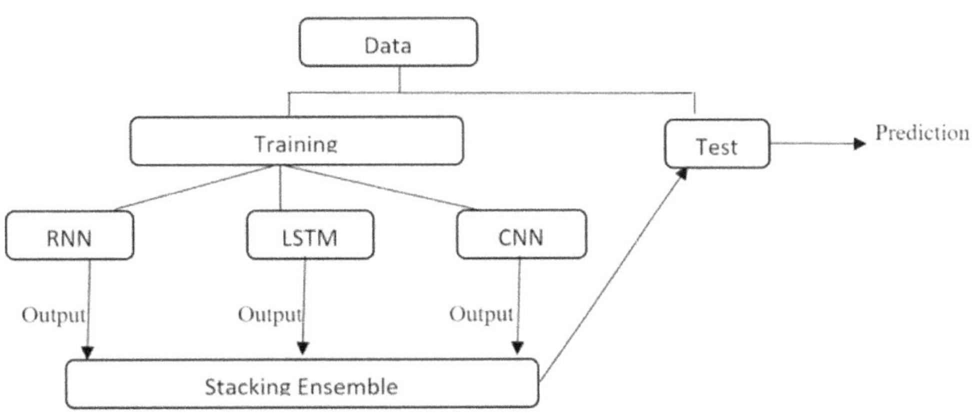

predictions to give a better output. In this chapter, we use the recurrent neural network, long short-term memory, and convolutional neural network models as the base learners; See Figure 4). The stacking-ensemble approach to time series forecasting is when heterogeneous base models are trained and tested on the dataset in parallel or stacking. The assumption is that prediction from the base models is weak.

In the following section, we will specify which learning algorithms have been used in the scheme we propose. We can define a stacking ensemble scheme more formally in the following way. Given a set of N different learning algorithms L_k, k=1,…,N N and the pair ⟨x,y⟩, with $x=(x_1,\ldots,x_w)$ representing the w recorded values and $y=(x_{w+1},\ldots,x_{w+h})$ the h values to predict. Let m_{kj}, k=1,…,N, j=1,…,h be the model induced by the learning algorithm L_k on x to predict x_{w+j}, and let f_j be the generalizer function responsible for combining the models for predicting such value. It is worth noting that f_j can be a generic function, such as the average, or a model induced by a learning algorithm. Then, the estimated $\hat{x}_{w+j}$ value is given by the expression

$$\hat{x}_{w+j} = f_j\left(m_{1j},\cdots,m_{Nj}\right). \tag{10}$$

Ensemble methods have been successfully applied to solving pattern classification, regression, and forecasting in time series problems (Jin and Dong, 2016). Adhikari (2015), for instance, presented a linear combination method for time series forecasting that uses a unique neural network topology to compute the combining weights. Bagnall et al. (2015) proposed a method using an ensemble of classifiers on different data transformations to improve the accuracy of time-series classification. The authors demonstrated that the simple combination of all classifiers in one ensemble obtained better performance than any of its components. To classify biomedical data, Jin and Dong (2016) developed an ensemble approach based on deep neural networks that incorporates filtering views, local views, distorted views, explicit and implicit training, sub-view prediction, and Simple Average. In particular, they used the Chinese Cardiovascular Disease cardiograms database.

In stacking, the algorithm takes the outputs of the base models as input and attempts to learn the best way of combining the heterogeneous predictions to give a better output. Predictions from these base models are combined using either the Linear Quantile regression Averaging (LQRA) or the Weighted Averaging methods. However, this chapter proposed a generalized additive model. A comparison of the performance of these models is done using the Key Performance Indicators KPIs). The KPIs are discussed at the end of this section. The meta-model combines predictions from these base models using either the stacking-ensemble approach. A comparison of the performance of these models is done using the Key Performance Indicators KPIs).

Selection of Error Metrics for Assessment

The selection of appropriate error metrics is crucial for evaluating the performance of forecasting models. Error metrics provide a quantitative measure of the accuracy of predictions and help assess how well the model is capturing the underlying patterns in the data. Here are some error metrics that are relevant to this topic

Mean Squared Error (MSE): MSE is a fundamental and extensively employed error metric in the realm of predictive modeling. It serves as a robust measure to quantify the overall quality of predictions by assessing the magnitude of the discrepancies between predicted values and actual observations. At its core, the MSE operates by calculating the average of the squared differences between the predicted values and the corresponding actual values. This squaring of differences magnifies larger errors, accentuating the impact of deviations between predictions and actual outcomes. By summing up these squared errors and dividing by the number of observations, the MSE provides a comprehensive perspective on the model's accuracy. Mathematically, the Mean Squared Error is given by

$$MSE = \frac{1}{n}\sum_{i=1}^{n}\left(y_i - \hat{y}_t\right)^2. \tag{11}$$

In model (5), n is the number of observations while y_i represents the actual observed value at time t and $\hat{y}_t$ represents the predicted value at time t.

Mean Absolute Error (MAE): MAE measures the average absolute difference between predicted and actual values. It is less sensitive to outliers compared to MSE and provides a more balanced view of prediction accuracy (Willmott and Matsuura, 2005). This test stands as a fundamental error metric that gauges the average absolute disparity between predicted values and actual observations. In contrast to Mean Squared Error (MSE), MAE's distinctive characteristic lies in its robustness against outliers and its ability to present a well-balanced evaluation of prediction accuracy. The equation of MAE is

$$MAE = \frac{1}{n}\sum_{i=1}^{n}\left|y_i - \hat{y}_t\right| \tag{12}$$

Symmetric Mean Absolute Percentage Error (SMAPE): SMAPE presents itself as an insightful error metric that builds upon the foundation of percentage-based evaluation while incorporating a symmetrical consideration of both overestimation and underestimation errors. This symmetry empowers SMAPE to offer a balanced and comprehensive appraisal of prediction accuracy. (Makridakis and Hibon, 2000). In mathematical form, this error metric is expressed as

$$SMAPE = \frac{100}{n}\sum_{i=1}^{n}\frac{\left|y_i - \hat{y}_t\right|}{\left(\left|y_i\right| + \left|\hat{y}_i\right|\right)/2} \tag{13}$$

SMAPE's symmetric nature ensures that no bias is introduced towards either overestimation or underestimation. This balanced assessment makes SMAPE an ideal choice when the objective is to comprehend prediction accuracy without favoring one direction of error over the other. It particularly suits scenarios where both overestimation and underestimation have equal consequences, such as demand forecasting or resource allocation.

Mean Forecast Error (MFE): MFE is the average difference between predicted and actual values. It provides a measure of the average error without considering the direction of errors (Tashman, 2000). Furthermore, MFE plays a pivotal role in assessing the performance of forecasting models. It quantifies the average difference between predicted values and their corresponding actual observations, shedding light on the average error magnitude, regardless of the direction of errors.

Unlike other error metrics that consider the direction of errors, such as overestimation or underestimation, MFE focuses solely on the magnitude of deviations. This characteristic makes MFE a direction-agnostic measure that offers insights into the general accuracy of predictions without discerning the tendencies of errors. The MFE statistic is expressed ab

$$MFE = \frac{1}{n}\sum_{i=1}^{n}\left(y_i - \hat{y}_t\right) \tag{14}$$

The interpretability of MFE is straightforward. A lower MFE value indicates that, on average, the predictions are closer to the actual values. Conversely, a higher MFE value implies a greater average discrepancy between predictions and observations. MFE is a valuable metric when the primary concern is to grasp the typical error magnitude without delving into the nuances of directional errors.

APPLICATION TO HOURLY ELECTRICITY DEMAND LOAD

The data used in this chapter is the hourly electricity demand load (Dispatchable Generation) by ESKOM for the period of 2019-04-01 to 2023-10-22. This makes a total of 39960 observations[1]. The data is split into two sets: training and testing sets. The stacking-ensemble algorithm coupled with Gumbel copulas is developed from the base learner models namely recurrent neural network (RNN), convolutional neural network (CNN) and long-short term memory (LSTM). Figure 5 presents a univariate data analysis for the variable of interest: the hourly electricity demand for South Africa. The normal histogram with kernel density (a nonparametric method used for estimating probability distributions), normal Q-Q, and box and whisker plot and time series are presented in this Figure.

The normal kernel density with a histogram plot shows that the hourly electricity demand data is not normally distributed. Normal Q-Q plots also show slight deviations from normality. Time series plot further show complex non-linear patterns.

Trend Analysis Results

Prior main analysis, the data underwent classical decomposition, and Figure 6 illustrates the resulting trend from decomposed hourly electricity demand data. The visual inspection of Figure 6 demonstrates that the extracted trend is decreasing for the sample period. Moreover, Mann-Kendall test statistic and Sen's slope estimator are used to analyze the long-term trends of hourly electricity demand load in South Africa. The Mann-Kendall test statistic and Sen's slope in Table 2 present these results. The outcome of the Mann-Kendall test results revealed that hourly electricity demand load has a significant monotonic decreasing long-term trends because it is negative. The Sen's slope value shows significant decreasing magnitudes of trends, which are corresponding with the Mann-Kendall test results.

We extract a nonlinear trend by fitting the cubic smoothing spline function with cubic smoothing spline constraints represented by the equation

$$\min_{g} p * \sum_{i=1}^{n}\left(y_i - g\left(x_i\right)\right)^2 + (1-p) * \int\left(g''\left(x_i\right)\right)^2 dx, p \in [0,1] \tag{15}$$

Figure 5. Hourly electricity demand

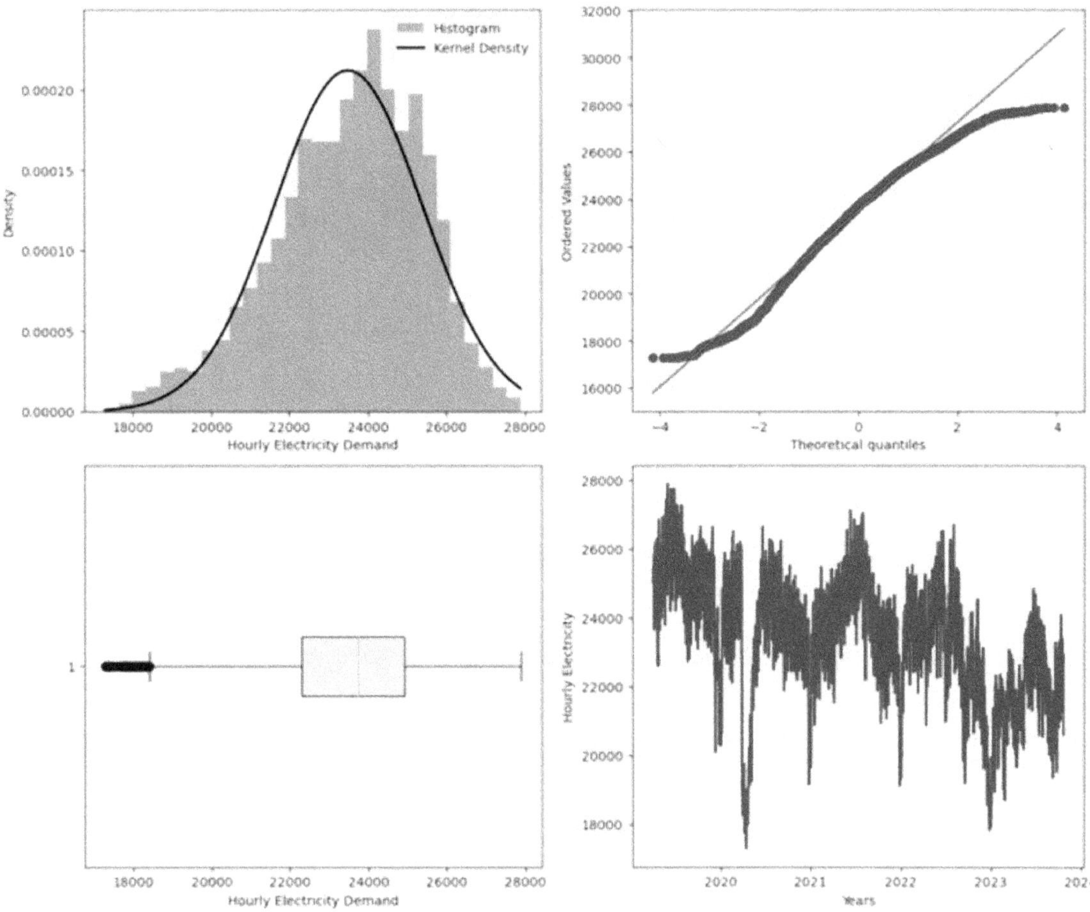

Table 1. Descriptive statistics

Min	25%	Mean	sd	50%	75%	Max	Kurtosis	Skewness
13797.94	21204.37	23513.52	2999	23519.93	25703.23	33065.91	-0.39198	0.057021

The nonlinear trend for the hourly contracted electricity demand time series data shows irregular cycles instead of regular seasonality. The peaks are sometimes smaller or larger. Thus, modelling this time series data requires the inclusion of the lag features to allow our forecasters to react dynamically to the changing conditions. This study includes lag variables as explanatory variables to allow varying amounts of the recent history into the forecast. The results of the extracted nonlinear trend are shown in Figure 7.

Figure 6. Trend for hourly electricity demand

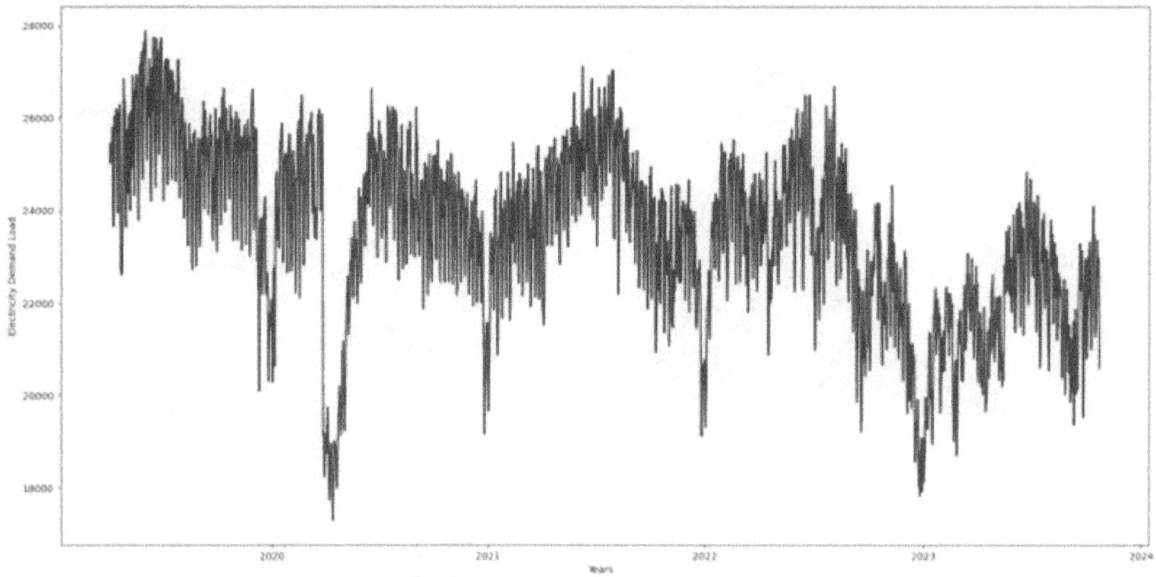

Table 2. Mann-Kendall test statistic and Sen's slope estimator

Variable	M-K Test Statistic	Kendall's Tau $\hat{\tau}$	p-Value	Sen's Slope
Electricity demand	-115.3355	-0.38466	0.001	-0.0882

Figure 7. Depersonalized hourly electricity demand superimposed with a nonlinear trend variable from a regression spline

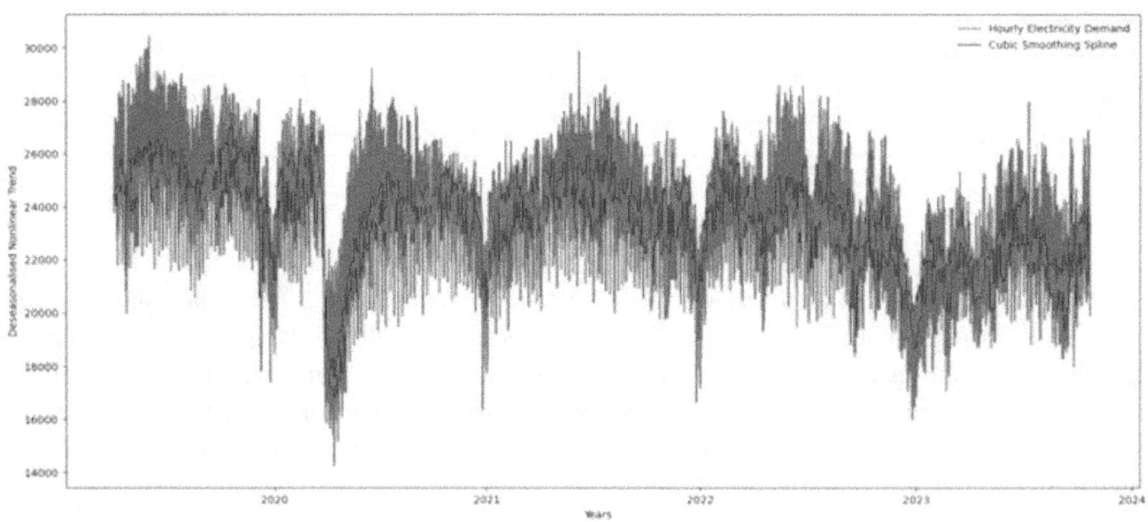

Prediction and Forecasting Performance of Hourly Electricity Demand Load

To train and test our base learners, we fixed the batch size to 32 and increased the number of epochs by a ratio of 0.2. Three criteria are used to perform a comparative analysis exercise which is done to find the model that best mimics the forecasting errors. These are namely mean absolute error (MAE), symmetric mean absolute percentage error (SMAPE), Mean Forecast Error (MFE), and mean square error for both in-sample and out-of-sample forecasting performance. The empirical results are shown in Table 3 from the perspectives of level forecasting and directional forecasting. Looking at the results reported in Table 3, the in-sample forecasts of the RNN give a small error when epochs are set at 50 this is because, out of the four metrics used, three have minimum values at this value of 50. However, in the out-of-sample forecasting, the model performed better when the epochs were increased to 250. The CNN has the minimum error when epochs are set at 250 in the in-sample forecasting but in out-of-sample forecasting, the model has minimum error when the epochs are at 10. The same analogy can be used for the LSTM. It is worth noting that our stacking ensemble did improve in our combined forecasting model. This indicates that as the number of epochs increases, both the base learners and meta-model improve the performance. Deep learning has become a prominent source in several sectors due to its superior performance, as far as we know. Thus, deep learning also produced positive results in this domain, as Table 3 illustrates. More recently, deep learning techniques have produced improved results, but in the process of modeling energy patterns, they are missing some crucial information, which indicates that model performance is not as good as it could be. We made this assumption throughout the literature review and developed an ensemble model to account for every feature, effortlessly model the energy pattern, and identify both spatial and spatiotemporal cues. We, therefore, suggest an ensemble deep learning technique and secure good results in comparison to other baseline methods because the performance of deep learning in the forecasting of time series issues is the inspiration for the construction of the ensemble framework. Figure 8 below shows the performance of the suggested system over hourly prediction usage.

In Figure 4 we present the plots of the training losses and validation losses for the RNN-LSTN-CNN model. These loss values give us an insight into how the learning performance changes over the number of epochs. The plots show that both the training loss and the validation loss decrease to a point of stability and there are small gaps between the plots. This shows that the RNN-LSTN-CNN model fits the observed data well.

After having our combined forecasting model, we use the residuals of this model and fitted a Gaussian copula. We use this copula because in our pre-processing of data for our base learners, we normalize the time series together with the hierarchies to best fit into the base learners. Figure 9 shows the visualization of our Gaussian copula. From this figure, it can be seen that only month and quarter that has high dependency with the week feature. There is a high positive dependency while half-year is negatively dependent of week feature. We also notice that quarter and month are positively dependent on each other. In addition, quarter and month are negatively dependent on and half-year feature. Table 4 presents the correlation matrix of the hierarchies using spearman rank correlation.

After we have found a few positive and negative dependencies in the established hierarchies, we now perform k-means and its two variants namely Fuzzy C-means clustering and Mini-Batch K-means clustering respectively. The choice of these two is based on their abilities. For instance, Mini-Batch K-Means is used to accelerate the convergence of the algorithm and reduce computational requirements, and it is suitable for large datasets such hourly electricity demand load used in this chapter. The algorithm uses a random subset (mini-batch) of the dataset for each iteration, updating the centroids based on this

Table 3. Forecasting performance

Epochs	Batch size	MSE	MAE	SMAPE	MFE	MSE	MAE	SMAPE	MFE
RNN-In-Sample Forecasting						**RNN-Out-of-Sample Forecasting**			
10	32	0.006	0.059	10.756	-0.004	0.012	0.086	15.185	0.001
50	32	0.003	0.044	7.826	0.004	0.009	0.072	12.892	-0.021
250	32	0.005	0.050	10.346	0.007	0.005	0.056	10.146	-0.011
CNN-In-Sample Forecasting						**CNN-Out-of-Sample Forecasting**			
10	32	0.0168	0.1024	1.0177	-0.0577	0.0074	0.0691	0.6879	0.0004
50	32	0.0120	0.0855	0.8503	-0.0484	0.0100	0.0824	0.8200	0.0501
250	32	0.0097	0.0821	0.8168	0.0496	0.0086	0.0757	0.7530	0.0299
LSTM-In-Sample Forecasting						**LSTM-Out-of-Sample Forecasting**			
10	32	0.005	0.055	0.540	-0.007	0.009	0.074	0.740	0.024
50	32	0.003	0.041	0.400	-0.003	0.008	0.069	0.690	-0.014
250	32	0.002	0.036	0.360	0.010	0.004	0.046	0.460	0.005
RNN-LSTM-CNN-In-Sample Forecasting						**RNN-LSTM-CNN-Out-of-Sample Forecasting**			
10	32	0.007	0.061	11.060	0.000	0.013	0.088	15.341	0.000
50	32	0.003	0.046	8.320	0.000	0.009	0.070	12.520	0.000
250	32	0.003	0.041	7.270	0.000	0.005	0.054	9.879	0.000

Figure 8. The model training and validation losses

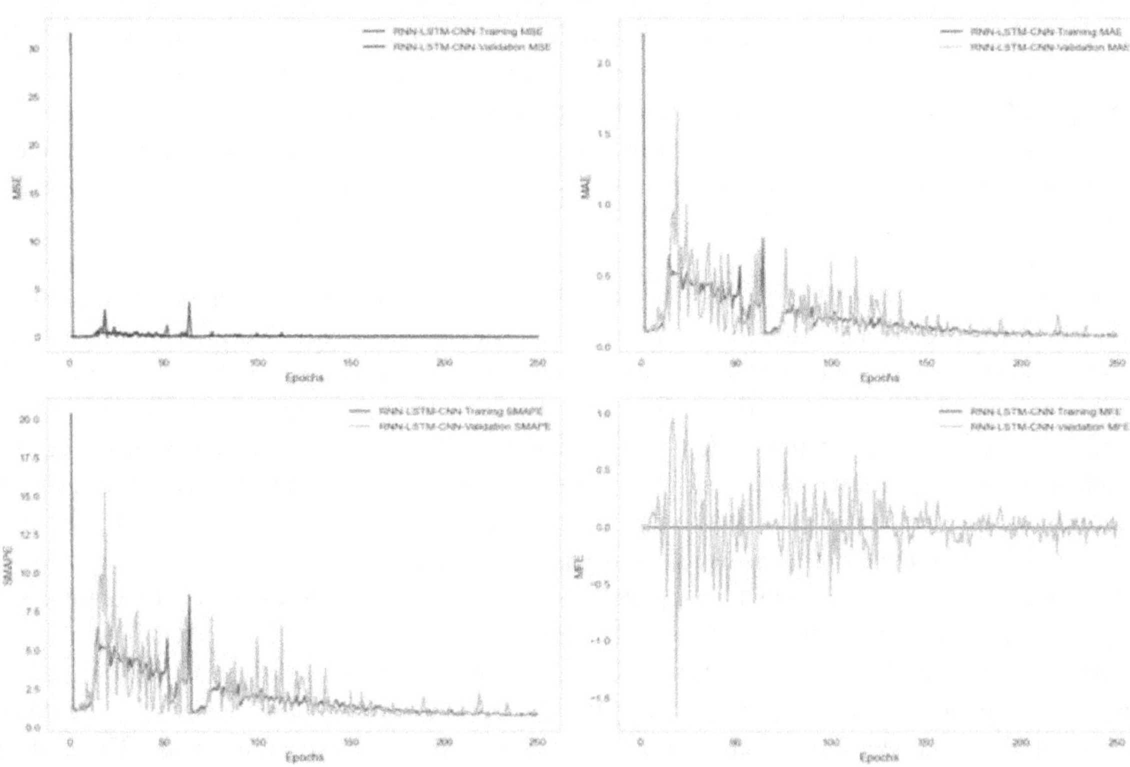

Figure 9. Gaussian copula visualization of hierarchies

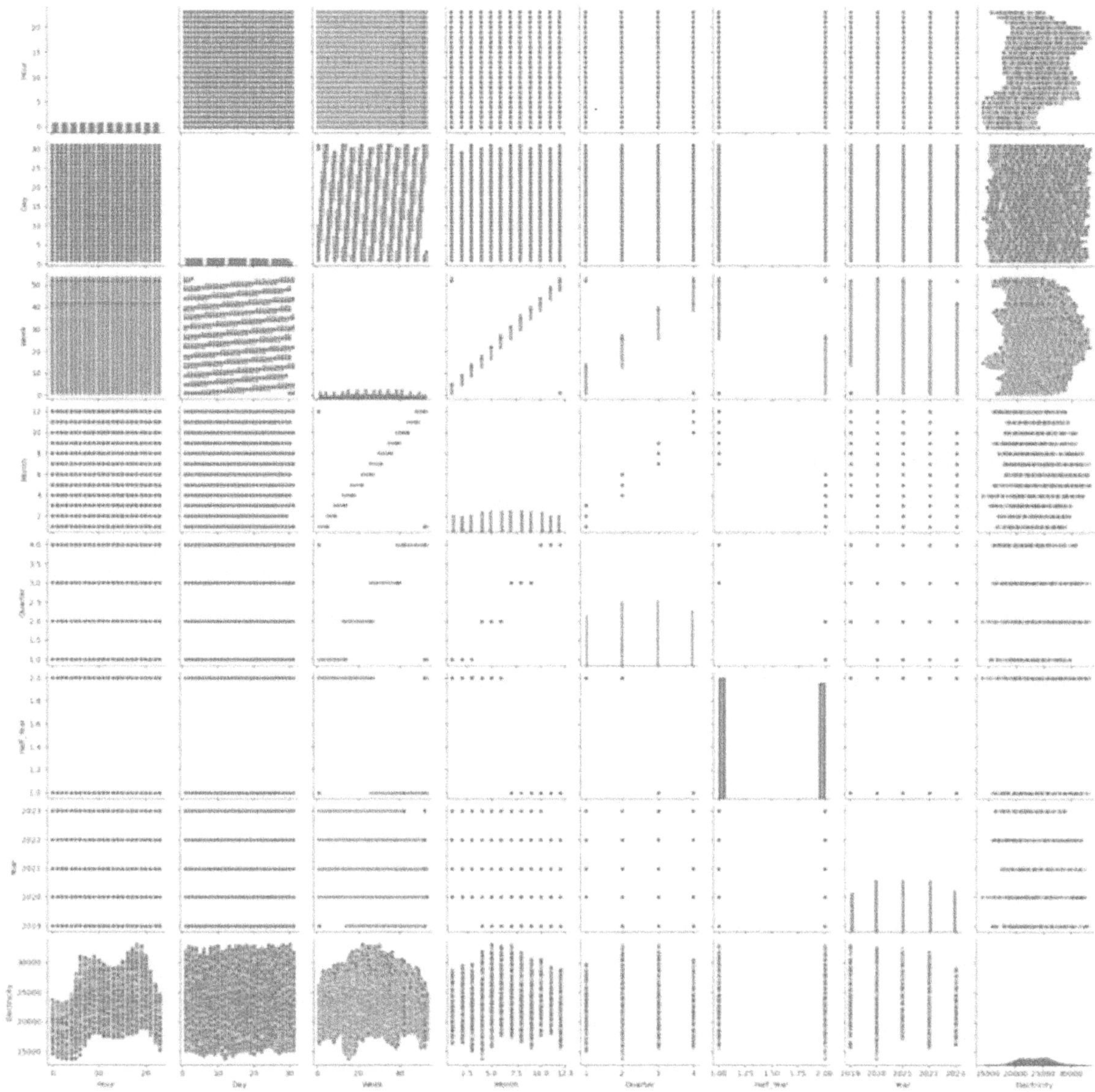

subset. This allows for faster convergence, especially when dealing with large datasets. On the other hand, Fuzzy C-Means (FCM) allow data points to belong to multiple clusters with varying degrees of membership. Fuzzy C-Means assigns each data point a membership value for each cluster, indicating the degree to which the point belongs to each cluster. This approach is more flexible than traditional k-means, especially when cluster assignments are ambiguous.

We discovered that two clusters is the ideal number after applying the elbow approach and Silhouette to determine the ideal number of clusters. Refer to the appendix's Figures 11 and 12. Table 5 presents the results of the three clustering techniques. It is evident that, in the case of k-means clustering, the Davies-Bouldin Index is given as 0.5011 (low), implying strong cluster separation, and the Silhouette

Table 4. Spearman rank correlation

	Hour	Day	Week	Month	Quarter	Half_year	Year
Hour	1.00						
Day	0.00	1.00					
Week	0.00	0.06	1.00				
Month	0.00	0.00	0.97	1.00			
Quarter	0.00	0.00	0.94	0.97	1.00		
Half_year	0.00	-0.01	-0.84	-0.86	-0.89	1.00	
Year	0.00	-0.01	-0.20	-0.20	-0.20	0.14	1.00

Score is 0.7284 (high), showing well-separated clusters. However, the Davies-Bouldin Index is 1.0289 (moderate), showing moderate cluster separation, and the Mini-Batch K-Means, Silhouette Score is determined to be 0.3805 (moderate), indicating moderate cluster separation as well.

Finally, with Fuzzy C-Means we found that Silhouette Score is found to 0.3734 (moderate), indicating moderate cluster separation while Davies-Bouldin Index is given 1.0554 (moderate), suggesting moderate cluster separation. All in all, K-Means appears to have the best performance among the three, with a higher Silhouette Score and lower Davies-Bouldin Index values. While Mini-Batch K-Means and Fuzzy C-Means show similar, moderate performance based on these metrics. Finally, our temporal dependencies for hierarchy in each cluster using k-means are given in Figure 10.

CONCLUSION AND FUTURE DIRECTION

Although accurate and reliable estimation of electricity/power consumption is extremely important, the field has despondently attracted less attention than other fields. But in the last few years, researchers have made some major strides and developed forecast models that are reliable, accurate, and effective by utilizing deep learning (DL) and artificial intelligence (AI). Due to their features and significance, these techniques—AI and DL—performed well and efficiently in handling non-linear problems with time-series data. In this chapter, we present a deep-learning ensemble method that uses chronological dependencies to anticipate and predict the hourly load of electricity demand. To guarantee model performance, our system first processes the data, cleaning, normalizing, and transforming it.

Furthermore, the preprocessed data is fed into an ensemble model to extract hybrid discriminative features by using RNN, LSTM, CNN and, stacking ensemble architectures. This will enable the utilities to plan resources and balance supply-demand, thereby ensuring stability and security of the power grid

Table 5. Performance of clustering algorithms

	Silhouette Score	Davies-Bouldin Index
K-means	0.728393737	0.501091441
Mini-Batch K-Means	0.380461786	1.028941784
Fuzzy C-Means	0.373376881	1.055433432

Figure 10. Clustering of electricity demand hierarchies

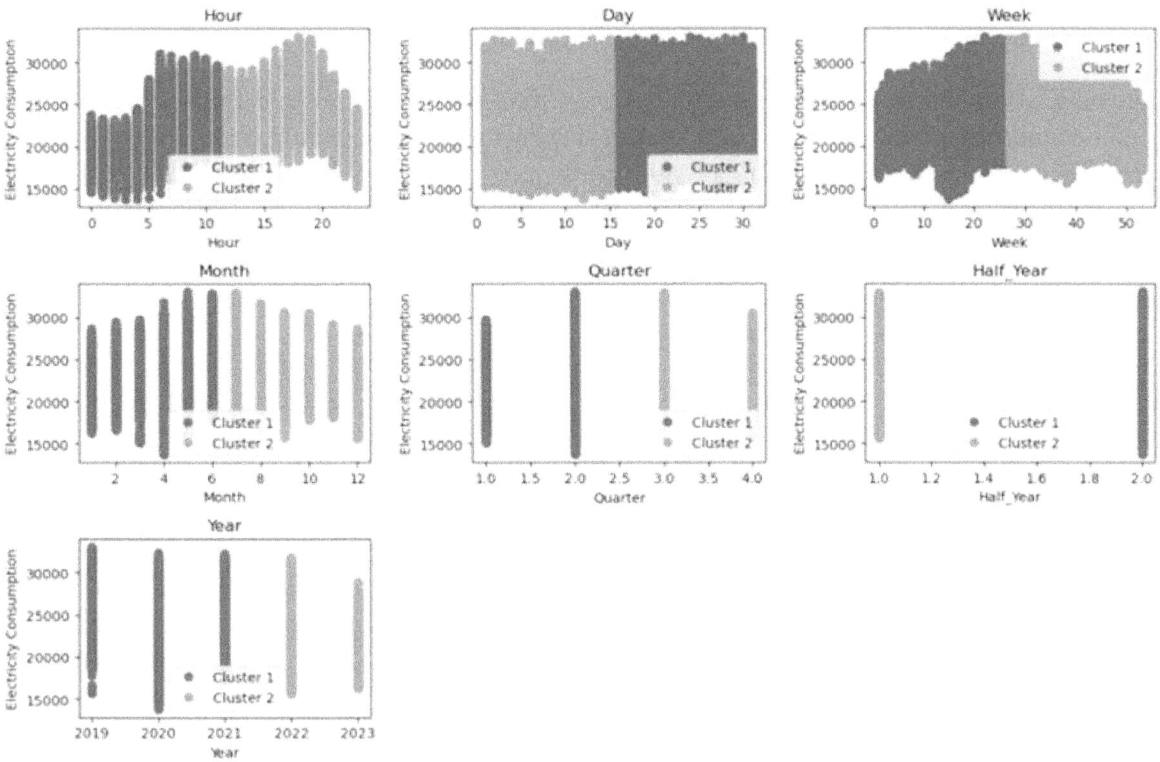

and reliability of service provisions. Using an hourly electrical demand load, we assessed the suggested system and compared the outcomes, which are displayed in Table 3. Targeting future effort, a variety of non-linear exogenous data structures, such as variables, monetary features, and weather/climate conditions or changes, can be investigated for the tendency analysis of power feasting patterns. Furthermore, a number of optimization techniques can be thought of and developed to raise the recommended system's forecast accuracy.

REFERENCES

Adhikari, R. J. N. (2015). A neural network based linear ensemble framework for time series forecasting. Academic Press.

Ahmadi, A., Daccache, A., Sadegh, M., & Snyder, R. L. (2023), Statistical and deep learning models for reference evapotranspiration time series forecasting: A comparison of accuracy, complexity, and data efficiency. Computers and Electronics in Agriculture, 215. doi:10.1016/j.compag.2023.108424

Aldhshan, S. R. S., Abdul Maulud, K. N., Wan Mohd Jaafar, W. S., Karim, O. A., & Pradhan, B. (2021). Energy Consumption and Spatial Assessment of Renewable Energy Penetration and Building Energy Efficiency in Malaysia: A Review. *Sustainability (Basel)*, *13*(16), 9244. doi:10.3390/su13169244

Bagnall, A., Lines, J., Hills, J., & Bostrom, A. (2015). Time-series classification with COTE: The collective of transformation-based ensembles. *IEEE Transactions on Knowledge and Data Engineering*, *27*(9), 2522–2535. doi:10.1109/TKDE.2015.2416723

Bandara, K., Bergmeir, C., & Smyl, S. (2020). Forecasting across time series databases using recurrent neural networks on groups of similar series: A clustering approach. *Expert Systems with Applications*, *140*, 112896. doi:10.1016/j.eswa.2019.112896

Cordeiro-Costas, M., Villanueva, D., Eguía-Oller, P., Martínez-Comesaña, M., & Ramos, S. (2023). Load Forecasting with Machine Learning and Deep Learning Methods. *Applied Sciences (Basel, Switzerland)*, *13*(13), 7933. doi:10.3390/app13137933

Ensafi, Y., Amin, S. H., Zhang, G., & Shah, B. (2022). Time-series forecasting of seasonal items sales using machine learning–a comparative analysis. *International Journal of Information Management Data Insights*, *2*(1), 100058. doi:10.1016/j.jjimei.2022.100058

Hagenau, M., Liebmann, M., & Neumann, D. (2013). Automated news reading: Stock price prediction based on financial news using context-capturing features. *Decision Support Systems*, *55*(3), 685–697. doi:10.1016/j.dss.2013.02.006

Hamed, M. M., Ali, H., & Abdelal, Q. (2022). Forecasting annual electric power consumption using a random parameters model with heterogeneity in means and variances. *Energy*, *255*, 124510. doi:10.1016/j.energy.2022.124510

Hochreiter S, & Schmidhuber J (1997). Long Short-Term Memory. *Neural Comput,* 9(8), 1735–1780. doi:10.1162/neco.1997.9.8.1735

Huang, S.-H., & Pan, Y.-C. (2014). Ergonomic job rotation strategy based on an automated RGB-D anthropometric measuring system. *Journal of Manufacturing Systems*, *33*(4), 699–710. doi:10.1016/j.jmsy.2014.02.005

Huang, S.-H., & Pan, Y.-C. (2015). Automated visual inspection in the semiconductor industry: A survey. *Computers in Industry*, *66*, 1–10. doi:10.1016/j.compind.2014.10.006

Jin, L., & Dong, J. (2016). Ensemble deep learning for biomedical time series classification. *Computational Intelligence and Neuroscience*. doi:10.1155/2016/6212684 PMID:27725828

Jogunola, O., Adebisi, B., Hoang, K. V., Tsado, Y., Popoola, S. I., Hammoudeh, M., & Nawaz, R. (2022). CBLSTM-AE: A Hybrid Deep Learning Framework for Predicting Energy Consumption. *Energies*, *15*(3), 810. doi:10.3390/en15030810

Kamal-Chaoui, L., & Robert, A. (2009). Competitive cities and climate change. Academic Press.

Kohzadi, N., Boyd, M. S., Kermanshahi, B., & Kaastra, I. (1996). A comparison of artificial neural network and time series models for forecasting commodity prices. *Neurocomputing*, *10*(2), 169–181. doi:10.1016/0925-2312(95)00020-8

Kranjc, J., Smailović, J., Podpečan, V., Grčar, M., Žnidaršič, M., & Lavrač, N. (2015). Active learning for sentiment analysis on data streams: Methodology and workflow implementation in the ClowdFlows platform. *Information Processing & Management*, *51*(2), 187–203. doi:10.1016/j.ipm.2014.04.001

Le, T., Vo, M. T., Vo, B., Hwang, E., Rho, S., & Baik, S. W. (2019). Improving electric energy consumption prediction using CNN and Bi-LSTM. *Applied Sciences (Basel, Switzerland)*, *9*(20), 4237. doi:10.3390/app9204237

Liu, T., Tan, Z., Xu, C., Chen, H., & Li, Z. (2020). Study on deep reinforcement learning techniques for building energy consumption forecasting. *Energy and Building*, *208*, 109675. doi:10.1016/j.enbuild.2019.109675

Lotfipoor, A., Patidar, S., & Jenkins, D. P. (2024). 'Deep neural network with empirical mode decomposition and Bayesian optimisation for residential load forecasting', Expert Systems with Applications, vol. 237, no. *Expert Systems with Applications*, *121355*. Advance online publication. doi:10.1016/j.eswa.2023.121355

Mariano-Hernández, D., Hernández-Callejo, L., García, F. S., Duque-Perez, O., & Zorita-Lamadrid, A. L. (2020). A Review of Energy Consumption Forecasting in Smart Buildings: Methods, Input Variables, Forecasting Horizon and Metrics. *Applied Sciences (Basel, Switzerland)*, *10*(23), 8323. doi:10.3390/app10238323

Rahman, M. M., Shakeri, M., Tiong, S. K., Khatun, F., Amin, N., Pasupuleti, J., & Hasan, M. K. (2021). Prospective methodologies in hybrid renewable energy systems for energy prediction using artificial neural networks. *Sustainability (Basel)*, *13*(4), 2393. doi:10.3390/su13042393

Razavi, R., Gharipour, A., Fleury, M., & Akpan, I. J. (2019). Occupancy detection of residential buildings using smart meter data: A large-scale study. *Energy and Building*, *183*, 195–208. doi:10.1016/j.enbuild.2018.11.025

Sharma, N., Jain, V., & Mishra, A. (2018). An analysis of convolutional neural networks for image classification. *Procedia Computer Science*, *132*, 377–384. doi:10.1016/j.procs.2018.05.198

Shiri, Perumal, Mustapha, & Mohamed. (n.d.). A Comprehensive Overview and Comparative Analysis on Deep Learning Models: CNN, RNN, LSTM, GRU. https://arxiv.org/ftp/arxiv/papers/2305/2305.17473.pdf

Singaravel, S., Suykens, J., & Geyer, P. (2018). Deep-learning neural-network architectures and methods: Using component-based models in building-design energy prediction. *Advanced Engineering Informatics*, *38*, 81–90. doi:10.1016/j.aei.2018.06.004

Slater, L. J., Arnal, L., Boucher, M., Chang, A., Moulds, S., Murphy, C., Nearing, G., Shalev, G., Shen, C., Speight, L., Villarini, G., Wilby, R. L., Wood, A., & Zappa, M. (2022). Hybrid forecasting: Using statistics and machine learning to integrate predictions from dynamical models. *Hydrology and Earth System Sciences*. Advance online publication. doi:10.5194/hess-2022-334

Somu, N., Raman, M. R. G., & Ramamritham, K. (2021). A deep learning framework for building energy consumption forecast. *Renewable & Sustainable Energy Reviews*, *137*, 110591. doi:10.1016/j.rser.2020.110591

Wang, Z., & Oates, T. (2015). Imaging time-series to improve classification and imputation. *arXiv preprint arXiv.00327*.

Wongrin, W., Chaisee, K., & Suphawan, K. (2023). Comparison of Statistical and Deep Learning Methods for Forecasting PM2.5 Concentration in Northern Thailand. *Polish Journal of Environmental Studies*, *32*(2), 1419–1431. doi:10.15244/pjoes/157072

World Energy Council. (2016). World Energy Council. World Energy Resources Solar.

ENDNOTE

[1] This data is available from https://www.eskom.co.za/dataportal/

APPENDIX

Figure 11. Elbow method for optimal number of clusters

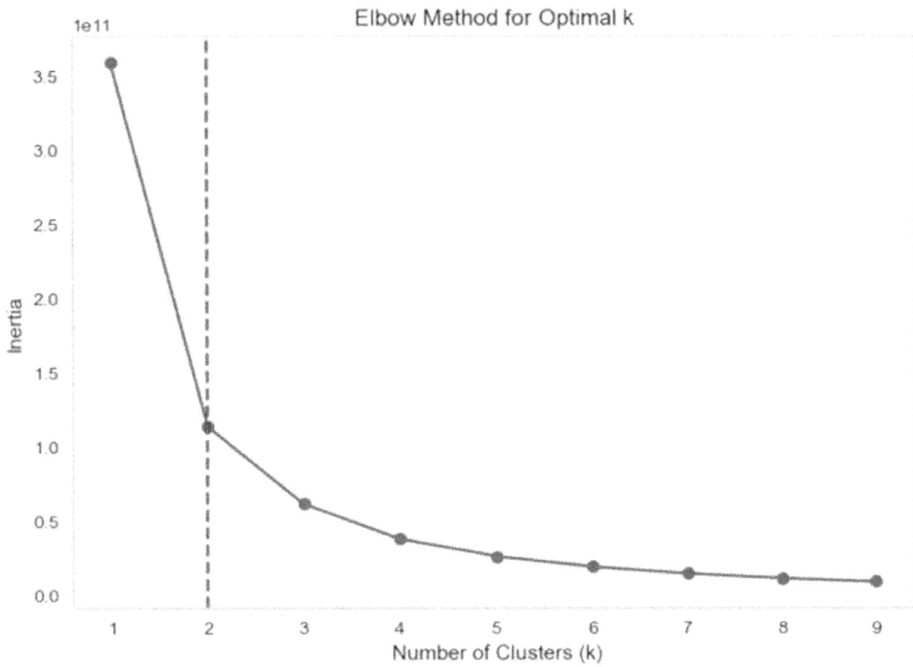

Figure 12. Silhouette method for optimal number of clusters

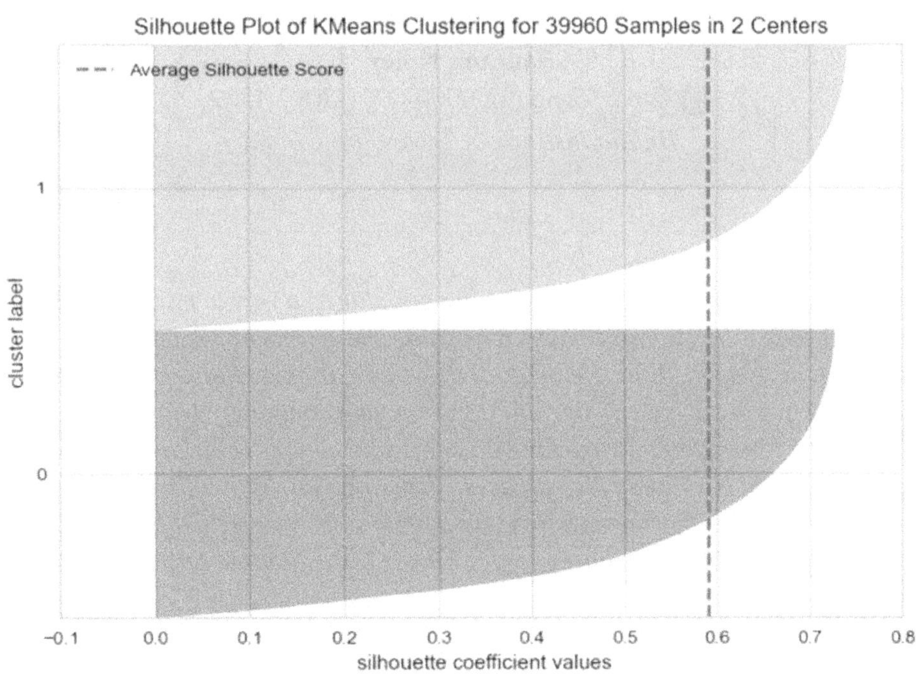

Chapter 4
A Study of Computer Vision, Deep Learning, and Machine Learning Techniques for Forecasting Solar Power and Renewable Energy

Jayeeta Majumder
Haldia Institute of Technology, India

Pinaki Pratim Acharjya
https://orcid.org/0000-0002-0305-2661
Haldia Institute of Technology, India

Subhabrata Barman
Haldia Institute of Technology, India

Santanu Koley
https://orcid.org/0000-0001-8082-1332
Haldia Institute of Technology, India

ABSTRACT

Utilising renewable energy sources is becoming more popular as a way to mitigate the effects of climate change and global warming. In an effort to make renewable energy more predictable, numerous prediction techniques have been developed. The objectives of this study are best illustrated by this chapter, which aims to provide a review and analysis of machine-learning and computer vision techniques in renewable solar energy projections. In addition to machine-learning and computer vision techniques for renewable solar energy projections, this chapter also focuses on the objective to deliver an optimized academic outcome, potentially necessary for the development of new solar energy fields. This could significantly contribute to the amplified usage of solar energy, which is a sustainable and cleaner energy source.

DOI: 10.4018/979-8-3693-2355-7.ch004

INTRODUCTION

Socio-economic development has led to progressive increase in energy demand that cannot be met with traditional technology with local resources only. Renewable energy is seen as a transformative solution to this increasing energy demand and economic challenges like rising coal prices. (Huang et. al., 2017; Jursa, 2007; Rohrig et. al., 2008) Further, renewable and natural sources are available easily, are in abundance and are also environment friendly. The nature of power generation from renewable energy plants is unpredictable because it is dependent on numerous variables, including location, wind speed, weather, and other climatic conditions. (M. Abdel-Nasser et al., 2019; Bouzerdoum et al, 2013) It is therefore pertinent to mention that integration of variable generation from solar and wind sources have become a real challenge for system operators in ensuring reliability and security of the power system network particularly with an exponentially increasing renewable energy share in our worlds energy mix.

For variable energy sources to be integrated into the grid, renewable energy forecasting is crucial because it allows power systems to manage the intermittent nature of the energy supply at different spatiotemporal scales. To predict the future impact of cloud displacements on the energy generated by solar facilities, traditional modelling approaches rely on numerical weather prediction or physical models. (Bacher et al, 2009; Bapai et al., 2019). These models find it difficult to pick up systematic biases and integrate cloud data. By combining surface measurements from multiple sources with real-time cloud cover observations via machine learning, a few of these restrictions are overcome for improving computer vision. With a focus on deep learning, (AlKandari et al., 2019; Alsharif et al., 2019, Aslam et al., 2019) which offers the theoretical groundwork required to create architectures that can extract relevant data from data produced by weather stations, satellites, ground-level sky cameras, and sensor networks, this review gives an overview of recent developments in solar forecasting from multisensory observations of

Figure 1. Renewable energy source

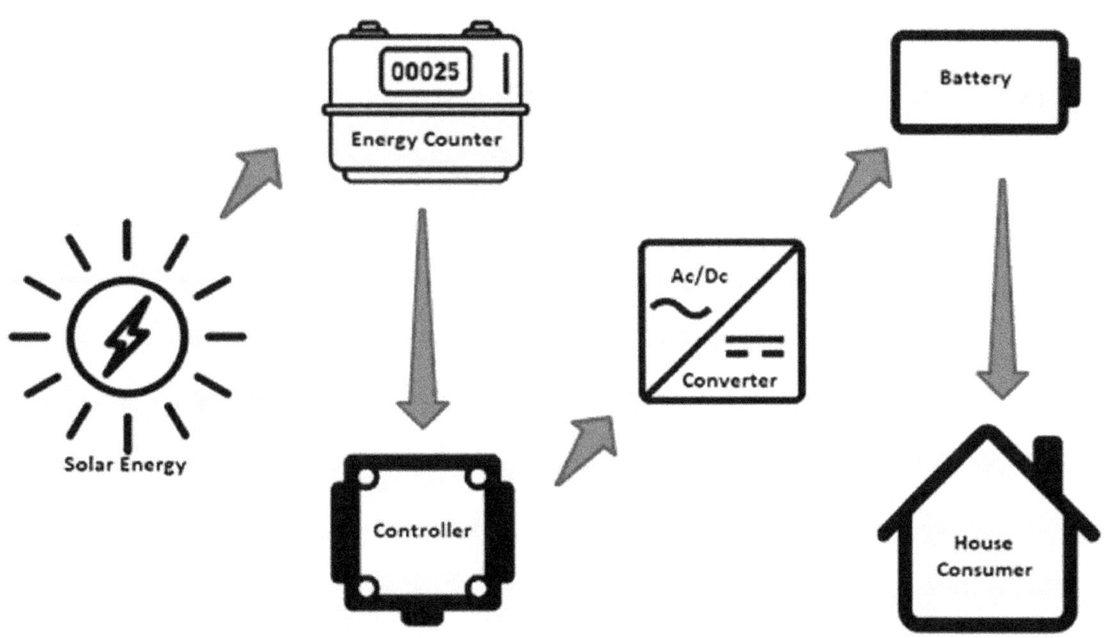

Earth. Taking everything into account, machine learning has enormous potential to improve the accuracy and robustness of solar energy meteorology. Nevertheless, further investigation is required to fully grasp this potential and tackle its constraints (Sheng, et al., 2018; Torres- Barran et al. 2019).

This chapter has been divided into various sections. The introduction and the purpose of the study were covered in section 1. In order to lessen the effects of solar intermittency, section 2 discusses solar power forecasting, which is the procedure for gathering and assessing data to predict the generation of solar power on different time horizons. The study's motivation and the list of current review articles are presented in section 3. Section 4 describes the computer vision of solar forecasting method. The various deep learning techniques utilised in vision-based solar energy modelling are covered in Section 5; the study's conclusion is presented in Section 6.

SOLAR FORECASTING

Among the best methods for reducing the unpredictability and variability of solar power brought on by atmospheric changes is solar forecasting. In order to estimate current meteorological conditions and predict future global solar irradiance (GSI) or power, ranging from seconds to years ahead, the majority of solar forecasting techniques use ground or remote sensing data as input. Known also as operational solar forecasting, solar forecasting has found widespread application in the power and energy sector. (Cellura et. al, 2008; Chen et. al, 2008; Dehghani et. al., 2019). Forecasting for the very short term, short term, medium term, and long term are the general categories into which solar forecasting can be divided based on various lead times and horizons.

Notwithstanding the differences in these definitions, forecasts of solar over every timeline are crucial for power system planning and operations. Solar forecasts for the long term are usually used in resource assessment, system planning, and energy trading. Energy markets and daily power system operations rely on short- and very-term solar forecasts, which are typically characterised as projections up to a few days

Figure 2. Solar forecasting

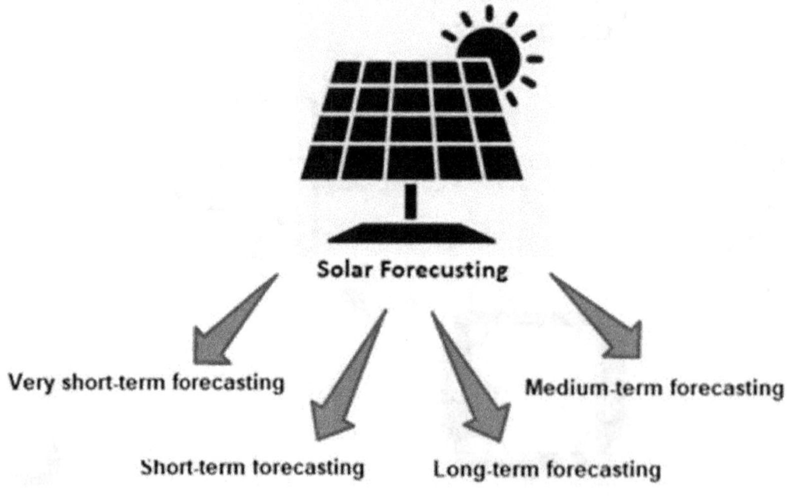

out. Power systems and markets use different definitions and jargon. (Han et. al., 2011; Harbola et. al., 2019; Hong et. al., 2019). Accurate solar forecasting can reduce costs and increase power system resilience and dependability for on-the-spot adjustments, system operators and market participants generally use very short-term solar forecasts, and for day-ahead bidding or preliminary generator scheduling, they typically use medium- to short-term solar forecasts. Solar forecasting techniques fall into one of three categories models that are physical, data-driven, or hybrid, depending on how closely they follow physical principles. Any kind of model can use data from remote and ground sensors as input. (Atique et al., 2019; Cali et. al, 2019). The most popular type of satellite photos are the data from remote sensing which can be used to generate solar forecasts through physical model processing, models with numerical weather prediction (NWP), for instance. Additionally, statistical models can be used directly to process satellite data and make solar forecasts. (Navarin et. al., 2020; Gutierrez et. al., 2020; Oakleaf et. al., 2007). The best ground-based sensors for measuring local surface irradiance are pyranometers and pyrheliometers. There are special chances for sensors to play important roles in solar forecasting, such as satellites and sky cameras because Compared to other target variables in energy forecasting, such as load, price, and wind forecasting, solar forecasting is more sensitive to cloud movement. Satellites and sky cameras cover a wide range of solar forecasting. Skycams and satellites' combined capacity are to give information in the form of pictures is a significant commonality. Advanced computer vision techniques can be used to process these cloud cover observations and model the variability of solar power. (Filik et. al., 2017; Fischetti et. al., 2019; Hombrados-Herrera et al., 2018).

LITERATURE REVIEW

1. Ibrahim et al. (2012) developed a linear regression model to forecast solar radiation in Perlis using information gathered from a weather station over a three-year period. The model fit the data well, as evidenced by the average daily solar radiation, average daily maximum and minimum temperatures, and the model's R-squared value of 0.954. The designers of the model concluded that Perlis could gain from applying it to solar radiation estimation. [Ibrahim et al., 2012]
2. Ekanayake et al. (2021) created power regression, multiple linear regression, and artificial neural networks (ANNs) to develop models to forecast wind power for a Sri Lankan wind farm. Average wind speed, average ambient temperature was used as input variables for their analysis in their modelling approach. (Khandakar et. al, 2019; Kim et. al., 2019; Kong et. al., 2015). The correlation coefficient was high, the RMSE was low, and the models were biased., indicating acceptable accuracy. Data on power generation gathered over a five-year period was used in their development. Though the ANN model was the most accurate The MLR and PR models offer insights for additional wind farms in the same region. [Ekanayake et al., 2021]
3. To forecast solar power, Mustafa et al. (2022) also analysed four regression models: logistic, elastic, Lasso, and linear regression. The findings demonstrated the effectiveness of all four models, with elastic regression outperforming the other two in predicting the solar power output at maximum point. Additionally, principal component analysis (PCA) was utilised, and the elastic regression model produced better results. The advantages and disadvantages of each solar power prediction model are the main topics of the study. [Mustafa et al., 2022]
4. In their 2020 study, Vassallo et al. examine optimal procedures for random forest (RF) modelling in speed of wind prediction of power. In this work, direct and multi-step prediction in reverse are

compared, the effect of training data availability is evaluated, and the use of random forest (RF) as an accurate measure is examined. The findings show that RF performs better when used as a tool for correcting errors in the persistence approach, and the direct forecasting strategy beats the recursive strategy by a small margin. The accuracy of forecasting increases with increased data availability. [Vassallo et al, 2020]

5. To further address weak reasoning, overfitting, and generalisation in neural network models during short-term wind power forecasting, (C.J. Camila et al, 2019; Cardona et. al, 2019; Mabel et. al., 2008). Shi et al. (2018) proposed a two-step feature selection process that is combined with a supervised random forest model. The suggested method eliminates pointless characteristics, chooses pertinent samples, and assesses each decision tree's effectiveness. A new external validation index that correlates with wind speed is presented in order to solve the shortcomings of the internal validation index. (Kou et al., 2014; S. Kumar et. Al., 2015; Zheng et al. 2009) Empirical examples and case studies show that the model performs better than other models in terms of robustness, efficiency, and accuracy—particularly with regard to wind power curtailment and noisy data. [Shi et al., 2018]
6. When a sizable amount of unstructured data needs to be analysed and the optimal target variable is still unknown, unsupervised learning algorithms are typically helpful. The clustering of meteorological data is a widely used application of unsupervised learning in renewable energy forecasting. For instance, J. Varanasi and M. Tripathi (2019) used K-means clustering to forecast short-term PV power generation by clustering the days of the year sunny, cloudy, and rainy based on resemblance. [J. Varanasi et al., 2019]
7. Renewable energy forecasting is one of reinforcement learning's many applications. Controlling the behaviour of energy systems is one way to apply RL to forecasts of renewable energy. For instance, utilising RL, Sierra-García J. and S. Matilde (2020) [Sierra-García J et al, 2020] created an advanced yaw control method for wind turbines. In contrast to the RL algorithm, which uses an ANN to maximise power generation and minimise mechanical loads this approach makes use of a particle swarm optimisation (PSO) and Pareto optimal front (PoF)-based algorithm to determine the best course of action that balances power gain and mechanical loads. Real wind data from Salt Lake City, Utah, FAST simulations, and the NREL 5-MW reference wind turbine all supported the plan. Artificial neural networks (ANNs) have been in use for a long time and have proven to be effective in forecasting renewable energy, including solar, wind, and multi-renewable energy. (Kalogirou et. al.,1999; Kamadinata et. al., 2019; Kechriotis et. al, 1994)

In Abha, Saudi Arabia, for example, Rehman and Mohandes estimated the amount of solar radiation by feeding relative humidity and air temperature into an artificial neural network. Meanwhile, to calculate the total solar radiation in Al-Madinah, Saudi Arabia, Benghanem et al. designed six ANN-based models. The model with the highest precision was determined to be based on air temperature and sunshine duration. [Benghanem et al, 2009]

8 Ozgoren et al. used the stepwise MNLR technique for identifying influential inputs and multiple input variables in a different study, wherein they developed an ANN model to forecast monthly global solar radiation in Turkey. It was found that the accuracy of the ANN model was satisfactory when compared to the measured values. Notwithstanding variations in the models and input variables employed, artificial neural networks (ANNs) have shown to be a trustworthy instrument

for forecasting global solar radiation in various locations. ANN was also used by S. Kumar and T. Kaur (2016) to forecast solar radiation for Himachal Pradesh's use in solar energy applications. Temperature, barometric pressure, humidity, sunshine hours, and precipitation were among the input variables used in the ANN model [Ozgoren et al,2012]. Following the development and testing of three models with three to five input parameters, the ANN-I5 model demonstrated the highest prediction accuracy was achieved with a mean absolute percentage error (MAPE) of 16.45%. This study demonstrated that, without the need for precise measurement tools, the method can be used to determine the solar energy potential of any location in the world. [S. Kumar et al, 2016].

9. Wind energy is another important area of research in terms of renewable energy forecasting. Jamii et al. (2022) proposed an ANN-based paradigm for forecasting wind power generation and load demand using meteorological parameters like wind speed, atmospheric pressure, and temperature. The outcomes demonstrated that the ANN produced high power forecasting accuracy and efficacy and outperformed four other machine learning techniques [Jamii et al., 2022]. Chen and K. Folly (2019) proposed an artificial neural network (ANN) model for estimating short-term wind power in small wind farms. They examine the effects of sample size, air pressure, temperature, relative humidity, wind direction, and wind speed at various elevations on the computational cost and forecasting accuracy of the model. The study looks into how sample size and input variables affect the model's cost computation and forecasting accuracy. Their research indicates that an ANN model with a large training sample size and all input features performs the best for forecasting [Chen et al, 2019].
10. Convolutional neural networks (CNNs) are a type of artificial neural networks (ANN) that excel at processing and interpreting data that is organised in a grid, such as pictures or videos. CNNs use convolutional layers to extract localised features from the input data, and they are inspired by the structure and function of the brain's visual cortex [Yao et al., 2019]. Convolutional, pooling, and fully connected layers are among the layers that make up a CNN. The core components of a CNN, called convolutional layers, use filters, also known as kernels, to extract localised features from the input data. [Akram et al., 2019].

All filters are made up of tiny matrices that traverse through the input data and multiply each filter by the corresponding input pixel. This process results in the creation of a feature map that depicts the sections of the input data that are most important for the current task. Pooling layers are commonly added to networks after convolutional layers to reduce the spatial dimensions of feature maps and increase computational efficiency. Max pooling selects the highest value within a defined region of the feature map. and average pooling selects the mean value within a defined region of the feature map. These are the two most popular forms of pooling [McCann, et al, 2017]. The pooling and convolutional layers' acquired features are used by the fully connected layers to generate the final predictions. These layers, which resemble those in a traditional neural network, forecast outcomes based on the current task using the learned features. The CNN weights are changed during training, just like those of other neural network architectures, to minimise the discrepancy between the expected and actual output. The weights are then updated in light of this information after the gradients of the loss function with respect to the weights are computed using the back-propagation algorithm [Ajit et al., 2020].

SOLAR FORECASTING USING COMPUTER VISION

When sky features are taken from spatial data captured by light-sensing apparatuses such as satellite imagery, ground-based imagers, or both, the method of solar forecasting using computer vision is referred to as image-based (Niall et al, Leron et al, 2021). On the other hand, while statistical techniques rely on a moving average and autoregressive covariate structure, physical approaches are based on the sun's position. Most intraday variability in irradiance is explained by the dynamics of cloud cover. Researchers have developed vision on computers techniques based on observations of clouds to better predict corresponding fluctuations in solar power. From past observations, the partially the spatial configuration can be inferred due to the deterministic nature of cloud motion of the cloud cover in the future (Quentin et al, 2022). In addition to the intrinsic stochastic component of cloud modelling. Predicting cloud displacement provides a major benefit over forecasting methods that rely solely on local meteorological data, which are unable to forecast future events with sufficient accuracy beyond the solar time series' short-term autocorrelation period. Consequently, in order to produce solar predictions, the forecasting approach is necessary, that are genuinely valuable at various spatiotemporal scales (Vishnu et al., 2020; Vincent et al., 2020). .

Satellites and sky cameras are the primary cloud observation sources for solar forecasting with a short time horizon. A fish-eye lens on a sky camera allows it to record the spatial arrangement of a few square kilometres of local cloud cover at a time, while a satellite provides maps of a continent's atmospheric radiation at a lower temporal resolution of five to fifteen minutes interval of less than a minute. Due to their differing technical features and perspectives on cloud cover, for solar forecasting, these two imaging methods are available right now. employed in tandem, even though it is anticipated that their spatiotemporal resolution will rise and eventually catch up to that of ground-level imagery. In reference to the forecast horizon, methods utilising sky cameras seek to offer forecasts of the local solar variability as far ahead as thirty minutes, contingent upon the local cloud velocity; in contrast, satellite imagery, with its broader field of view, offers longer-range forecasts, reaching several hours in advance (Gaétan et al., 2019). Sky cams are most appropriate for places with an elevated solar power density, like solar or hybrid power plants, unless they are part of a network arrangement that covers a wider area. On the other hand, predictions of the nation's solar output can be obtained through satellite-based methods, which can be advantageous for energy trading and grid balancing. Some of the comparable methods utilised in conventional vision of computer techniques Cloud segmentation, cloud localization, cloud properties estimation, cloud tracking, and cloud motion modelling are all based on sky or satellite images (Gaétan et al., 2019; Dragomir et al, 2018).

For sky images, the cloud mask is often generated using techniques for thresholding based on statistics of pixel values. In addition to comparable thresholding analyses, the effective cloud albedo, also known as the cloud index (0: no cloud, 1: dense cloud), can also be used to improve cloud segmentation in satellite imagery. This is calculated based on the percentage of light reflected by a surface, the percentage of light reflected by an observation, and the highest empirical percentage of light reflected by a surface. The cloud transmittance is indicated by this continuous variable, which can also be estimated for sky camera applications taking advantage of a clear-sky model and a pyrheliometer. Next, multiple sky cameras in stereovision mode are used to carry out cloud geolocation. Subsequently, to predict the future deterministic or probabilistic effects of cloud cover dynamics on solar output, cloud velocity vectors can be estimated using block matching techniques or optical flow (Vignesh et al., 2019; Ahmad et al, 2019). Forecasts of local or regional irradiance from the cloud's future location can be obtained by

Figure 3. Traditional computer vision approaches based on sky images or satellite images

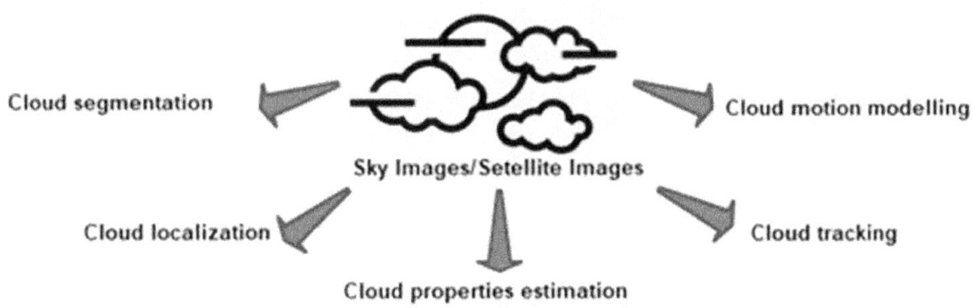

using estimates of the physical properties of the clouds. Using the recent advancements Several data-driven methods have been developed to tackle this computer vision task, thanks to advancements in deep learning and the growing availability of large datasets. The conventional method to forecast solar variability is to train a neural network to identify spatiotemporal elements in a series of historical satellite or sky photos. Like traditional computer vision techniques, this class of techniques can be used to obtain both regional predictions, like GSI or cloud-index maps, and local values, like photovoltaic power output or GSI. Even with significant Despite deep learning models' performance gains as measured by average metrics such as mean absolute error (MAE) or root mean square error (RMSE), have a number of shortcomings, such as the inability to break through the persistence barrier, forecast significant occurrences in the future, or associate a picture—especially one taken from space—with its matching solar value (Huiyu et al., 2022, Bharath et al., 2015). As such, certain models rely on external data, like solar elevation or GSI readings the solar community has taken an interest in solar forecasting based on computer vision, as seen by the increase in papers published on the topic in online journals in recent years. (Chatwin et. al., 2019; Feng et. al, 2018; Hodge et. al., 2017). Similar percentages of forecasting the sun using images of the sky and satellite data have been developed recently.

VISION-BASED SOLAR ENERGY USING DEEP LEARNING TECHNIQUES

In this section, various deep learning techniques for vision-based solar energy modelling are presented. These methodologies concentrate on different facets of general modelling, quality assurance, data processing, and data augmentation.

Many architectural design decisions, such as the number of layers, convolutional filter size, and neuron count, as well as alternative learning configurations, such as batch size and learning rate, are frequently involved in the implementation of deep learning models. (Lansberry et al., 1992; Lauret et.al., 2015; Leholo et. al, 2019) It takes work to adjust these hyperparameters so that deep learning models perform better. Generally, there is a defined range of values for every hyperparameter. Due to time and computational resource limitations, a comprehensive search (also known as a grid search) incorporating all possible combinations of these parameters is frequently unfeasible. Grid search is ineffective for finding hyperparameters since hyperparameters can also vary depending on the dataset. Put differently, the hyperparameter values that work best for one dataset might not work well for another. Sampling techniques are usually used in light of these issues. Rather than designating a distinct range of values

Figure 4. Vision-based solar energy modelling

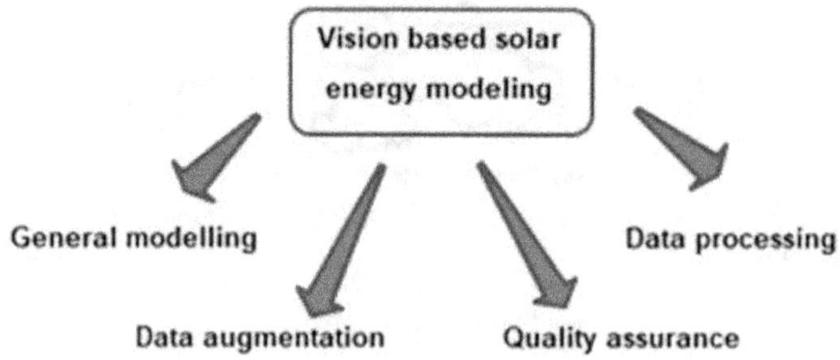

for every hyperparameter, a statistical distribution is established, from which values can be selected at random. Not all hyperparameters have the same importance. While grid search can focus on finding the best values for the crucial hyperparameters, random search has more exploratory power and may spend less time looking into unimportant parameters. Besides random search, Bayesian optimisation has recently gained strength as a hyperparameter tuning tool. (Rodriguez et. al, 2018; Salcedo-Sanz et. al, 2014; Saloux et. al, 2014). The optimal hyperparameter settings can be found automatically using a machine learning algorithm called Bayesian optimisation. This is in contrast to random search, which requires numerous separate but concurrent experiments. Several iterations are used by the algorithm to determine the ideal hyperparameter settings. Preliminary probability estimation of model performance throughout the hyperparameter range is a common application of the Gaussian process. The sampling strategy for the subsequent iteration can be chosen based on the outcomes of the preceding iterations, utilising an acquisition function until an optimal solution is reached. (Wang et al., 2017; Welch et. al, 2009; Wen et. al, 2019).

Together with model-centric methods that concentrate on enhancing other works attempt to facilitate modelling by improving the data, such as the model architecture or training plan. These data-centric methods include data augmentation and feature engineering, among others. Feature engineering seeks to create new features from unprocessed data by applying domain expertise. One common pre-processing step in vision-based solar forecasting, for example, is to unwrap fish-eye lens images of the sky, which severely distort the scene. Clouds can be projected onto a rectangular grid, assuming, (Leva et. al, 2017; Li S, 2003; Yang et al, 2018). to achieve this. This technique was used in deep learning studies to help predict distant cloud displacement or to give object trajectory consistency. On the other hand, additionally, it was shown that the original lens distortion enhanced by adding more information to the area in the image centre where the sun and surrounding clouds are visible during high solar flux, solar predictions can be made.

When, by creating altered versions of pre-existing samples, data augmentation seeks to artificially expand a dataset. By utilising existing data more effectively, data augmentation attempts to increase a model's generalizability without the need for costly or impractical additional data collection. In actual use, synthetic data can be produced by applying visual transformations like noise addition, cropping, rotation, or flipping to pre-existing samples or by using generative models like GANs. (Nazaripouya et al., 2018; Meng et. al, 2016; Mohandes et al, 2004) For vision-based solar forecasting, a variety of data augmentation techniques have been used. To perform noise injection, for example, pixel values were subjected to the addition of Gaussian noise an alternative method of colour space transformation is to independently adjust each of the three RGB channels' intensity level at random. An alternative is to simultaneously change the values of each of the three channels' pixels by a constant amount (Wang et. al, 2018; Yu et. al., 2011; Zhen et. al, 2015). When used for solar modelling from sky images, these methods have been demonstrated to help with the nowcasting task of matching a picture to a solar value; however, they have little effect on the forecasting task. Combining images is another type of data augmentation technique. One of the k-nearest neighbours or a random sample can be combined with the original images. It was shown that mixing images made the greatest difference in nowcasting predictions, much like colour space transformation techniques. (Will et. al, 2011; Williams et. al, 1989; Yousif et. al, 2019).

Deep learning methods for solar energy based on vision also require careful consideration of the sampling strategy (Lin et. al., 2018; Liu et al., 2018; Zhai et. al., 2019) . Sky image datasets and satellite observations are uneven in various ways, just like a lot of real-world datasets. For example, the distribution of cloud types at a given geographic location is highly dependent on the local weather. As a result, certain cloud patterns in the original dataset were either overrepresented or underrepresented. (Mujeeb et al., 2019; Kubby et. al, 2017; Paoli et. al, 2010) uniformly sampled mini-batches from a training set that is indicative of the dataset will teach the model a site-specific distribution, which may affect how well it performs in various climates.

A dataset's quality, in addition to its size, can significantly affect the model's performance and generalisation qualities. Incorrectly labelled samples or mis measured irradiance, for example, can cause an incorrect supervision signal during model training, and they can also negatively impact the model's exploitation if it is used in untested environments. These factors make quality control a crucial component of machine learning initiatives (Coimbra et. al, 2018; Qing et. al, 2018; Rana et al., 2020). The causes of anomalies in vision-based solar power monitoring include operational disturbances like dust, snow, dew, and temperature response, as well as equipment failures like misaligned shade rings, spectral selectivity, and temperature response. Rain, snow, animals, and soiling are just a few of the external hazards that sky cameras are subject to, which can hinder the collection of cloud cover observations. (Yu et. al, 2019; Zambrano et. al, 2020). One common method to filter these anomalies in a dataset is to inspect the outliers. Outliers can be defined, for instance, as samples that are significantly limited in the large original RGB feature space due to the curse of dimensionality. One way to accomplish this is to gauge the separation between samples within a feature space. To circumvent this issue a filtering method based on the CVAE model was developed (Sapitang et. al, 2010; Senthil et. al., 2019; Shimray et. al, 2017). An autoencoder was trained to reconstruct sky images using a feature space with low dimensions. Outliers that result from this low-dimensional representation are then manually y verified to facilitate the separation of real anomalies from visually uncommon sky images. however, this filtering technique may miss common anomalies, like rain or dust in sandy areas. Visual tools were created more recently to complement the different approaches to quality control (Zhang et. al, 2016; Peng et. al, 2017).

CONCLUSION

Computer vision-based solar energy meteorology is crucial for addressing the variations in solar radiation caused by changing cloud cover and how well they integrate into the power system. In this situation, machine learning offers an effective modelling framework that can analyse and integrate enormous volumes of complex data from ground-based sky imagers and in situ measurement instruments are used to observe the Earth. This chapter describes the resulting deep learning methods and computer vision tasks for solar forecasting are reviewed and studied. Promising research areas and easily available resources are also fully described. This thorough investigation demonstrates how computer vision and machine learning can enhance solar energy generation modelling.

REFERENCES

Abdel-Nasser & Mahmoud. (2019). Accurate photovoltaic power forecasting models using deep LSTM-RNN. Neural Computing and Applications, 31, 2727–2740.

Ahmad, S. T., Samarth, B., & Shivkumar, K. (2019). A deep learning approach to solar-irradiance forecasting in sky-videos. In IEEE winter conference on applications of computer vision (WACV). IEEE.

Ajit, A., Acharya, K., & Samanta, A. (2020). A Review of Convolutional Neural Networks. Proceedings of the 2020 International Conference on Emerging Trends in Information Technology and Engineering (ic-ETITE 2020), 1–5. 10.1109/ic-ETITE47903.2020.049

Akram, M. W., Li, G., Jin, Y., Chen, X., Zhu, C., Zhao, X., Khaliq, A., Faheem, M., & Ahmad, A. (2019). CNN based automatic detection of photovoltaic cell defects in electroluminescence images. *Energy*, *189*, 116319. doi:10.1016/j.energy.2019.116319

AlKandari & Ahmad. (2019). Solar power generation forecasting using ensemble approach based on deep learning and statistical methods. Appl. Comput. Inform.

Alsharif, M. H., Younes, M. K., & Kim, J. (2019). Time series ARIMA model for prediction of daily and monthly average global solar radiation: The case study of Seoul, South Korea. *Symmetry*, *11*(2), 240. doi:10.3390/sym11020240

Aslam, M., Lee, J.-M., Kim, H.-S., Lee, S.-J., & Hong, S. (2019). Deep learning models for long-term solar radiation forecasting considering microgrid installation: A comparative study. *Energies*, *13*(1), 147. doi:10.3390/en13010147

Atique, S., Noureen, S., Roy, V., Subburaj, V., Bayne, S., & Macfie, J. (2019). Forecasting of total daily solar energy generation using ARIMA: A case study. *Proceedings of the 2019 IEEE 9th Annual Computing and Communication Workshop and Conference (CCWC)*, 114–119. 10.1109/CCWC.2019.8666481

Bacher, P., Madsen, H., & Nielsen, H. A. (2009). Online short-term solar power forecasting. *Solar Energy*, *83*(10), 1772–1783. doi:10.1016/j.solener.2009.05.016

Bapai, A., & Duchon, M. (2019). A hybrid approach of solar power forecasting using machine learning. *Proceedings of the 3rd International Conference on Smart Grid and Smart Cities*.

Benghanem, M., Mellit, A., & Alamri, S. N. (2009). A. Mellit, S. Alamri, ANN-based modelling and estimation of daily global solar radiation data: A case study. *Energy Conversion and Management*, *50*(7), 1644–1655. doi:10.1016/j.enconman.2009.03.035

Bharath, H., Pablo, A., Ross, G., & Jitendra, M. (2015). Hypercolumns for object segmentation and fine-grained localization. *Proceedings of the IEEE conference on computer vision and pattern recognition*, 447–56.

Bouzerdoum, Mellit, & Massi. (2013). A hybrid model (SARIMA-SVM) for short-term power forecasting of a small-scale grid-connected photovoltaic plant. Sol. Energy, 98, 226–235.

Cali, U., & Sharma, V. (2019). Short-term wind power forecasting using long-short term memory based recurrent neural network model and variable selection. *Int. J. Smart Grid Clean Energy*, *8*, 103–110. doi:10.12720/sgce.8.2.103-110

Camila, C. J., Cardemil, J. M., Droguett, E. L., & Behzad, M. (2020). Assessment of deep learning techniques for prognosis of solar thermal systems. *Renewable Energy*, *145*, 2178–2191. doi:10.1016/j.renene.2019.07.100

Cardona, J. L., Howland, M. F., & Dabiri, J. O. (2019). Seeing the wind: Visual wind speed prediction with a coupled convolutional and recurrent neural network. arXiv 2019, arXiv:1905.13290

Carolin Mabel, M., & Fernandez, E. (2008). Analysis of wind power generation and prediction using ANN: A case study. *Renewable Energy*, *33*(5), 986–992. doi:10.1016/j.renene.2007.06.013

Cellura, M., Cirrincione, G., Marvuglia, A., & Miraoui, A. (2008). Wind speed spatial estimation for energy planning in {Sicily}: {A} neural kriging application. *Renewable Energy*, *33*(6), 1251–1266. doi:10.1016/j.renene.2007.08.013

Chen, J., Yu, J., Song, M., & Valdmanis, V. (2019). Factor decomposition and prediction of solar energy consumption in the United States. *Journal of Cleaner Production*, *234*, 1210–1220. doi:10.1016/j.jclepro.2019.06.173

Chen, Q., & Folly, K. (2019). Effect of Input Features on the Performance of the ANN-based Wind Power Forecasting. Proceedings of the Southern African Universities Power Engineering Conference/Robotics and Mechatronics/Pattern Recognition Association of South Africa (SAUPEC/RobMech/PRASA), 673–678.

Damousis, I. G., Alexiadis, M. C., Theocharis, J. B., & Dokopoulos, P. S. (2004). A fuzzy model for wind speed prediction and power generation in wind parks using spatial correlation. *IEEE Transactions on Energy Conversion*, *19*(2), 352–361. doi:10.1109/TEC.2003.821865

Danfei, X., Dragomir, A., & Jain, P. A. (2018). Deep sensor fusion for 3d bounding box estimation. *Proceedings of the IEEE conference on computer vision and pattern recognition*, 244–53.

Dehghani, M., Riahi-Madvar, H., Hooshyaripor, F., Mosavi, A., Shamshirband, S., Zavadskas, E. K., & Chau, K.-W. (2019). Prediction of hydropower generation using Grey wolf optimization adaptive neuro-fuzzy inference system. *Energies*, *12*(2), 289. doi:10.3390/en12020289

Dorado-Moreno, M., Navarin, N., Gutierrez, P. A., Prieto, L., Sperduti, A., Salcedo-Sanz, S., & Hervas-Martinez, C. (2020a). Multi-task learning for the prediction of wind power ramp events with deep neural networks. *Neural Networks*, *123*, 401–411. doi:10.1016/j.neunet.2019.12.017 PMID:31926464

Dorado-Moreno, M., Navarin, N., Gutierrez, P. A., Prieto, L., Sperduti, A., Salcedo-Sanz, S., & Hervás-Martínez, C. (2020b). Multi-task learning for the prediction of wind power ramp events with deep neural networks. *Neural Networks*, *123*, 401–411. doi:10.1016/j.neunet.2019.12.017 PMID:31926464

Ekanayake, P., Peiris, A. T., Jayasinghe, J. M. J. W., & Rathnayake, U. (2021). Development of Wind Power Prediction Models for Pawan Danavi Wind Farm in Sri Lanka. *Mathematical Problems in Engineering*, *2021*, 4893713. doi:10.1155/2021/4893713

Ernst, B., Oakleaf, B., Ahlstrom, M. L., Lange, M., Moehrlen, C., Lange, B., Focken, U., & Rohrig, K. (2007). Predicting the wind. *IEEE Power & Energy Magazine*, *5*(6), 78–89. doi:10.1109/MPE.2007.906306

Erten, M. Y., & Aydilek, H. (2022). Solar Power Prediction using Regression Models. *Int. J. Eng. Res. Dev.*, *14*, s333–s342.

Eze, E. C., & Chatwin, C. R. (2019). Enhanced recurrent neural network for short-term wind farm power output prediction. *J. Appl. Sci.*, *5*, 28–35.

Feng, C., Cui, M., Hodge, B. M., & Zhang, J. (2017). A data-driven multi-model methodology with deep feature selection for short-term wind forecasting. *Applied Energy*, *190*, 1245–1257. doi:10.1016/j.apenergy.2017.01.043

Feng, C., & Zhang, J. (2018). Hourly-similarity based solar forecasting using multi-model machine learning blending. arXiv 2018, arXiv:1803.03623 doi:10.1109/PESGM.2018.8586091

Filik, Ü. B., & Filik, T. (2017). Wind speed prediction using artificial neural networks based on multiple local measurements in Eskisehir. *Energy Procedia*, *107*, 264–269. doi:10.1016/j.egypro.2016.12.147

Fischetti, M., & Fraccaro, M. (2019). Machine learning meets mathematical optimization to predict the optimal production of offshore wind parks. *Computers & Operations Research*, *106*, 289–297. doi:10.1016/j.cor.2018.04.006

Gaétan, B., Lionel, D., Matthieu, M., & Florent, L. (2019). Low-power neural networks for semantic segmentation of satellite images. *Proceedings of the IEEE/CVF international conference on computer vision workshops.*

García-Hinde, O., Terrén-Serrano, G., Hombrados-Herrera, M. A., Gómez-Verdejo, V., Jiménez-Fernández, S., Casanova-Mateo, C., Sanz-Justo, J., Martínez-Ramón, M., & Salcedo-Sanz, S. (2018). Evaluation of dimensionality reduction methods applied to numerical weather models for solar radiation forecasting. *Engineering Applications of Artificial Intelligence*, *69*, 157–167. doi:10.1016/j.engappai.2017.12.003

Han, S., Li, J., & Liu, Y. (2011). Tabu search algorithm optimized ANN model for wind power prediction with NWP. *Energy Procedia*, *12*, 733–740. doi:10.1016/j.egypro.2011.10.099

Harbola, S., & Coors, V. (2019). One dimensional convolutional neural network architectures for wind prediction. *Energy Conversion and Management*, *195*, 70–75. doi:10.1016/j.enconman.2019.05.007

Hong, W. C. (2008). Rainfall forecasting by technological machine learning models. *Applied Mathematics and Computation*, *200*(1), 41–57. doi:10.1016/j.amc.2007.10.046

Huang, G. B., Zhu, Q. Y., & Siew, C.-K. (2006). Extreme learning machine: Theory and applications. *Neurocomputing*, *70*(1-3), 489–501. doi:10.1016/j.neucom.2005.12.126

Huiyu, G., & Miaomiao, L. (2022). Short-term solar irradiance prediction from sky images with a clear sky model. *Proceedings of the IEEE/CVF winter conference on applications of computer vision*, 2475–83.

Ibrahim, S., Daut, I., Irwan, Y., Irwanto, M., Gomesh, N., & Farhana, Z. (2012). Linear Regression Model in Estimating Solar Radiation in Perlis. *Energy Procedia*, *18*, 1402–1412. doi:10.1016/j.egypro.2012.05.156

Jamii, J., Mansouri, M., Trabelsi, M., Mimouni, M. F., & Shatanawi, W. (2022). Effective artificial neural network-based wind power generation and load demand forecasting for optimum energy management. *Frontiers in Energy Research*, *10*, 898413. doi:10.3389/fenrg.2022.898413

Jiang, Y., & Huang, G. (2017). Short-term wind speed prediction: Hybrid of ensemble empirical mode decomposition, feature selection and error correction. *Energy Conversion and Management*, *144*, 340–350. doi:10.1016/j.enconman.2017.04.064

Jordan, M. I. (1987). Attractor dynamics and parallelism in a connectionist sequential machine. *Proceeding of 8th Annual Conference of the Cognitive Science Society (CogSci)*, 112–127.

Jursa, R. (2007). Wind power prediction with different artificial intelligence models. Proceedings of the 2007 European Wind Energy Conference and Exhibition (EWEC), 1–10.

Jursa, R., & Rohrig, K. (2008). Short-term wind power forecasting using evolutionary algorithms for the automated specification of artificial intelligence models. *International Journal of Forecasting*, *24*(4), 694–709. doi:10.1016/j.ijforecast.2008.08.007

Kalogirou, S. A., Panteliou, S., & Dentsoras, A. (1999). Artificial neural networks used for the performance prediction of a thermosiphon solar water heater. *Renewable Energy*, *18*(1), 87–99. doi:10.1016/S0960-1481(98)00787-3

Kamadinata, O. J., Ken, T. L., & Suwa, T. (2019). Sky image-based solar irradiance prediction methodologies using artificial neural networks. *Renewable Energy*, *134*, 837–845. doi:10.1016/j.renene.2018.11.056

Kechriotis, G., Zervas, E., & Manolakos, E. S. (1994). Using recurrent neural networks for adaptive communication channel equalization. *IEEE Transactions on Neural Networks*, *5*(2), 267–278. doi:10.1109/72.279190 PMID:18267796

Khandakar, A., Chowdhury, M. E. H., Khoda Kazi, M., Benhmed, K., Touati, F., Al-Hitmi, M., & Gonzales, J. S. (2019). Machine learning based photovoltaics (PV) power prediction using different environmental parameters of Qatar. *Energies*, *12*, 2782. doi:10.3390/en12142782

Kim, S. G., Jung, J. Y., & Sim, M. K. (2019). A two-step approach to solar power generation prediction based on weather data using machine learning. *Sustainability (Basel)*, *11*(5), 1501. doi:10.3390/su11051501

Kong, X., Liu, X., Shi, R., & Lee, K. Y. (2015). Wind speed prediction using reduced support vector machines with feature selection. *Neurocomputing*, *169*, 449–456. doi:10.1016/j.neucom.2014.09.090

Kou, P., Liang, D., Gao, F., & Gao, L. (2014). Probabilistic wind power forecasting with online model selection and warped Gaussian process. *Energy Conversion and Management*, *84*, 649–663. doi:10.1016/j.enconman.2014.04.051

Kumar, S., & Kaur, T. (2016). Development of ANN Based Model for Solar Potential Assessment Using Various Meteorological Parameters. *Energy Procedia*, *90*, 587–592. doi:10.1016/j.egypro.2016.11.227

Kumar, S., & Lopez, D. (2015). Feature selection used for wind speed forecasting with data driven approaches. *J. Eng. Sci. Technol. Rev.*, *8*(5), 124–127. doi:10.25103/jestr.085.17

Kusiak, A., Zheng, H., & Song, Z. (2009). Short-term prediction of wind farm power: A data mining approach. *IEEE Transactions on Energy Conversion*, *24*(1), 125–136. doi:10.1109/TEC.2008.2006552

Lansberry, J. E., Wozniak, L., & Goldberg, D. E. (1992). Optimal hydro generator governor tuning with a genetic algorithm. *IEEE Transactions on Energy Conversion*, *7*(4), 623–630. doi:10.1109/60.182643

Lauret, P., Voyant, C., Soubdhan, T., David, M., & Poggi, P. (2015). A benchmarking of machine learning techniques for solar radiation forecasting in an insular context. *Solar Energy*, *112*, 446–457. doi:10.1016/j.solener.2014.12.014

Leholo, S., Owolawi, P., & Akindeji, K. (2019). Solar energy potential forecasting and optimization using artificial neural network- South Africa case study. *Proceedings of the Amity International Conference on Artificial Intelligence*. 10.1109/AICAI.2019.8701372

Leron, J., & Sankaranarayanan, C. A. (2021). Precise forecasting of sky images using spatial warping. *Proceedings of the IEEE/CVF international conference on computer vision*, 1136–44.

Leva, S., Dolara, A., Grimaccia, F., Mussetta, M., & Sahin, E. (2017). Analysis and validation of 24 hours ahead neural network forecasting for photovoltaic output power. *Mathematics and Computers in Simulation*, *131*, 88–100. doi:10.1016/j.matcom.2015.05.010

Li, C., Lin, S., Xu, F., Liu, D., & Liu, J. (2018). Short-term wind power prediction based on data mining technology and improved support vector machine method: A case study in Northwest China. *Journal of Cleaner Production*, *205*, 205. doi:10.1016/j.jclepro.2018.09.143

Li, S. (2003). Wind power prediction using recurrent multilayer perceptron neural networks. *Proceedings of the 2003 IEEE Power Engineering Society General Meeting*, 2325–2330. 10.1109/PES.2003.1270992

Li, Y., Yang, P., & Wang, H. (2018). Short-term wind speed forecasting based on improved ant colony algorithm for LSSVM. *Cluster Computing*, *22*(S5), 11575–11581. doi:10.1007/s10586-017-1422-2

Liu, H., Mi, X., & Li, Y. (2018). Wind speed forecasting method based on deep learning strategy using empirical wavelet transform, long short term memory neural network and Elman neural network. *Energy Conversion and Management*, *156*, 498–514. doi:10.1016/j.enconman.2017.11.053

Ma, Y.-J., & Zhai, M.-Y. (2011). A dual-step integrated machine learning model for 24h- ahead wind energy generation prediction based on actual measurement data and environmental factors. *Energy Procedia*, *12*, 733–740.

Majidpour, M., Nazaripouya, H., Chu, P., Pota, H., & Gadh, R. (2018). Fast univariate time series prediction of solar power for real-time control of energy storage system. *Forecasting*, *1*(1), 8. doi:10.3390/forecast1010008

McCann, M. T., Jin, K. H., & Unser, M. (2017). Convolutional Neural Networks for Inverse problems in imaging. *IEEE Signal Processing Magazine*, *34*(6), 85–95. doi:10.1109/MSP.2017.2739299

Meng, A., Ge, J., Yin, H., & Chen, S. (2016). Wind speed forecasting based on wavelet packet decomposition and artificial neural networks trained by crisscross optimization algorithm. *Energy Conversion and Management*, *114*, 75–88. doi:10.1016/j.enconman.2016.02.013

Mohandes, M. A., Halawani, T. O., Rehman, S., & Hussain, A. A. (2004). Support vector machines for wind speed prediction. *Renewable Energy*, *29*(6), 939–947. doi:10.1016/j.renene.2003.11.009

Mujeeb, S., Alghamdi, T. A., Ullah, S., Fatima, A., Javaid, N., & Saba, T. (2019). Exploiting deep learning for wind power forecasting based on big data analytics. *Applied Sciences (Basel, Switzerland)*, *9*(20), 4417. doi:10.3390/app9204417

Niall, O., Sean, C., Anderson, C., Suman, H., Gustavo, V. H., & Lenka, K. (2020). Deep learning vs. traditional computer vision. In A. Kohei & K. Supriya (Eds.), *Advances in computer vision* (pp. 128–144). Springer International Publishing.

O'Leary, D., & Kubby, J. (2017). Feature selection and ANN solar power prediction. *Journal of Renewable Energy*, *2017*, 1–7. doi:10.1155/2017/2437387

Ozgoren, M., Bilgili, M., & Sahin, B. (2012). Estimation of global solar radiation using ANN over Turkey. *Expert Systems with Applications*, *39*(5), 5043–5051. doi:10.1016/j.eswa.2011.11.036

Paoli, C., Voyant, C., Muselli, M., & Nivet, M.-L. (2010). Forecasting of preprocessed daily solar radiation time series using neural networks. *Solar Energy*, *84*(12), 2146–2160. doi:10.1016/j.solener.2010.08.011

Pedro, H. T. C., & Coimbra, C. F. M. (2012). Assessment of forecasting techniques for solar power production with no exogenous inputs. *Solar Energy*, *86*(7), 2017–2028. doi:10.1016/j.solener.2012.04.004

Pedro, H. T. C., Coimbra, C. F. M., David, M., & Lauret, P. (2018). Assessment of machine learning techniques for deterministic and probabilistic intra-hour solar forecasts. *Renewable Energy*, *123*, 191–203. doi:10.1016/j.renene.2018.02.006

Qing, X., & Niu, Y. (2018). Hourly day-ahead solar irradiance prediction using weather forecasts by LSTM. *Energy*, *148*, 461–468. doi:10.1016/j.energy.2018.01.177

Quentin, P., Anthony, H., Guillaume, A., Philippe, B., & Joan, L. (2022). SPIN: Simplifying polar invariance for neural networks application to vision-based irradiance forecasting. Proceedings of the IEEE/CVF conference on computer vision and pattern recognition workshops, 5182–91.

Rana, M., Koprinska, I., & Agelidis, V. G. (2016). Univariate and multivariate methods for very short-term solar photovoltaic power forecasting. *Energy Conversion and Management*, *121*, 380–390. doi:10.1016/j.enconman.2016.05.025

Rana, M., & Rahman, A. (2020). Multiple steps ahead solar photovoltaic power forecasting based on univariate machine learning models and data re-sampling. *Sustain. Energy Grids Netw.*, *21*, 100286. doi:10.1016/j.segan.2019.100286

Rodriguez, F., Fleetwood, A., Galarza, A., & Fontán, L. (2018). Predicting solar energy generation through artificial neural networks using weather forecasts for microgrid control. *Renewable Energy*, *126*, 855–864. doi:10.1016/j.renene.2018.03.070

Salcedo-Sanz, S., Pastor-Sanchez, A., Prieto, L., Blanco-Aguilera, A., & Garcia-Herrera, R. (2014). Feature selection in wind speed prediction systems based on a hybrid coral reefs optimization-extreme learning machine approach. *Energy Conversion and Management*, *87*, 10–18. doi:10.1016/j.enconman.2014.06.041

Saloux, E., & Candanedo, J. A. (2018). Forecasting district heating demand using machine learning algorithms. *Energy Procedia*, *149*, 59–68. doi:10.1016/j.egypro.2018.08.169

Sapitang, M., Ridwan, W. M., Kushiar, K. F., Ahmed, A. N., & El-Shafie, A. (2020). Machine learning application in reservoir water level forecasting for sustainable hydropower generation strategy. *Sustainability*, *12*, 6121. doi:10.3390/su12156121

Senthil, K. P. (2019). Improved prediction of wind speed using machine learning. *EAI Endorsed Trans. Energy Web*, *6*(23), 1–7.

Sheng, Xiao, Cheng, Ni, & Wang. (2018). Short-term solar power forecasting based on weighted Gaussian process regression. *IEEE Trans. Ind. Electron.*

Shi, K., Qiao, Y., Zhao, W., Wang, Q., Liu, M., & Lu, Z. (2018). An improved random forest model of short-term wind-power forecasting to enhance accuracy, efficiency, and robustness. *Wind Energy (Chichester, England)*, *21*(12), 1383–1394. doi:10.1002/we.2261

Shimray, B. A., Singh, K., Khelchandra, T., & Mehta, R. (2017). Ranking of sites for installation of hydropower plant using MLP neural network trained with GA: A MADM approach. *Computational Intelligence and Neuroscience*, *2017*, 2017. doi:10.1155/2017/4152140 PMID:28331490

Sierra-García, J. E., & Santos, M. (2020). Exploring Reward Strategies for Wind Turbine Pitch Control by Reinforcement Learning. *Applied Sciences (Basel, Switzerland)*, *10*(21), 7462. doi:10.3390/app10217462

Sun, S., Wang, S., Zhang, G., & Zheng, J. (2018). A decomposition-clustering-ensemble learning approach for solar radiation forecasting. *Solar Energy*, *163*, 189–199. doi:10.1016/j.solener.2018.02.006

Torres, J. F., Troncoso, A., Koprinska, I., Wang, Z., & Martínez-Álvarez, F. (2019). Big- data solar power forecasting based on deep learning and multiple data sources. *Applied Energy*, *238*, 1312–1326.

Torres-Barran, A., Alonso, Á., & Dorronsoro, J. R. (2019). Regression tree ensembles for wind energy and solar radiation prediction. *Neurocomputing*, *326*, 151–160. doi:10.1016/j.neucom.2017.05.104

Varanasi, J., & Tripathi, M. M. (2019). K-means clustering based photo voltaic power forecasting using artificial neural network, particle swarm optimization and support vector regression. *J. Inf. Optim. Sci.*, *40*(2), 309–328. doi:10.1080/02522667.2019.1578091

Vassallo, D., Krishnamurthy, R., Sherman, T., & Fernando, H. J. S. (2020). Analysis of Random Forest Modeling Strategies for Multi-Step Wind Speed Forecasting. *Energies*, *13*(20), 5488. doi:10.3390/en13205488

Vignesh, V., Yuchi, S., & Brandt, R. A. (2019). Short-term solar PV forecasting using computer vision: The search for optimal CNN architectures for incorporating sky images and PV generation history. *Journal of Renewable and Sustainable Energy*, *11*(6), 066102. doi:10.1063/1.5122796

Vincent, L. G., & Nicolas, T. (2020). A deep physical model for solar irradiance forecasting with fisheye images. *Proceedings of the IEEE/CVF conference on computer vision and pattern recognition workshops*, 630–1.

Vishnu, S., Anirudh, J., Burak, U., & Stefano, E. (2020). Cloud removal from satellite images using spatiotemporal generator networks. *Proceedings of the IEEE/CVF winter conference on applications of computer vision*, 1796–805.

Wang, F., Zhen, Z., Mi, Z., Sun, H., Su, S., & Yang, G. (2015). Solar irradiance feature extraction and support vector machines based weather status pattern recognition model for short-term photovoltaic power forecasting. *Energy and Building*, *86*, 427–438. doi:10.1016/j.enbuild.2014.10.002

Wang, F., Zhen, Z., Wang, B., & Mi, Z. (2017). Comparative study on KNN and SVM based weather classification models for day ahead short term solar PV power forecasting. *Applied Sciences (Basel, Switzerland)*, *8*(1), 28. doi:10.3390/app8010028

Wang, S., Tang, L., & Yu, L. (2011). SD-LSSVR-based decomposition-and-ensemble methodology with application to hydropower consumption forecasting. *Proceedings of the 4th International Joint Conference on Computational Sciences and Optimization (CSO)*, 603–607. 10.1109/CSO.2011.303

Welch, R. L., Ruffing, S. M., & Venayagamoorthy, G. K. (2009). Comparison of feedforward and feedback neural network architectures for short term wind speed prediction. *Proceedings of International Joint Conference on Neural Networks*, 3335–3340. doi:10.1109/IJCNN.2009.5179034

Wen, L. L., Zhou, K. L., Yang, S. L., & Lu, X. H. (2019). Optimal load dispatch of community micro grid with deep learning based solar power and load forecasting. *Energy*, *171*, 1053–1065. doi:10.1016/j.energy.2019.01.075

Will, A., Bustos, J., Bocco, M., Gotaya, J., & Lamelas, C. (2011). On the use of niching genetic algorithms for variable selection in solar radiation estimation. *Renewable Energy*, *50*, 168–176. doi:10.1016/j.renene.2012.06.039

Williams, R., & Zipser, D. (1989). A learning algorithm for continually running fully recurrent neural networks. *Neural Computation*, *1*(2), 270–280. doi:10.1162/neco.1989.1.2.270

Yao, G., Lei, T., & Zhong, J. (2019). A review of Convolutional-Neural-Network-based action recognition. *Pattern Recognition Letters*, *118*, 14–22. doi:10.1016/j.patrec.2018.05.018

Yousif, J. H., Kazem, H. A., Alattar, N. N., & Elhassan, I. I. (2019). A comparison study based on artificial neural network for assessing PV/T solar energy production. *Case Studies in Thermal Engineering*, *13*, 1–13. doi:10.1016/j.csite.2019.100407

Yu, R., Gao, J., Yu, M., Lu, W., Xu, T., Zhao, M., Zhang, J., Zhang, R., & Zhang, Z. (2019). LSTM-EFG for wind power forecasting based on sequential correlation features. *Future Generation Computer Systems*, *93*, 33–42. doi:10.1016/j.future.2018.09.054

Zambrano, A. F., & Giraldo, L. F. (2020). Solar-irradiance forecasting models without onsite training measurements. *Renewable Energy*, *152*, 557–566. doi:10.1016/j.renene.2020.01.092

Zhang, C., Wei, H., Zhao, J., Liu, T., Zhu, T., & Zhang, K. (2016). Short-term wind speed forecasting using empirical mode decomposition and feature selection. *Renewable Energy*, *96*, 727–737. doi:10.1016/j.renene.2016.05.023

Zhang, X., Peng, Y., Xu, W., & Wang, B. (2017). An optimal operation model for hydropower stations considering inflow forecasts with different Lead-times. *Water Resources Management*, *33*(1), 173–188. doi:10.1007/s11269-018-2095-1

Chapter 5
AI Approach Towards Optimal Finding of Renewable Sources of Energy and Their Classification

Pampa Saha
Brainware University, India

Anshit Mukherjee
https://orcid.org/0009-0001-7930-401X
Abacus Institute of Engineering and Management, India

Gunjan Mukherjee
https://orcid.org/0000-0002-3959-3718
Brainware University, India

ABSTRACT

The smart energy domain poses substantial challenges in future research, necessitating advanced investigation into optimizing smart customizable networks using artificial intelligence (AI) and machine learning (ML). With renewable energy (RE) being pivotal for global development amid climate change, AI introduces new paradigms to reshape activities, demanding revamped energy infrastructure and RE deployment strategies. The chapter explores the adoption of AI in future smart cities research with considerable economic benefits. Moreover, in the field of environmental science and engineering (ESE), the ML's potential in revolutionizing ESE by addressing complex problems and outlining crucial components for successful ML implementation in ESE, correct model development, proper interpretation, and applicability analysis has been done. The renewable source of energy like solar and wind energy can be generated in place where they are found in plenty. The ML application of energy source tracing and the prediction of energy type with multiple models of efficiencies has been highlighted.

DOI: 10.4018/979-8-3693-2355-7.ch005

INTRODUCTION

The integration of advanced artificial intelligence (AI) into Smart Energy systems necessitates a comprehensive understanding of computational, economic, and societal implications. Establishing this socio-technical platform requires defining the domain and specifying research problems cohesively. Sustainable development challenges post- industrial society, urging academia, industry, and society to seek tangible solutions. The allure of renewable energy (RE) emerged during the 1970s energy crisis when the risk of depleting conventional fuels became apparent, prompting the development of RE and resource conservation. Subsequent concerns about pollution, global warming, and resource scarcity in the 1980s highlighted the imperative to mitigate environmental damage. Integrating renewable into the energy sector significantly mitigates greenhouse gas emissions and import reliance, emphasizing the imperative of environmental protection. AI's evolution permeates various sectors, influencing global productivity, equality, environmental outcomes, and more. Machine learning(ML) enables machines to learn from data, enhancing performance and predicting outputs autonomously. Note worthy ML algorithms in photovoltaic (PV) applications include support vector machines (SVM), neural networks (NNs), decision trees (DT), and deep learning (DL) techniques like convolution neural networks (CNNs), long-short-term memory (LSTM), and generative adversarial networks (GANs). DL, a subset of ML, eliminates manual feature extraction, allowing algorithms to autonomously extract features from raw data, marking a novel advancement in neural networks.

In addition to supervised learning, recent advancements have been introduced unsupervised ML approaches. In unsupervised methods, the computer algorithm autonomously identifies patterns in the data, operating in an "unsupervised" manner without user intervention. The primary application of unsupervised ML involves automatically categorizing data into distinct groups or "clusters" that exhibit similar characteristics. Unsupervised learning algorithm avoids the training with labeled and classified datasets. The time of training procedure is very optimal in unsupervised learning algorithm.

Within the realm of environmental studies, recent applications of unsupervised ML, such as t-distributed stochastic neighbor embedding (t-SNE) or k-means clustering, have been employed to categorize the carbon–fluorine bond dissociation energies of per- and polyfluoroalkyl substances (PFAS). These applications aim to comprehend bond dissociation energies, enabling the visualization of high-dimensional data as two-dimensional "clusters" where data points within a cluster share comparable characteristics. It is crucial to emphasize that these clusters are automatically identified by unsupervised ML algorithms, devoid of human intervention. Consequently, these results underscore the utility of such algorithms in automatically classifying and elucidating chemical trends in environmental contaminants, which would be challenging to detect manually.

Artificial intelligence (AI) heralds a new epoch in the knowledge society, reshaping global development through information and knowledge. Its evolution signifies a shift in traditional production factors and organizational paradigms to confront modern challenges. Recent strides in neural networks, cognitive computing, and machine learning bolster energy management, monitoring, optimization capabilities and identification of the energy availability trends. Integrating AI with 5G and sensor networks lays the groundwork for future-generation smart city services. Application of IOT technology also provides the degree of improvement in accurate identification of the energy followed by the appropriate optimization strategy using AI techniques. In the sphere of smart energy grids, the fusion of advanced machine learning and cognitive computing within AI presents unparalleled efficiency prospects. Our contribution to Future Smart Cities research introduces AI-driven renewable energy (RE) services, promising significant social impact and economic efficiency. Acknowledging the complexity of this domain, our study serves as an inaugural step toward initiating a robust research dialogue.RE, abundant in remote areas, challenges the development of technologies for energy production, integration, and control, aiming to enhance conversion efficiency and diminish dependence on fluctuating fossil fuel prices. So the economic connectivity is being indirectly hit by the unique application of machine learning. However, managing the variability of renewable energy sources demands predictive methods to optimize energy production, storage, and usage, ensuring reliability and efficiency. Intermittent power generation in RE grids poses challenges, risking reliance on backup fossil fuels if improperly managed. Accurate forecasting becomes pivotal for stable power grid management, minimizing deviations and associated costs while meeting demand fluctuations. The security at the grid point poses a major issue as the grid is much likely to succumb to security threat. Aligning supply with demand remains critical for grid stability, urging advancements in forecasting accuracy to mitigate disruptions and operational costs. The integration of AI in energy grids underscores the necessity of precise variable resource forecasting, crucial for grid stability, cost reduction, and meeting consumer demands amidst evolving power supply dynamics. There exist several RESs, such as hydropower, bioenergy, wind energy, solar thermal, solar photovoltaic (PV) energy, and many more. Identification of the renewable energy source has become the most significant challenge in order to get access to the source properly. Identification of the fake source of energy has become a threat sometimes as it fails to contribute anything to the renewability of energy mitigating the crucial and acute demand. The AI enabled approach combined with the core principle of machine learning ahs helped a lot in tracking the real renewable energy source in some optimized manner saving both the time and cost connected to the search options.

LITERATURE REVIEWS

The research trends in the identification of renewable source of energy using the machine learning principle has been of great demand. Number of research works has been performed producing overall good outcomes. Many research work are going on right now producing many different directions of new possibilities with robust future prospects. Some of the work and their discussions along with their score on the achieved accuracy values has been covered in the following part of the section.

(Zhu et al.,2020) employed ML techniques to analyze key variables affecting the performance of Cu(In1−x Gax)Se2 (CIGS) solar cells and the underlying correlations. They selected 15 input features, including CGI, GGI ratio, CIGS layer thickness, highest substrate temperature during fabrication, fabrication method, alkali treatment, substrate type, among others. Random Forest (RF) performed the best in predicting CIGS solar cell efficiency. According to RF results, alkali post-deposition treatment (PDT), CIGS thickness, buffer layer thickness, CGI, and GGI gradient were identified as the most significant features influencing device performance. The study recommended a co-evaporation fabrication method for CIGS with CdS as the buffer layer. It also suggested an ideal CIGS thickness of 2-2.3 micrometers, a substrate temperature range of 540-600oC, and CGI and GGI ratios adjusted within specified ranges for highly efficient CIGS solar cells. (Salman et al.,2020) developed an Artificial Neural Network (ANN)-based ML model to extract Cu doping profiles resulting from the diffusion annealing and cool-down process in the fabrication sequence of CdTe solar cells. Using a dataset generated by the PVRD-FASP solver, they aimed to mimic the Cu diffusion process with the ANN ML model. The study successfully replicated the Cu diffusion process, showing good agreement between PVRD-FASP solver simulation

Table 1. Literature review in this domain

Author Name	Purpose	Accuracy
(Budak et al., 2019)	To develop and implement a systematic approach to assess renewable energy and identify the energy alternatives for a city using the analytic hierarchy process	High accuracy
(Lagarrigue et al., 2021)	To explore how artificial intelligence can help manage decentralized grids and optimize energy use and storage	Considerable accuracy
(Singh et al.,2020)	To present a computationally efficient technique to find the most feasible values of different control variables of the power system having distributed renewable energy sources	99.8%
(Kaur et al., 2023)	To provide a review of AI explainability and governance in smart energy systems and discover the parameters or themes for AI governance in energy	Considerable accuracy
(Liu et al.,2020)	To propose a novel deep reinforcement learning framework for optimal energy management of microgrids with renewable energy sources and energy storage systems	97.6%
(Wang et al.,2019)	To develop a hybrid intelligent optimization algorithm based on artificial bee colony and particle swarm optimization for optimal power flow with renewable energy integration	98.2%
(Zhang et al.,2019)	To design a deep neural network based approach for short-term wind power forecasting using spatio-temporal data	95.4%
(Li et al.,2019)	To apply a machine learning method for solar power prediction using satellite images and numerical weather prediction data	94.7%
(Al-Sumaiti et al.,2018)	To present a comprehensive survey on the applications of artificial intelligence techniques for optimal planning and operation of hybrid renewable energy systems	Considerable accuracy
(Siano et al., 2014)	To analyze the main challenges and opportunities for effective integration of renewable energy sources and demand response programs by using artificial intelligence tools	Considerable accuracy value
(AlRashidi and El-Naggar, 2010)	To employ artificial neural networks for long term electric load forecasting in Kuwait with consideration of renewable energy sources	93.1%
(Mohandes et al.,2004)	To demonstrate the use of artificial neural networks for wind speed and power prediction in Saudi Arabia	92.8%
(Mellit and Kalogirou, 2008)	To compare different artificial intelligence techniques for daily global solar radiation prediction in Algeria	96.3%
(Kusiak et al., 2010)	To propose a data-driven approach based on data mining and machine learning for optimization of wind turbine performance	99.2%
(Mohammadi-Ivatloo et al., 2013)	To solve the stochastic security-constrained unit commitment problem with high penetration of wind power using artificial bee colony algorithm	97.9%
(Abbas et al.,2018)	To solve the problem, the genetic algorithm have been used for the optimization purpose .	95%
(Touati et al.,2017)	To predict the output power from the photo voltaic cell subject to different atmospheric conditions	94%
(Eltamly et al..2020)	To provide the efficiency of the cyber physical system based wind energy system	94%

results and ANN predictions. Additionally, the ANN algorithm demonstrated an almost instantaneous process for creating Cu doping profiles compared to the lengthy procedure with the PVRD-FASP solver. This makes the ANN algorithm a valuable tool for efficiently developing optimal doping profiles for CdTe solar cells.(W.B Xiao et al.,2020) applied ANN-based ML models to predict the Power Conversion Efficiency (PCE) of three types of silicon solar cells: mono-crystalline, multicrystalline, and amorphous crystalline. Training data were obtained by characterizing experimental solar cell samples under various conditions of incident light intensity and ambient temperature. Prediction results were evaluated based on correlation coefficient and mean square error (MSE). The study found that prediction results aligned with traditional physics-based studies of solar cells. The authors also suggested optimal numbers of hidden layers in the ANN architecture for each of the three solar cell technologies and discussed the significance of the number of hidden layers in determining prediction quality. According to their analysis, multi- and amorphous crystalline cells utilizing 3 or 4 hidden layer units demonstrated optimal results. Regarding single crystalline cells, the optimal results were achieved with eight hidden layer units, resulting in a high correlation coefficient and low MSE.Anti-reflection coatings (ARCs) are employed on solar cell surfaces to minimize light reflection and enhance light absorption, leading to increased efficiency and power output (Shivam et al., 2020). In a recent study, (Shivam et al.,2020) suggested utilizing Artificial Neural Networks (ANNs) to predict the optimal thickness of ARC coatings based on solar cell-generated voltage and current. The authors utilized PC1D, a 1-D numerical simulation tool, to establish the training database for their ANN model. Results indicated that augmenting the ARC coating's thickness can contribute to higher power output from the solar cell. (Kaya and Hajimirza, 2020) crafted a Neural Network (NN) surrogate model to forecast the optical performance of thin film multi-layered amorphous-silicon-based solar cells using Finite-Difference Time-Domain (FDTD) simulation methods for dataset creation. The NN-based approach exhibited accuracy and increased speed compared to FDTD simulations in predicting optical absorption concerning incident light wavelength and cell geometry. The study explored the fundamental reasons behind the ability of an ML-based surrogate model to approximate device physics-based simulation model results. The optimization solution suggested a substantial improvement in external quantum efficiency compared to bare silicon and a randomly designed structure. (Wasmer et al.,2018) proposed an ensemble of decision trees with feature sub-sampling as the ML model to comprehend the Power Conversion Efficiency (PCE) of mass-produced Q. ANTUM solar cells based on p-type Czochralski Silicon (Cz-Si) wafer. Their objectives were to understand the global hourly time trends of solar cell efficiency in a 1-week dataset and the impact of different features of solar cell production on the temporal evolution of performance. The training data, collected from 500,000 solar cells through inline measurements over one week, encompassed 329 features grouped using hierarchical clustering. The ML model demonstrated high prediction accuracy for solar cell efficiency over time, indicating its potential application in real production lines. (Dawood et al.,2020) has shown that the hydrohgen energy source is much efficient and free from carbon di oxide and so could have been used as the most conducive and perfect source of energy for sustainable renewable energy supply. (Yue et al.,2021) has explored and unearthed the underlying problem related to the renewable source of energy and the integration of hydrogen into the energy systems. He has also reported the work related to the cost effective economy concerned to the alternative energy source (Parra et al.,2019) has performed the hydrogen based power systems and inferred the consensus that the energy derived can be used in decarburizing sectors such as transport, power and heat by using them in much economically viable way. (Carrillo et al.,2019) has reported the unique method of alternative source of energy by harnessing the power of sun. The concentrated solar power (CSP) has been reported to be the conducive source of alternative energy for the generation of electricity. (Angulo et al.2020) has reported the role of bubbles in the transfer of energy and shown the influence of bubbles in transfer of energy through the electrochemical reactions.

It is evident that from the above discussion that a large progress of work has been found in the research towards finding the optimal approach of finding the alternative energy sources. During the era till the recent year number publications has been done in this trend . In the future year to come, many more research has been planned to take place to escalate the value of the alternative source of energy to the greater height to the human society for more perfect sustainable environment.

INTRODUCTION TO SOLAR POWER AND NEED FOR ITS FORECASTING

Solar power, derived from sunlight through photovoltaic (PV) or concentrated solar power systems, standing as a sustainable and competitive renewable energy source. Solar photovoltaic (PV) energy specifically is highly regarded for its sustainability. However, a key challenge lies in its intermittency due to weather conditions, notably variations in temperature and irradiance, significantly affecting electric power production quality. Accurate short-term forecasting of PV module power output becomes crucial for efficiently managing daily/hourly power grid production, delivery, and storage. Moreover, such forecasting holds significance in decision-making within energy markets. The ability to predict the power output of PV modules aids in optimizing grid operations and enables better resource allocation, enhancing the reliability and efficiency of solar power utilization.

Solar photovoltaic (PV) systems have witnessed substantial growth in electricity generation, marked by a rise in the number of installations. These systems are now commonly found in open spaces, rooftops, integrated into buildings and vehicles, and organized into expansive arrays within solar power plants. Despite the positive impacts of solar PV on diversifying power generation and addressing global warming, the swift adoption of this technology brings along certain challenges. As such, it varies based on site-specific elements as well as weather trends. Power instability is a risk that increases as renewable energy sources become more and more prevalent in the energy mix, which emphasizes how important it is to control the flow of renewable energy. The practice of projecting how much electricity a solar photovoltaic (PV) system will produce over a given period of time is known as solar PV forecasting. This entails predicting a PV system's output based on a variety of input characteristics using mathematical models. Differences between the available power generation and the desired load may arise from the ambiguity of the load consumption as well as the intermittent power output of the solar PV production. Therefore, it is crucial to predict or forecast solar PV power generation capacity accurately .Precise solar photovoltaic forecasting is crucial for multiple reasons. Initially, it enables independent systems or grid operators to predict variations in the electrical supply and modify the distribution of power in line with that. This supports the upkeep of the power grid's stability and dependability, as well as that of standalone systems that are disconnected from it. Second, precise projections can minimize the requirement for backup power sources, maximize the usage of renewable energy sources, and eventually lower customer electricity costs.

Solar energy production is a complex and dynamic process, making projecting solar PV a tough challenge. Inaccuracies in forecasting can give rise to a mismatch between the available power generation and the demand, potentially causing power outages or other disruptions. Moreover, inaccurate forecasts may result in the underutilization of solar PV systems, restricting the integration of renewable energy into the power grid and augmenting reliance on fossil fuel-based sources of electricity. Temperature, humidity, and cloud cover are a few examples of variables that might impact a PV system's output. As such, it varies based on site-specific elements as well as weather trends. Power instability is more likely to occur as renewable energy sources become more prevalent in the energy mix which emphasizes the need of efficiently managing the flow of renewable power. In 1954, a team led by Pearson at US Bell Labs pioneered the development of the initial crystalline silicon solar cell, boasting a power conversion efficiency (PCE) of 4.5%. This event marked a crucial breakthrough in the utilization of solar electricity. The PCE of contemporary mono-crystalline/poly-crystalline silicon solar cells in industrial applications has now surpassed 20%. However, these silicon-based solar cells are comparatively costly to produce, delicate, prone to damage, environmentally impactful, and exhibit limited temperature tolerance. A unit photovoltaic cell harvests photons from incident solar radiation and creates electron-hole pairs (EHP), which result in the production of photo-electric current as an output. PV cells are joined in series to form solar panels, which can be used alone or in conjunction with other panels to create larger power plants.

The three primary components of designing a typical solar cell are selecting appropriate PV materials, refining the device structure, and creating manufacturing processes. A unit PV cell captures photons from the sunlight incident, upon it and generates electron-hole pairs (EHP), the result of which is photo-electric current. PV cells are joined in series to form solar panels, which can be used alone or in conjunction with other panels to create larger power plants. The three primary components of designing a typical solar cell are selecting appropriate PV materials, refining the device structure, and creating manufacturing processes.

Solar PV systems demand continuous monitoring due to the vulnerability of solar radiation to unpredictable fluctuations and uncertainties. Consequently, it is crucial to address the challenges associated with the operation of PV infrastructure. Machine learning (ML) algorithms have surfaced as alternatives to traditional approaches to tackle these challenges, aiming to improve the efficiency, reliability, and cost-effectiveness of PV systems, akin to advancements in the fabrication process. Various application areas of ML in both aspects have been extensively researched and are exemplified in the following contexts.

PHYSICAL MODELS (WHITE BOX)

Physical models are grounded in the fundamental physical principles that govern the behavior of solar PV systems. These models consider the essential physical properties of the solar PV system, such as panel efficiency, the angle of incidence of the sun's rays, and panel shading. Additionally, environmental factors influencing solar energy production, including temperature, cloud cover, and humidity, are taken into account. Typically formulated through mathematical equations, these models describe the physical behavior of the solar PV system. However, the complexity of these equations necessitates a deep understanding of the underlying physics and system characteristics (installation details, electrical configuration, geographic location, and technical specifications) of the modelled PV system. Moreover, their reliance on weather conditions for forecasting makes their performance heavily dependent on the accuracy of weather forecast data

DATA DRIVEN MODELS

Data-driven forecasting models leverage historical data analysis to derive insights, patterns, and relationships, enabling the prediction of solar PV system outputs. These models, grounded in statistics and algorithms, learn patterns and relationships from data to make accurate predictions. Effective implementation requires substantial amounts of high-quality data, and prediction accuracy can be influenced by biases or incomplete data. Data-driven models can be further categorized into statistical and machine learning models. The distinction between statistical modelling and machine learning can be blurry, with some researchers considering them interchangeable depending on the specific context and user perspective. As a normal trend of classification, the statistics takes a major and significant role in the machine learning based analysis.

A: Statistical Models

Statistical models rely on the statistical analysis of historical data to identify trends and patterns for forecasting future solar PV power output. These models employ statistical techniques and mathematical algorithms to fit historical data and establish relationships between various variables, such as solar irradiance, weather conditions, and energy production. Numerous types of statistical models are used in solar PV forecasting, including autoregressive models (AR), moving average models (MA), and Autoregressive Integrated Moving Average (ARIMA). ARIMA models, commonly employed for time series analysis and forecasting, assume that the current value of a time series depends on its past values. The common tasks involved in the analysis of such model is the resource integration and their proper classification based on the model prepared. Various associated metrics produces relevant outputs for the further analysis. The statistical features derived can also be used as the inputs to different machine learning procedures to analyse the outputs further to greater precision level.

B: Machine Learning Models (Black Box)

Contrasting with statistical models, machine learning models rely on more intricate algorithms that enable them to autonomously learn patterns and relationships within the data. These models excel in handling complex and nonlinear connections between input and output variables, often yielding more precise forecasts compared to statistical models. Consequently, they have demonstrated efficacy in producing accurate PV forecasting. Noteworthy machine learning models applied in solar PV forecasting encompass artificial neural networks (ANN), support vector machines (SVM), and random forests (RF).

A fundamental distinction between statistical models and machine learning models lies in the level of human involvement required. Statistical models typically demand more human input concerning feature selection and engineering, designing the model architecture, and tuning hyper parameters. Conversely, machine learning models can be trained on raw data with minimal preprocessing, and the model architecture, along with hyper parameters, can often be automatically optimized using techniques like grid search or Bayesian optimization. Another contrasting aspect between the two approaches is their capacity to handle uncertainty. Statistical models often offer more transparency and can furnish estimates of uncertainty in their forecasts. In contrast, machine learning models are often more opaque or "black-box," providing limited information on how predictions were derived. However, techniques such as Monte Carlo simulation can be employed to generate probabilistic forecasts from machine learning models. The machine learning approach prompts the classification on larger and potential scale of precision. The training of model is very crucial and equally follow for supervised and non-supervised technique as well. In case of energy classification some relevant attributes can be used such as the characteristics of source of energy, energy flow trends, intensity of energy flow, direction of the energy etc. Such features of energy can be utilized properly in carrying out the training of models.

HYBRID MODELS

Hybrid models are those that amalgamate one or more forecasting techniques derived from either physical or data-driven methods, aiming to overcome the limitations of individual approaches while capitalizing on the strengths of each method. However, this amalgamation does incur a trade-off in the form of increased computational time. By utilizing a blend of these methodologies and continually enhancing the precision of solar PV forecasting, it becomes feasible to uphold system reliability, optimize the integration of renewable energy, and ultimately progress towards a more sustainable and carbon-free energy system. In the context of this paper, the primary scope and objective revolve around addressing accurate solar PV power forecasting as a pivotal tool for ensuring system reliability and maximizing the integration of renewable energy. The goal is to present a comprehensive and comparative examination of existing Machine Learning (ML) based approaches in PV power generation. While there are related literature surveys on solar PV power generation forecasting, most do not conduct simulation comparisons for short-term forecasts. Consequently, the presented work stands out by placing emphasis on the comparison of short-term horizons and conducting a comprehensive review with simulations. The relevant factors impacting solar PV power forecasting methods encompass the forecasting horizon and prevailing weather conditions. The regional weather condition has been strongly being influenced by the proper forecasting of PV power as it can vary from one time instance to another in a single day.

Forecasting Horizons: Numerous methods in the literature undertake forecasts for solar PV power generation, classified based on temporal, spatial, or data-centric aspects. Key temporal components include forecast interval (fi), forecast resolution (fr), and forecast horizon (fh). The forecast horizon represents the temporal span between the current time (t) and the subsequent prediction time. Forecast resolution denotes the frequency of forecasts, while forecast interval signifies the time span of projections. These aspects categorize PV power generation forecasting based on forecast horizons.

Different applications align with distinct forecasting horizons, such as short-term forecasts for day-ahead energy markets, energy storage, load scheduling, and energy market participation. Neglecting the impact of the forecast horizon on accuracy can lead to poor forecasting performance, affecting activities reliant on the information, Studies, such as (Al-Saedi et al., 2018), emphasize variations in forecasting accuracy relative to the forecast horizon's length, emphasizing the significant impact of this factor on forecasting performance.

Weather Conditions: In solar energy conversion, Photovoltaic (PV) transducers convert solar irradiance into electrical energy. Factors influencing PV efficiency include solar irradiance, ambient temperature, geographical location, panel design, and electrical components. Weather conditions, particularly variations in solar irradiation reaching the Earth, introduce uncertainty and variability in PV output power. These variations consist of deterministic (e.g., diurnal and annual cycles) and stochastic (meteorological or weather-related) components. Weather forecasting and forecasted data, considering factors like cloud cover, humidity, wind speed, temperature, and air pressure, contribute to improved solar PV systems power forecasting. Different meteorological variables may exhibit varying correlation factors with PV power based on the PV plant's location (Kaya et al., 2020).

SOLAR PV POWER FORECASTING MODELS

Reliable predictions of short and long-term production capabilities for solar PV units are essential for utilities heavily dependent on PV reliability. Two main approaches exist: the direct approach provides PV power as an immediate output, while the indirect approach predicts solar irradiation first, using meteorological or weather forecast data as input for PV system models to generate the predicted PV power harvest. The indirect approach is favoured when historical PV power data is lacking or when weather forecast data serves other applications.

Various techniques are explored in the literature for solar PV power forecasting, catering to different forecasting horizons and conditions. Some prominent techniques include machine learning, statistical models, and hybrid approaches, each offering distinct advantages and applications

MACHINE LEARNING MODELS

Efficient ML algorithms are employed for both regression and classification tasks across various aspects of solar cell technology. In the fabrication stages, predicting parameters like Power Conversion Efficiency (PCE), bandgap, conduction band, valence band, doping concentration, and the selection of optimal materials holds significant importance. ML approaches such as Random Forest (RF), Decision Tree (DT), XGBoost (XGB), and Gradient Boosted Decision Tree (GBDT) are commonly applied for these tasks. The utilization of deep learning is constrained by the limited availability of a large dataset relative to the numerous features involved.

Conversely, in the implementation stage of PV systems, models based on Deep Learning, such as Convolutional Neural Network (CNN), Long Short-Term Memory (LSTM), and hybrid architectures, are more prevalent owing to the abundance of large training samples. This section provides a summary of some of the frequently employed ML techniques in the PV research literature.

A. Decision Tree (DT)

The Decision Tree (DT) stands out as one of the most straightforward machine learning algorithms. While it is predominantly known for its application in classification tasks, it finds extensive use in regression as well (Kausiak et al., 2011). A decision tree comprises roots, branches, and leaves, with each branch making decisions about the truthfulness of a statement. The final output is derived from the leaves. A critical requirement for DT involves determining the optimal tree structure. This is accomplished by selecting a splitting criterion in the root and branches that effectively separates the input samples. This process is also referred to as attribute selection. The selection measure, often referred to as ASM, is commonly evaluated using methods such as information gain and the Gini index, calculated as outlined in equations (1) and (3) respectively. In the context of a node, the entropy before splitting is represented as Entropy(S), where S denotes the number of samples at the current node, and Sv is the number of samples belonging to class v. The entropy for N features is determined by equation (2), where Pj represents the probability of randomly selecting an element of class j. In the case of decision trees (DT) based on information gain, the objective is to maximize this value, whereas Gini scores are aimed at minimization.

$$\text{Information gain=Entropy(S)-} \sum \left(\frac{Sv}{S} \text{Entropy}(Sv)\right) \tag{1}$$

$$\text{where, Entropy(S)= -} \sum_{j=1}^{N} -P_j \times \log_2 (P_j) \tag{2}$$

$$\text{gini}=1 - \sum_{j=1}^{N} P_j^2 \tag{3}$$

B. Random Forest (RF)

Random Forest (RF) is a widely adopted machine learning technique designed to address the high variance associated with Decision Trees (DT). It functions as an ensemble of numerous unpruned independent decision trees, each fitted on training data using the bagging method to capture non-linear relationships effectively (Sing et al., 2019). The learning process initiates by creating a bootstrapped dataset through random sampling from the original one, allowing each observation to be selected multiple times. Typically, about two-thirds of the training data constitutes the bootstrapped dataset, while the remaining portion forms an ''out-of-bag'' (OOB) set utilized for evaluation purposes.

When dealing with an input featuring a total of m features, each decision tree is constructed by randomly selecting a subset of k from the m features (where k belongs to m). Each tree is allowed to grow to its full extent. In the context of regression analysis, the final output ^y(x) represents the average prediction across all trees, as indicated in equation (4), where K denotes the number of trees, and hi(x) signifies the output prediction for the ith tree. In the case of classification tasks, RF employs the majority vote from all trees as its ultimate output.

$$\hat{y}(x)\frac{1}{K}\times\sum_{i=1}^{k} h_i(x) \tag{4}$$

C. Gradient Boosted Decision Tree (GBDT)

Gradient Boosted Decision Tree (GBDT) is an additional ensemble machine learning algorithm based on trees, employing hundreds of Decision Trees (DTs) as base learners (Kaya et al., 2020). These trees are constructed sequentially, with each tree giving more emphasis to the misclassifications made by the preceding one. GBDT is a highly popular method extensively utilized in both regression and classification tasks. Given a set of n input observations, (xi, yi) where i ranges from 1 to n, and a differential loss function L, GBDT begins by making an initial prediction, as depicted in equation (5). This equation aims to determine an initial value γ that minimizes the sum of the loss function across all observations before commencing the training iterations.

$$F_0(x) = argmin_{\gamma} \sum_{i=1}^{N} \mathcal{L}(y_i, \gamma_i) \tag{5}$$

$$r_{sm} = -\left[\frac{\partial\mathcal{L}(y_i, F(x_i))}{\partial F(x_i)}\right] \int_x = \int_{m-1}(x) \tag{6}$$

$$\gamma j_{m=} \ argmin_{\gamma} \sum_{x_i \epsilon R_{ij}} \mathcal{L}(y_i, F_{m-1}(x_i) + \gamma) \tag{7}$$

$$F_m(x) = F_{m-1}(x) + \upsilon \sum_{i-1}^{j} r_{jm} I(x \in R_{jm}) \tag{8}$$

Let M represent the number of trees. For any tree "m" in the training sequence, pseudo residuals, denoted as "rim," are computed using equation (6) across all input samples. Following this, the tree is tailored to predict these residuals, with the tree structure potentially restricted based on the user-defined parameter known as the maximum depth. The output of each leaf in the tree is determined by equation (7). The prediction output of the current tree undergoes an update based on the output of the preceding tree and the learning rate ν multiplied by the leaf output, as illustrated in equation (8). This iterative process continues until the convergence conditions are satisfied.

D. XGBoost (XGB)

XGBoost (XGB) (Abadi et al., 2016) stands out as a robust machine learning model comprising an ensemble of decision trees. The architecture of these trees is determined using a quality score, akin to the impurity score employed by other tree-based models through greedy algorithms. Operating as a boosting algorithm, XGBoost constructs each tree based on the performance of the preceding tree, minimizing the objective function outlined in equation (9). This iterative process leads to successive trees addressing smaller residuals.

$$\mathcal{L}(\theta) = \sum_{i=1}^{n} \mathcal{L}(y_i, \hat{yi}) + \sum_{j=1}^{T} \Omega(f_j) \tag{9}$$

$$\Omega(f) = \gamma T + \frac{1}{2} \lambda \sum_{j=1}^{T} w_j^2 \tag{10}$$

In equation (9), L represents the loss function, assessing the disparity between the predicted value ˆyi and the actual label yi, where fj signifies the j-th tree in the ensemble. What sets XGBoost (XGB) apart from traditional gradient tree boosting techniques is its incorporation of diverse regularization techniques, denoted by the term □. Regularization aims to diminish model variance by introducing slight bias, reducing reliance on observations and enhancing the accuracy of long-term predictions. The number of leaf nodes is indicated by T.

Within equation (10), γ facilitates pruning, diminishing complexity in ensemble trees, wj denotes the leaf weight, and λ is another user-defined parameter governing the magnitude of the regularization-induced penalty. The ultimate output of the XGB model is determined by aggregating the outputs of all trees. XGBoost has gained popularity for its swift performance and proficiency in handling a substantial number of features, missing values, and noisy data. The authors have also introduced features like sparsity-aware split finding, parallel learning, etc., further optimizing its performance, especially on larger datasets.

The fundamental unit of an MLP is the artificial neuron, also known as a perceptron. Each perceptron within a layer receives inputs from the preceding layer, conducts a weighted sum of these inputs, and then employs an activation function to produce its output. The activation function introduces non-linearity, enabling the modeling of intricate relationships between inputs and outputs. The design parameters of an MLP encompass the number of neurons in each layer and the overall number of layers, which can be adjusted to regulate the network's capacity. The MLP's parameters, including the weights and biases of each neuron, are acquired from the training data through an optimization algorithm such as back propagation. This algorithm utilizes stochastic gradient descent to minimize a loss function, gauging the disparity between the network's predicted outputs and the actual outputs.

$$p = \phi\left(\sum_j w_j x_j + b\right) \tag{11}$$

Equation (11) presents the mathematical depiction of a single perceptron unit "p," where the inputs to the unit, weights, and biases are symbolized by xj, wj, and b, respectively. The activation function is denoted by φ. Extending this concept to include L hidden layers, the MLP network is expressed in equation (12), where x and y represent the input and output of the model. The output of any layer l is represented by h(l), and W(l) signifies the weight matrix of that specific layer. Various layers within the network can employ distinct activation functions, dxenoted by φ(l).

$$h^{(i)} = \phi^{(i)}\left(W^{(i)}x + b^{(i)}\right)$$

$$h^{l} = \phi^{(l)}(W^{(l)}x + b^{(i)})$$

$$y = \phi^{(f)}\left(W^{(f)}h^{(l)} + b^{(f)}\right) \tag{12}$$

E. Long Short-Term Memory (LSTM)

Long Short-Term Memory (LSTM) represents a distinctive form of recurrent neural network (RNN) architecture, specifically crafted to tackle the challenge of vanishing gradients, encountered in conventional RNNs. The issue of vanishing gradients can impede the learning of long-term dependencies within sequential data. LSTMs introduce a memory cell state, enabling them to retain information from distant past instances across numerous timesteps. Functioning with three gating mechanisms, LSTMs selectively manage the forgetting, remembering, or updating of information from preceding time steps. Equations (13)-(18) provide a comprehensive depiction of an LSTM unit for a singular timestep.

$$f_t = \sigma\left(W_f\left[h_{t-1}, x_t\right] + b_f\right) \tag{13}$$

$$i_t = \sigma\left(W_i\left[h_{t-1}, x_t\right] + b_i\right) \tag{14}$$

$$\widehat{C} = \tanh\left(W_c\left[h_{t-1}, x_t\right] + b_c\right) \tag{15}$$

$$C_t = f_t * C_{t-1} + i_t * \tilde{C} \tag{16}$$

$$o_t = \sigma\left(W_o\left[h_{t-1}, x_t\right] + b_o\right) \tag{17}$$

$$h_t = o_t * \tanh\left(C_t\right) \tag{18}$$

where xt and ht−1 denote the input sequence component of the current timestep and the hidden state from the previous timestep, respectively. The forgetting, input, and output gates are denoted by ft, it, and Ot, respectively. These gates' outputs are determined through the application of a sigmoid activation function, yielding values between 0 and 1. The forget gate (ft) governs the extent to which the previous cell state should be disregarded. Subsequently, the current cell state (Ct) undergoes an update involving it and ˜C, where ˜Ct represents the candidate cell state. The weight matrices, denoted as Wf, Wi, Wc, and Wo, optimize the behavior of these gates and are learned during the LSTM training process on input samples. Additionally, bf, bi, and bo are referred to as biases.

$$C_{1,j} = ReLU\left(Z^{*} h_{1.j} + b_{1,j}\right) \tag{19}$$

Here, (*) signifies the convolution operation. The result feature map of the initial convolutional layer is represented as C1,j, where h1,j denotes the Jth convolution kernel, and b1,j is the bias term. The equation includes a rectified linear unit (ReLU) as the activation function, introducing non-linearity to the input. The ensuing feature map is subsequently input into the following convolutional layer, undergoing additional processing to extract more advanced features. Consequently, the Ith convolutional layer CI, generating the output feature space Jth, can be expressed as depicted in the equation.

$$C_{I,j} = ReLU\left(C^{*}_{I-1,j} h_{I.j} + b_{I,j}\right) \tag{20}$$

In recent years, Perovskite Solar Cells (PSCs) have garnered significant attention due to their noteworthy attributes such as low cost, high absorption coefficient, tunable bandgap, long carrier lifetime and diffusion length, high defect tolerance, low exciton energy, and high carrier mobility. The configuration of PSCs comprises an absorber layer, often referred to as the perovskite layer, positioned between the hole transport layer (HTL) and the electron transport layer (ETL). Substantial advancements in PSC power conversion efficiency (PCE) have been achieved, increasing from 3.8% to 25.7% through extensive research efforts capitalizing on the superior properties of these cells. This rapid progress positions PSCs as a practical and appealing alternative to traditional solar technology (Figure 1 illustrates the structure of PSC). While PSCs have demonstrated high efficiency within a relatively short timeframe, they still confront several challenges before becoming commercially viable. The perovskite material possesses the general molecular structure of ABX3, where A represents an organic or inorganic monovalent cation, B is a divalent metal ion, and X is a halogen anion. The crystal structure of perovskite material is outlined in Figure 1, with commonly used elements such as Cesium (Cs+), Rubidium (Rb+), Methylammonium (MA+/CH3NH3+), and Formamidinium (FA+/NH2CHNH2+) at the A site; lead (Pb+), Tin (Sn+) typically used at the B site; and Iodide (I−), Bromide (Br−), Chloride (Cl−) at the X site. Essentially, there exist infinite combinations and possibilities for perovskites in solar cell applications.

A notable characteristic of perovskite materials is their tunable bandgap, allowing them to absorb a broad range of photons. Altering the perovskite's composition enables the adjustment of its bandgap, facilitating the tuning of the material's absorption properties. Achieving optimal efficiency in PSCs requires meticulous attention to various aspects of fabrication, including the composition of perovskite materials, interface engineering, and device architecture.

Identifying a perovskite material with an ideal bandgap matching the solar spectrum is crucial for the efficiency of solar cells. Perovskites exhibit the capacity to fine-tune their bandgap within the range of 1.5 eV to 3.2 eV, with the most efficient bandgap for perovskite solar cells being approximately 1.6 eV. This specific bandgap optimally absorbs a diverse range of photons present in the solar spectrum (Cozzi et al., 2020). Consequently, tuning the composition of perovskites is essential to optimize their properties. However, relying solely on trial and error for such a intricate scenario is impractical and illogical.

Structural instability poses a significant challenge to the commercialization of Perovskite Solar Cells (PSCs) as it can lead to cell degradation over time. Factors such as temperature, humidity, and light exposure contribute to this instability. The deterioration of perovskite materials negatively impacts the efficiency and lifespan of solar cells. Additionally, the use of Pb, a toxic substance, in PSCs is an environmental concern as it can leach out over time and contaminate the surroundings. To address these issues, researchers are actively developing new perovskite materials that are more stable and less toxic. Strategies include creating lead-free perovskite materials, utilizing alternative materials to stabilize the perovskite structure, and employing encapsulation methods to shield solar cells from environmental factors.

In recent years, Machine Learning (ML) techniques have been employed to predict the bandgap, Power Conversion Efficiency (PCE), and stability of PSCs with high accuracy, significantly expediting the development and optimization of PSCs. Figure 1 illustrates the implementation of ML in the fabrication process of PSCs. Various studies have demonstrated the predictive capabilities of ML methods in effectively estimating the bandgap, PCE, and stability of PSCs. This section provides a succinct overview of the application of ML techniques in the fabrication of PSCs.

Role of Solar Energy in the Sustainable Development

Sustainable energy distribution strategies (Radovanovic et al., 2012) is the main stream energy distribution model consisting of the generation, distribution and proper utilization of energy in the optimized fashion. The impact of such energy generation on the environment is very less as it entertains the reduction of emission (Salvarli et al., 2020) in the atmosphere. The best outcome of such infrastructure is the cleaner and more reliable energy access and its utilization. Some of the pathways for such renewable energies are grid computing, energy security and control policy of energy pricing. The overall demand-side response can help meet the flexibility requirements in electricity generation. The integration of the renewable energy source has managed to ride over the sudden demand peak controlling the cost and the carbon di oxide emission. The infrastructural (Cozzi et al., 2020) upliftment also helps in enhancing the overall economic condition along with the proper framework of consistency in the trade line. The grid based system can easily assist in the immediate medical crisis and therefore the improvement of economy based on the advanced mode of trades. So the sustainable development of the alternative source of energy has influence on the environment as well as the economy.

ML IN THE DEVELOPMENT OF TRADITIONAL SOLAR CELLS

Abadi et al. (Balle et al., 2018) introduced an innovative approach to predict device performance and unveil structure-property correlations in organic photovoltaics (OPVs) by combining Machine Learning (ML) and Taguchi Design Experiments (TODE) (Wasmer et al., 2018). Their database comprised 240 small-molecule organic solar cells with 220 diverse donor materials blended with two fullerene-based acceptors (PC61BM and PC71BM). Using ten quantum chemical features related to energy conversion as descriptors, the model employed TODE and Artificial Neural Network (ANN) algorithms to derive design guidelines. The predictions indicated that reducing the optical band gap enhances organic solar cell performance by increasing photon absorption. Additionally, the model identified optimal values for 1L (energetic difference of LUMO and LUMO+1) and 1H (energetic difference of HOMO and HOMO-1) corresponding to each optical band gap. Lower values of 1L and 1H were associated with higher Power Conversion Efficiency (PCE) in materials with a low optical band gap, and the model suggested that having 1L and 1H nearly equal to 0 leads to good PCE values.

The Role of Internet of Things in the Renewable Source of Energy

The internet of things (IOT) plays a significant role in the monitoring, protection and overall control mechanism for accessing any energy sources. The concept of automation along with the network has made the roadways much conducive to make the assistance to the energy control much flexible and cost effective (Syed et al., 2018). The IOT system has been appointed for data acquisition, processing, transmission, and storage. The grid architecture has become such an instance which provides lot of security aspects of the energy source. The IOT system comprises different sensors which helps in capturing data and send them to different control centers. The decision strategy has been executed by means of control commands. The communication goes on the control systems and some micro grids. The whole IOT process has been broken into umber of layers. The outcome of the entire system is the sum total of the outcomes of the integrated IOT components.

ALGORITHMS FOR ENERGY SOURCE IDENTIFICATION

The real challenge lies in the identification of alternative source of energy. The above discussion provides an overview of different types of models and their proper utilization in identification of various sources. A unique algorithm has been devised up to identify the source with greater precision. The algorithm can be presented in the section given below

- Let X be the matrix of features of the energy sources, with n rows and m columns, where n is the number of energy sources and m is the number of features.
- Let y be the vector of labels of the energy sources, with n elements, where each element is either “solar”, “wind”, “hydro”, or “biomass”.
- Let k be the number of clusters, which is 4 in this case.
- Let C be the vector of cluster labels of the energy sources, with n elements, where each element is an integer from 0 to k−1.
- Let f be the function that maps the energy types to numbers, such that f(“solar”) =0, f(“wind”)=1, f(“hydro”)=2, and f(“biomass”)=3.
- Let y′ be the vector of numerical labels of the energy sources, with n elements, where each element is $f(y_i)$, where y_i is the i-th element of y.
- Let g be the linear regression model that predicts the energy type based on the cluster, such that $g(C_i)=\beta C_i+\alpha$, where C_i is the i-th element of C, β is the coefficient, and α is the intercept.
- Let y′′ be the vector of predicted numerical labels of the energy sources, with n elements, where each element is $g(C_i)$, where C_i is the i-th element of C.
- Let E be the mean squared error of the predictions, such that $E=\sum_{i=1}^{n}(yi'-yi'')2$, where $y_{i'}$is the i-th element of y′, and $y_{i''}$is the i-th element of y′′.

The respective mathematical symbols provide their usual meanings. The KMeans clustering algorithm has been applied to classify different concerned sources of energy and the regression analysis has been carried out to predict the energy type based on the cluster. The following summarized version provides the steps of operation in details.

- Cluster the energy sources based on their features, such that C=KMeans(X,k), where KMeans is the clustering algorithm.
- Assign a numerical value to each energy type, such that y′=f(y), where f is the mapping function.
- Train a linear regression model to predict the energy type based on the cluster, such that g=LinearRegression(C,y′), where LinearRegression is the linear regression algorithm.
- Predict the energy type based on the cluster, such that y′′=g(C), where g is the linear regression model.
- Calculate the mean squared error of the predictions, such that E=MSE(y′,y′′), where MSE is the mean squared error function.

The efficiency of respective algorithms has been measured on executing the same over a duration of 1 hour. The maximum efficiency of all such algorithms have been measured and the variation of the same has been plotted in the graph as shown in the fig -1.The efficiency assessment with respect to the time plot provides the instantaneous variation of the efficiency. The corresponding peaks of the graph correspond to the maximum efficiency of the classifier attached with the optimizing algorithm.

It has been found that the maximum peak of the curve has been obtained for the particle swarm optimization algorithm which confirms also attaining the maximum efficiency in comparison to the others. The maximum efficiency values have been shown in Table 2.

FUTURE SCOPE FOR THE STUDY

The AI ML approach (Zadeh et al., 2016) is the recent platform for development of many algorithms for much accuracy and precision. The predictive approach for any experiment can easily be done by employing such AI ML concept. The future prospect of such paradigm in tracking different energy sources is also huge. The optimal approach for algorithm must fetch much bigger responses. The application of different optimal algorithms with various features relevant to the sources can be the big pathways for achieving precision result. The application of IOT sensor integrated with bigdata support can also solve problem of sensing various sources of energies. The highly efficient network

Figure 1. The variation of efficiency ofthe different optimized models with respect to time stamps

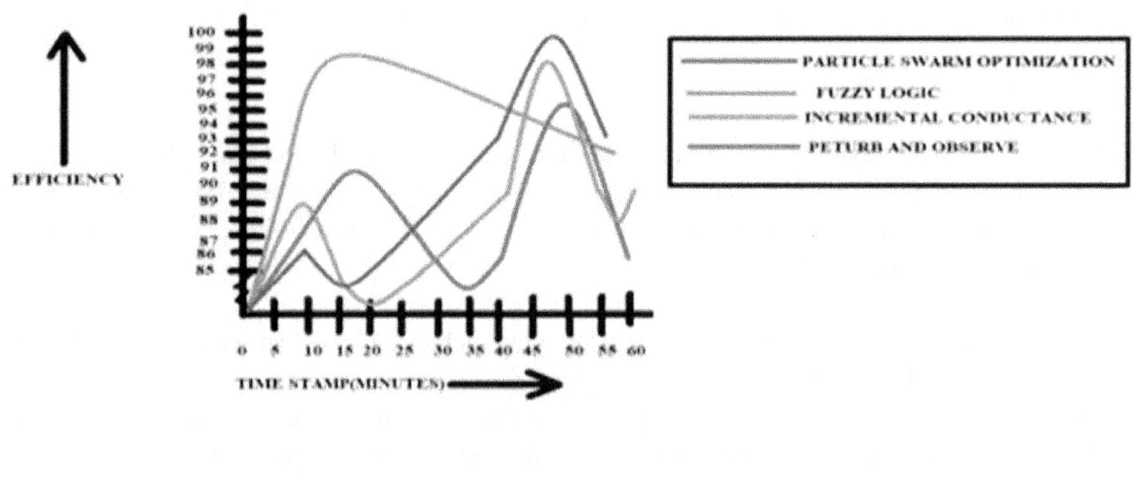

Table 2. Maximum efficiency values corresponding to different optimizing algorithm for solar renewable energy

Algorithm	Maximum Efficiency
Perturb and Observe (P & O)	96%
Incremental Conductance (INC)	98%
Fuzzy Logic	99%
Particle Swarm Optimization (PS0)	99.5%

integrated with proper security can also serve for finding the potential information related to the source of renewable source of energies.

CONCLUSION

The renewable source of energy in the current paradigm of energy crisis has become the great relief path towards finding direction of sources. As the natural source of energy are gradually dwindling with time and increasing number of populations, the thought for alternative means has been realized. The renewable source of energy can fulfill the demands and easily overcome such problem of shortfall. Out of many different renewable sources of energies, the solar power energy has become the great and reliable source of energy and can easily become the alternative mean for all other sources of energies. The source of such energy has been found with the help of different models. The machine learning model out of them has gained much potential response in assessing the source of such energies using different types of optimizing algorithms. The detailed analytical approach has been done in the specified chapter. The regression analysis has also been done to develop a predictive model for precision prediction of the types of different energy sources. The concept of such machine learning modeling can be extended also in future to tap some potential energy sources other than the conventional ones with greater possibility of usage and utilization.

REFERENCES

Abadi, M., Chu, A., Goodfellow, I., McMahan, H. B., Mironov, I., Talwar, K., & Zhang, L. (2016). Deep learning with differential privacy. In *Proceedings of the 2016 ACM SIGSAC Conference on Computer and Communications Security* (pp. 308-318). 10.1145/2976749.2978318

Abbas, F., Habib, S., Feng, D., & Yan, Z. (2018). Optimizing Generation Capacities Incorporating Renewable Energy with Storage Systems Using Genetic Algorithms. *Electronics (Basel)*, *7*(7), 100. doi:10.3390/electronics7070100

Ahmed, M. A., Eltamaly, A. M., Alotaibi, M. A., Alolah, A. I., & Kim, Y.-C. (2020). Wireless Network Architecture for Cyber Physical Wind Energy System. *IEEE Access : Practical Innovations, Open Solutions*, *8*, 40180–40197. doi:10.1109/ACCESS.2020.2976742

Al-Saedi, W., Lachowicz, S., Habibi, D., & Bass, O. (2018). A review of algorithms for control and optimization for energy systems. *Journal of Renewable and Sustainable Energy*, *10*(5), 053502. doi:10.1063/1.5032146

Al-Sumaiti, A., Salama, M., & Al-Hinai, A. (2018). A review of optimal planning and operation of hybrid renewable energy systems. *International Journal of Energy Research*, *42*(10), 3179–3204.

AlRashidi, M. R., & El-Naggar, K. M. (2010). Long-term electric load forecasting based on particle swarm optimization. *Applied Energy*, *87*(1), 320–326. doi:10.1016/j.apenergy.2009.04.024

Angulo, A., van der Linde, P., Gardeniers, H., Modestino, M., & Rivas, D. F. (2020). Influence of Bubbles on the Energy Conversion Efficiency of Electrochemical Reactors. [Google Scholar] [CrossRef] [Green Version]. *Joule*, *4*(3), 555–579. doi:10.1016/j.joule.2020.01.005

Balle, B., Bell, J., Gascon, A., & Nissim, K. (2018). The privacy blanket of the shuffle model. In *Proceedings of the 2018 ACM SIGSAC Conference on Computer and Communications Security* (pp. 638-650). ACM.

Budak, G., Budak, E., & Kaya, I. (2019). A systematic approach to assess renewable energy and identify the energy alternatives for a city. *Energy*, *189*, 116233.

Carrillo, A. J., González-Aguilar, J., Romero, M., & Coronado, J. M. (2019). Solar Energy on Demand: A Review on High Temperature Thermochemical Heat Storage Systems and Materials. *Chemical Reviews*, *119*(7), 4777–4816. doi:10.1021/acs.chemrev.8b00315 PMID:30869873

Cozzi, L., Gould, T., & Bouckart, S. (2020). *World Energy Outlook 2020*. International Energy Agency.

Dawood, F., Anda, M., & Shafiullah, G. M. (2020). Hydrogen Production for Energy: An Overview. *International Journal of Hydrogen Energy*, *45*(7), 3847–3869. doi:10.1016/j.ijhydene.2019.12.059

Kaya, M., & Hajimirza, S. (2020). Neural network surrogate model for optical performance of thin film multi-layered amorphous-silicon-based solar cells. *Optics Express*, *28*(4), 4946–4957.

Kusiak, A., & Li, W. (2011). The prediction and diagnosis of wind turbine faults. *Renewable Energy*, *36*(1), 16–23. doi:10.1016/j.renene.2010.05.014

Kusiak, A., Zheng, H., & Song, Z. (2010). Wind turbine performance assessment with data-mining models. *Renewable Energy*, *35*(12), 2734–2741.

Lagarrigue, C. (2021). How artificial intelligence can optimize renewable energy. *Forbes*.

Li, X., Zhang, J., Wang, J., & Zhao, H. (2019). A machine learning method for solar power prediction using satellite images and numerical weather prediction data. *Applied Energy*, *253*, 113596.

Liu, Z., Wang, J., Li, Y., & Wu, J. (2020). A deep reinforcement learning framework for optimal energy management of microgrids with renewable energy sources and energy storage systems. *Applied Energy*, *279*, 115836.

Mellit, A., & Kalogirou, S. A. (2008). Artificial intelligence techniques for photovoltaic applications: A review. *Progress in Energy and Combustion Science*, *34*(5), 574–632. doi:10.1016/j.pecs.2008.01.001

Mohammadi-Ivatloo, B., Rabiee, A., Soroudi, A., & Ehsan, M. (2013). Iterative solution of multi-objective security-constrained unit commitment problem with high penetration of wind power using artificial bee colony algorithm. *International Journal of Electrical Power & Energy Systems*, *53*, 318–329.

Mohandes, M., Rehman, S., & Halawani, T. O. (2004). A neural networks approach for wind speed prediction. *Renewable Energy*, *29*(6), 939–947. doi:10.1016/j.renene.2003.11.009

Parra, D., Valverde, L., Pino, F. J., & Patel, M. K. (2019). A review on the role, cost and value of hydrogen energy systems for deep decarbonisation. *Renewable & Sustainable Energy Reviews*, *101*, 279–294. doi:10.1016/j.rser.2018.11.010

Radovanovic, M., Popov, S., & Dodic, S. (2012). *Sustainable Energy Management*. Academic Press.

Raut, H. K., Ganesh, V. A., Nair, A. S., & Ramakrishna, S. (2011). Anti-reflective coatings: A critical, in-depth review. *Energy & Environmental Science*, *4*(10), 3779–3804. doi:10.1039/c1ee01297e

Salman, S., Kuciauskas, D., & Dippo, P. (2020). Machine learning for extracting Cu doping profiles in CdTe solar cells. *Solar Energy Materials and Solar Cells*, *209*, 110527.

Salvarli, M. S., & Salvarli, H. (2020). For sustainable development: future trends in renewable energy and enabling technologies. In Renewable Energy-Resources, Challenges and Applications. London: IntechOpen. doi:10.5772/intechopen.91842

Shivam, S., Kumar, A., & Singh, S. (2020). Artificial neural network based prediction of optimum thickness of anti-reflection coating for solar cell. *Materials Today: Proceedings*, *21*, 1628–1631.

Siano, P. (2014). Demand response and smart grids—A survey. *Renewable & Sustainable Energy Reviews*, *30*, 461–478. doi:10.1016/j.rser.2013.10.022

Singh, A., Singh, S., & Kumar, A. (2020). Artificial intelligence based optimal control of renewable energy integrated power system. *International Journal of Electrical Power & Energy Systems, 118*, 105794.

Singh, K., & Bean, J. D. (2019). *Optimizing India's electricity grid for renewables using AI and machine learning applications*. Center for Strategic and International Studies.

Syed, S. A., Al-Turjman, F., & Alghamdi, S. A. (2018). IoT-Based Energy Management System for Wind-Powered Microgrids. *Energies*, *11*, 2625.

Touati, F., Chowdhury, N. A., Benhmed, K., Gonzales, A. J. S. P., Al-Hitmi, M. A., Benammar, M., Gastli, A., & Ben-Brahim, L. (2017). Long-term performance analysis and power prediction of PV technology in the State of Qatar. *Renewable Energy*, *113*, 952–965. doi:10.1016/j.renene.2017.06.078

Wang, L., Zhang, J., Zhou, Y., & Li, G. (2019). A hybrid intelligent optimization algorithm for optimal power flow with renewable energy integration. *Applied Energy*, *242*, 1634–1647.

Wasmer, S., Rajsrima, N., Geisemeyer, I., Fertig, F., Greulich, J. M., & Rein, S. (2018). Analytical modeling of the temporal evolution of hot spot temperatures in silicon solar cells. *arXiv preprint arXiv:1803.02463*.

Xiao, W. B., Li, X. F., Zhang, C., Yang, G. Q., & Wang, X. D. (2020). Power conversion efficiency prediction of silicon solar cells using artificial neural network. *Journal of Semiconductors*, *41*(2), 022101.

Yue, M., Lambert, H., Pahon, E., Roche, R., Jemei, S., & Hissel, D. (2021). Hydrogen energy systems: A critical review of technologies, applications, trends and challenges. *Renewable & Sustainable Energy Reviews*, *146*, 111180. doi:10.1016/j.rser.2021.111180

Zadeh, P. A., & Jadid, S. (2016). A systematic approach for assessment of renewable energy using analytic hierarchy process: A case study of Iran. *Energy, Sustainability and Society*, *6*(1), 1–14.

Zhang, Y., Wang, J., Wang, J., & Zhao, H. (2019). A deep neural network-based approach for short-term wind power forecasting using spatial-temporal data. *Applied Energy*, *236*, 220–231.

Zhu, Y., Zhang, J., Wang, X., & Zhang, Y. (2020). Machine learning for key variables affecting the performance of CIGS solar cells. *Journal of Alloys and Compounds*, *827*, 154226.

Chapter 6
An Approach of Renewable Energy Based on Machine Learning:
OCR-Based Handwritten Character Datasets

Sumanta Kuila
https://orcid.org/0000-0002-6088-6806
Haldia Institute of Technology, India

Angana Chakraborty
Haldia Institute of Technology, India

Subhankar Joardar
Haldia Institute of Technology, India

Manasi Jana
https://orcid.org/0000-0002-7076-1775
Haldia Institute of Technology, India

Shradhanwita Mukherjee
https://orcid.org/0009-0006-1772-6395
Haldia Institute of Technology, India

ABSTRACT

Renewable energy comes from natural resources that replenish more quickly than they are used up. Compared with burning fossil fuels, producing renewable energy produces a lot fewer emissions. The key to solving the climate catastrophe is shifting away from fossil fuels, which presently provide the majority of emissions, and toward renewable energy. The facility has global impact and recent computer vision based on machine learning creates the business opportunity within all countries and regions globally. So there should be a communication between all countries and regions to exchange information between different global languages. In this study, an analytical survey was done to create the communication between countries and regions. For this purpose, different languages and datasets of the languages are studied where OCR-based character recognition can be done.

DOI: 10.4018/979-8-3693-2355-7.ch006

1. INTRODUCTION

Energy derived from renewable resources that replenish naturally within a human timescale is referred to as low-carbon, green, or renewable energy. Light, wind, river flow, and geothermal heat are examples of renewable resources. More electrification is frequently implemented in tandem with renewable energy sources since electricity is clean at the time of use and can convey heat or items with efficiency. Renewable energy currently accounts for more than 20% of the energy supply in many countries worldwide, with some producing more than half of their electricity this way. There are a few nations that use renewable energy to produce all of their electricity. The 2020s and beyond are expected to see significant growth in the national markets for renewable energy. There are major economic, climate change mitigation, and energy security benefits from the deployment of renewable energy and energy efficiency technology. The growing popularity of renewable energy sources is being propelled by concerns about climate change as well as the ongoing decline in the price of certain renewable energy technology, such solar panels and wind turbines. These types of information should be communicated globally through different leading languages globally. In this study we made a comparative study between different languages and its existing databases. These existing databases will support the common communication environments between different languages. Making handwritten text machine readable is known as handwritten character recognition, or HCR. Handwriting styles vary greatly among writers, which poses a significant challenge to handwritten character recognition (HCR) systems. An approachable computer-assisted character representation is the goal of the handwritten character recognition system. The successful extraction of characters from handwritten documents and the digitization and translation of the handwritten text into machine-readable language are made possible by the character representation. In Devanagari script the upper part of each character features a horizontal line called Shirorekha, which means headline. Characters in English don't possess this quality. English may be derived from these scripts to this unique characteristic (Nawaz et al., 2018). In sequential handwriting, the shirorekha of one character links to the shirorekha of the preceding or following letter of the same word when written from left to right. Characters are recognized in the online character recognition system while they are still being created. The handwritten papers in the off-line character recognition system are created scanned, saved in a computer and subsequently identified with the use of character analysis. Here the handwriting recognition detects the characters in a photograph. Character recognition is the process of recognizing, differentiating and identifying characters in a picture. Computers are able to read handwritten characters but here human handwriting recognition is superior. Because the handwritten characters can have many various flavors and are not always perfect, the machine finds this work difficult. This difficulty can be solved using handwritten character recognition which uses an image of characters to identify the characters that are there. Character recognition is a technique for identifying, detecting and segmenting characters (alphabets) in images. The goal of character recognition in research is to make computers more proficient readers so they can interact with text in a way that is similar to that of humans. These are the key issues in machine learning in handwritten character recognition (HCR). The purpose of its use has been to test new machine learning algorithms with different train and test datasets (Dalal & Triggs, 2005; Newell & Griffin, 2011). This methodologies to work with handwritten character recognition aims to accurately recognize each handwritten character on a computer using an image input. The Convolutional neural network (CNN) model is an leading module that can recognize the English alphabet's images which makes the sizable chunk of the dataset that will be used to train the model. Handwritten characters or the English alphabet from A to Z will be developed in this machine learning project. Several popular artificial neural networks used for image recognition and processing is the neural networks were developed primarily to analyze pixel input. Unsolved computer vision challenge of great difficulty to read text from photos. Relevant data are read from various many real-world applications. Finding and identifying characters in an incoming digital image and converting them to a comparable machine-editable format is a crucial challenge for handwritten character recognition (HCR). Pattern recognition and image processing both see significant increase from it. The interpretation of data presents significant hurdles ranging from bank checks and language recognition to the conversion of handwritten documents into structured text formats. A neural network-based soft computing technique which has been the subject of extensive research with numerous theories and implemented algorithms is used by the handwritten character recognition system. Character recognition feature extraction is accomplished by utilizing a novel method called diagonal-based feature extraction. Preprocessing is done once the scanned text document image has been obtained. The paths to obtain the input photos and store the output variables have now been set and all variables and counters have been initialized. Unwanted outliers and noise have all been eliminated. Its skew is adjusted, and the picture is made regular. Moreover the binary picture is obtained by converting the RGB image to a grayscale image. Utilizing the project profiles a pre-processed binary image is obtained and further segmented to extract lines, words, and characters. In order to identify and find every line in the bounding boxes, the image is first segmented using a horizontal projection profile. This is carried out because text lines have a higher density of black pixels than the spaces between neighboring lines. After this type of line identification, a vertical projection profile is used to identify and recognize words within bounding boxes. The geometric features of a character are retrieved after a character geometrical structure consumes a number of black pixels, each of which is represented by a corresponding value. Research into newer strategies that could assist improve recognition accuracy is still ongoing in the off-line handwriting recognition field. It's because a number of applications such as postal address recognition, bank processing, document reading, and mail sorting need offline handwriting recognition systems. In this research hand written datasets of different languages are discussed and its features are studied. The popular languages like English, Hindi, Bengali, Arabic etc. have existing datasets which are used for research, academic and commercial purposes. This study aims to deliver a comparative study among these languages their properties, features also merits and demerits of these datasets.

2. RELATED STUDY

Several existing dataset for used in different research publications. These datasets are used for feature extraction and classification. In this study the properties of the existing datasets are discussed with their impact in the machine learning and deep learning classifications. N. Dalal et al. (2005) discusses a variety of phases, including pre-processing, segmentation, classification, and post-processing, make up the suggested system for handwriting recognition. This research describes an innovative offline character recognition system based on neural networks. S. B. Khan et al. (2018) works with a unique type of network structure utilized for HCR is called a region proposal network. The primary goals of this approach are to preserve accuracy while simultaneously accelerating text recognition.

T. Watanabe et al. (2009) describes a CNN architecture that can handle digital penmanship without requiring its conversion to a picture. Character and broad vocabulary word recognition tasks demonstrate the efficacy of the suggested CNN features. The whole new design is known as DAG, a network segmented into numerous routes which was made possible by the basic CNN concept. M Altuwaijri et al. (1994) focuses on a network segmented into several routes that was made possible by the foundational CNN paradigm. Here three sections comprise the framework: CNN models, voting, and sample creation. C. Bahlmann et al. (2002) focuses the Research on applying artificial intelligence for handwriting recognition (HWR) of the Comenia script. The image-based tasks like OCR and pattern recognition have been shown to be feasible and more accurately completed via deep learning. et al. To identify the Handwritten Mathematical Expression (HME) written on the paper, the paper is scanned, and the image is sent to the identification system. An alternate method of training a subset of CNN layers is suggested in order to regularize the neural network. ATR-CNN attains cutting-edge performance on handwritten digit datasets from the MNIST and ICDAR'13. C Nadal et al. (1990) describes a special profile normalization technique that was used to word and line images, and random perturbations on a regular grid were employed to improve pre-existing text images. These methods work regardless of the network and can be applied to enhance the functionality of various HWR networks and strategies. CL Liu et al. (2003) identifies the pooling function fully connected, and convolutional layers for feature extraction, image representation, and dimensionality reduction. The distance between separate feature values in the model for regression and classification is determined using the KNN technique, which is based on geometric measurement. M. Shi et al. (2002) describes the results of the simulation trials, it is evident that the CNN algorithm achieves a higher recognition rate than the KNN approach. It is very important what kind of features are used in the digit recognition system. A k-nearest neighbor classifier is used to classify the digits, and a Euclidean minimum distance criterion is applied to find minimum distances. The works used different existing hand written datasets.

3. OPTICAL CHARACTER RECOGNITION

The computer vision technology known as optical character recognition, or OCR, is used to identify characters like numbers, letters, signs and other characters that are frequently encountered in daily life. Character identification can be done in accordance with specific needs. Through the process of character recognition by the software using computer can identify letters, numbers or symbols and can convert them into a digital format that people can use for different purposes. Systems for recycling forms and processing bank checks can both benefit from handwritten character recognition. One popular but challenging area of study is character recognition. A paperless environment is produced via character recognition in the womb. Using knowledge of recent data on a particular topic the machine learning seeks to extract hidden information from the data. One of machine learning's main uses is pattern recognition. To identify the current trends the large picture data sets are usually used for typical optical recognition. Handwriting recognition by pattern recognition in a picture is a common practice. According to research character recognition techniques have developed successively where the initial research focused on identifying individual handwritten characters and eventually the entire words were examined. A few publications have lately tackled free handwritten recognition which is free from domain-specific limitations and big vocabularies. Such a system still needs to be improved because of its low recognition rate. Using such concepts we could train computers to read characters and numbers from any language that appears in a picture. The character's images make up a sizable chunk of the dataset that will be used to train our model. Sometimes the characters in handwriting are so similar that the reader cannot tell them apart without context. To differentiate between characters that are so similar in nature, it is necessary to recognize the little variances between them. Because different language writers write characters in different proportions, one of the main issues with accomplishing this for handwritten characters is that they do not appear at the same relative place of the letter. Not everyone writes letters in the same proportions or even by the same person every time (Khan et al., 2018; Watanabe et al., 2009).

The figure (Figure 1) represents sample dataset of English numeric digits. Here the OCR dataset represents the English numeric digit. The dataset is the sample dataset of MNIST which is the subset of the standard NIST dataset. The figure shows the hand written English numeric digits from 0 to 9. The raw hand written digits are collected and from there the datasets are created. In today's digital age automated English character identification is a novel, visible and demanding field that has developed through the fusion of artificial intelligence, pattern recognition, machine learning and data mining principles. Comparing Indic scripts to non-English languages, such as Devanagari, Chinese, Japanese, Korean, German, etc., OCR systems are already advanced.

4. EXISTING HANDWRITTEN DATASETS

There are several existing publicly available datasets are used for feature extraction classification of character recognition. In this study the characteristics and features of the datasets of the mostly used datasets are discussed.

4.1 NIST Database

NIST is a publicly available dataset used for the training and testing of machine learning classification. NIST's whole training material corpus for hand printed document where the character recognition is stored in Special Database 19. 10,000 character images extracted from their forms and the ground truth classifications for those images, reference forms for additional data gathering. The software utilities for

Figure 1. English character dataset of MNIST dataset

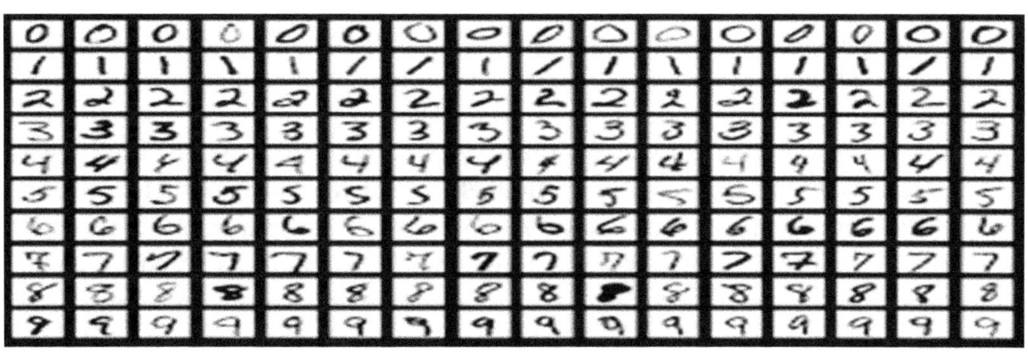

image storage and handling are all published together with hand printed sample forms from 3600 writers. It replaces Special Databases 3 and 7 of NIST. Each form's fields were divided into separate characters. It can greatly enhance the interface between a machine and a human in a variety of applications and greatly advances automated processes. New approaches and strategies that would shorten processing times while increasing recognition accuracy have been the subject of several research projects. An automatic classification was given to each of the resultant images. Every referenced character image has been examined by hand, and any errors found have resulted in the image's class being altered. The process of verifying was carried out twice by distinct individuals each time. The checking process's guiding principle was to keep examples near class boundaries by avoiding deleting or altering anything that might have been the intended class—even if, in retrospect, it seemed more like an example of a different class (Al-Yousefi & Udpa, 1988).

4.2 MNIST Database

A training set of 60,000 samples and a test set of 10,000 examples comprise the MNIST database of handwritten digits. This is part of a bigger set that NIST is making public. In a fixed-size image the digits have been centered and size-normalized. With little effort required for formatting and preprocessing, it's a useful database for those wishing to test learning strategies and pattern recognition algorithms on actual data. NIST's Special Databases 3 and 1, which include binary images of handwritten numbers, were used to build the MNIST database. At first, NIST used SD-3 as a training set and SD-1 as a test set. Nonetheless, SD-3 is far more recognizable and cleaner than SD-1. The distinction between the SD-3 and SD-1 surveys—which were conducted among Census Bureau employees and high school students, respectively—explains why this is the case. In order to make sense of learning studies, the outcome must be unaffected by the training set and test that are selected from the entire set of samples. As a result, combining several NIST datasets to create a new database was required. Thirty thousand SD-1 patterns and thirty thousand SD-3 patterns make up the MNIST training set. Five thousand patterns each from SD-1 and SD-3 made up our test set. Examples from roughly 250 writers were included in the 60,000-pattern training set (Badi & Shimura, 1982).

4.3 Devanagari Character

India's most widely used language and script is Devanagari, whereas Bangla is the second most widely used language and script in the Indian subcontinent and the fifth most widely used language worldwide. Significant portions of the Indian populace also utilize a number of other scripts, including Tamil, Telegu, Kanada, Malayalam, and a few more. There has been ongoing research on OCR systems for scanned printed texts in Indian script for a considerable amount of time. However not much study has been done on handwriting recognition. There is literature on deteriorated document analysis of Indian scripts or recognition of scene texts. Due to the high degree of diversity in their samples some printed OCR technology is unfortunately unable to recognize such characters, photos or documents. Currently available research on handwriting or scene text identification of Indian scripts primarily rely on limited or non-standard databases that were gathered in lab settings. In order to alter this situation several sizable databases were created to contain handwritten or scene characters from significant Indic script(s). ISIDDI is an image database of degraded Bangla documents which is the newest database added to this group. Either these databases have already been made freely available to academic researchers, or the release process is in progress (Altuwaijri & Bayoumi, 1994).

4.4 Mathematics Expressions

The dataset which combines the data sets from three CROHME competitions offers more than 10,000 handwritten statements from hundreds of writers from various nations. From a corpus of utterances the writers were instructed to replicate printed expressions. The corpus was created to accommodate the variety that the various tasks offered and it was selected from pre-existing math corpora as well as expressions that were contained in different research articles. Different scales and resolutions have been employed because different devices (such as tablets with reasonable screens, white-board input devices and digital pen technologies) have been used. The only online signal provided by the dataset are INKML and MATHML standards where the CROHME dataset contains the segmentation, labeling and layout of every mathematical expression. The two tiers of ground truth data at both the expression and symbol levels that are manually executed. Additionally the file now includes some generic information about the image data of the characters read.

4.5 Chinese Characters

Studies pertaining to language acquisition and processing require the use of lexical databases. Little is known about reading materials for children and how lexical features from these materials affect children's reading comprehension. The majority of prior Chinese lexical databases have concentrated on materials for adults. The current study included the first extensive database of 2999 Chinese characters and 2182 words which were gathered from the official textbooks which are recently released by the People's Republic of China's Ministry of Education (MOE) for the majority of Mainland China's elementary schools along with norms from both adults and children of school age. From these lexical units, the database includes important orthographic, phonological and semantic components. To find out how these elements affected adults' and kids' word processing and character, word-naming tasks were employed.

4.6 Arabic Printed Text

Certain features of the Arabic script contribute to how difficult it is to recognize. Its cursive writing style and the way its character forms change depending on where they are in the word give it these qualities. Arabic characters alter in form based on the calligraphic style that is being utilized as well as their placement within the word. Arabic calligraphy comes in a variety of forms including Neskhi, Thoulthi, and Diwani. Using the training and test sets of the character image dataset, as well as the K-nearest neighbor classifier with K=1, the topological feature was selected to assess the statistical and structural aspects. The Arabic printed text recognition system with an open vocabulary can be tested using this database. The large-scale benchmarking of Arabic text recognition algorithms with multiple fonts, sizes, styles and open vocabularies are the aim of this database. A careful examination of the printed Arabic papers shows that the variety and quantity of font sizes and styles utilized are significant. Arabic optical character recognition (OCR) systems show that their identification rates can be highly sensitive to variations in font size and/or style. An explanation of the Arabic printed text database with multiple fonts, styles, and sizes in detail.

4.7 IAM Online Document

There are text blocks, lists, tables, formulas, diagrams, and illustrations in this work. These contents of these elements have been arranged arbitrarily on every page. The writers have been instructed to duplicate the template's material onto a piece of paper throughout the acquisition phase. Using a digital pen the writing was captured online. In all cases one thousand templates are made up of text from the brown corpus and images, charts and equations from several research datasets commonly used by the respective language. This result delivers the need of 200 distinct drawings 200 distinct formulas and 200 distinct diagrams. Random text was chosen from the Brown corpus's eleven categories for the text blocks, lists and tables. Following the completion of a document, the writer annotates it using a limited range of marking elements, such as underlining text, enclosing text in square brackets, marking text on one or more lines with an angle on the top left as a start mark and an angle on the bottom right as an end mark, encircling, underlining, and marking text on the document. Tiny text labels for these marks' annotation, as well as lines joining these text labels to the markings. Approximately 10 of these marking components are present in 839 of the total components used. The recorder has eliminated sample points that form a straight line. A timestamp in milliseconds generates a pressure value ranging from 0 to 255 where the coordinates of each sample point on the sheet are provided. A stroke or trace is formed by grouping together the sample points that are recorded from the starting point at which the pen contacts the paper until it is lifted once more (Beigi, 1993).

4.8 Street View Text

Google Street View provided the dataset known as Street View Text (SVT). This data's image text frequently generates the low resolution that shows significant variability. Initially the image text is frequently taken from company signage and then the business names are readily accessible through local business searches. The SVT set is especially well-suited for word spotting in the wild because of these factors where the objective is to recognize words from local businesses given a street view image. Lexicon-driven word recognition with cropping and word detection and recognition with full image is the major objective for this scenario. In order to facilitate bounding box image annotation the Alex Sorokin's annotation toolkit was used. Using the search nearby tool a list of nearby business names for every image was received at the address of the image in Google maps. The SVT data collection describes briefly to made up the pictures that were gathered from Google Street View and labeled with bounding boxes encircling phrases from nearby businesses.

4.9 Street View House Numbers (SVHN)

With little need for data cleaning or formatting the SVHN offers a real-world image dataset for creating object recognition and machine learning algorithms. There are 26032 digits for testing where 73257 digits for training and 531131 more comparatively easier samples to use be as additional training data. The file includes a digit structural material file that can be loaded with Matlab along with the original photos in png format. The fact is that at the character level presented this dataset works similar to MNIST dataset. Each and every digit has been scaled to a consistent 32 by 32 pixel resolution. The character bounding boxes are appropriately expanded to form square windows, preventing aspect ratio distortions when enlarging them to 32 by 32 pixels. However, this preprocessing adds a few distracting numbers on the sides of the relevant digit. A challenging computer vision challenge at the heart of many developing applications is the detection and reading of text from natural images. Scholars in the fields of computer vision and machine learning have extensively researched character recognition where the problem has essentially been addressed for real world uses like handwritten digit reading (Netzer et al., n.d.).

4.10 Natural Environment OCR (NEOCR)

The NEOCR dataset includes 659 real-world photos with 5238 text fields which is also known as bounding boxes annotated on them. Since multiple individuals took separate shots of the photographs the dataset encompasses a wide range of features that set real-world images apart from scanned documents. The dataset encompasses a wide range of features that differentiate real world photos from scanned documents which works as the images were shot separately by multiple individuals. Every image contains an annotation for every text that is human readable. Once there were at least 100 text fields in the dataset for each metadata dimension the dataset construction procedure was terminated. The visible text is included in the ground truth along with distortion quadrangles that encompass it far more precisely than bounding boxes. Using a modified version of the LabelMe annotation tool the annotation was manually created. Every image has an annotation next to every sentence that is readable and recognizable by people. We have expanded the character recognition schema to meet the needs and the annotation is provided in XML. As part of the dataset the extended XML schema is also given. Global and local metadata are also provided to support the schema. The image's height, width, depth, contrast, brightness and filename are all included in the global image metadata. Information about texture, brightness, contrast, inversion, resolution, noise, and blur is an example of an optical feature. Image characters were used to compute the remaining data automatically. The texture, noise and inversion were manually annotated. Geometric properties include information about occlusion, rotation, distortion, and character layout were used to select the features (Nagy et al., 2011a; Nagy et al., 2011b).

4.11 Scene Text

The KAIST scene text dataset consists of 3000 photos taken in various settings. This includes scenes both inside and outside in a variety of lighting settings such as bright artificial lights, nighttime and clear days. Either a high-resolution digital camera or a low-resolution cell phone camera was used to take the pictures. Text recognition on any shaped scene-text can be benchmarked using the Scene Text dataset. Natural photos with arbitrary shapes for scene text require models to do text recognition. With the help of about one million high-quality word annotations, Scene Text enables end-to-end reasoning to be applied to downstream activities like image captioning and visual question answering. Every image has had its size adjusted to 640x480. Language used to collect the scene text is used to categorize the KAIST scene text database (Jung et al., 2011).

4.12 MSRA Text Detection

Developed and made available to the public the MSRA Text Detection 500 Database (MSRA-TD500) serves as a benchmark for assessing text detection algorithms. It is intended to monitor the latest developments in the field of text detection in natural image particularly the progress made in identifying texts with variable orientations. 500 natural photos from interior (office and mall) and outdoor (street) settings were captured with a pocket camera and are included in the MSRA Text Detection 500 Database (MSRA-TD500). The training set and test set comprise the two halves of the dataset for the character set of this language. Out of the original dataset 300 photographs were randomly picked to form the training set, and the remaining 200 images were designated as the test set (Yao et al., 2012).

4.13 Stanford OCR

A challenging computer vision challenge at the heart of many developing applications is the detection and reading of text from natural images. For practical applications like reading handwritten numbers the related problems like document character recognition have been extensively researched by computer vision and machine learning experts and are essentially solved. It is much harder to reliably identify characters in more complicated settings such as photographs which has the impact on the advanced techniques fall far short of human performance in these tasks. Unsolved computer vision reads these data that can be used to create maps which are more accurate to enhance navigational services. The challenge of great difficulty is that it has to read text from photos which is relevant for many real-world applications. Hand engineered representation based systems fare worse that the average human have to face this challenge. The effect of learnt representations on the performance is an end-to-end system solution which has their persistent advantage for this job. Future researchers working on this benchmark which will be immediately benefitted from improvements in terms of overall system performance (Nadal et al., 1990).

4.14 Chars74K Data

The objective is to investigate techniques for segmented handwritten character recognition from matching image files. The 64 classes (a-z,A-Z,0-9) in the Chars74K dataset are made up of 7705 characters that were taken from natural photos and 3410 characters that were hand-drawn on a tablet PC and 62992 characters that were substituted using computer typefaces. This yields more than 74K photos in total which accounts for the dataset's name. Hindu-Arabic numbers and Latin letters use sans accents which are utilized in the English language. This is referred to as the English character set for simplicity. This dataset attempts to address three main sub-problems i.e., pre-processing photos, obtaining pertinent characteristics from the images and training an artificial neural network model to categorize the data points into classes (Lee et al., 2010).

4.15 CEDAR Database

NCI funds the Cancer Epitope Database and Analysis Resource (CEDAR), a publicly accessible resource used for research and academic purposes. In the context of cancer disease it compiles experimental data on antibody and T cell epitopes investigated in humans, non-human primates and other animal species. Additionally, the CEDAR provides resources to help with cancer epitope analysis and prediction. Complementing the Immune Epitope Database (IEDB) which focuses on infectious breaks the pixels of the image of the photo CEDAR which are freely available for extensive collection of cancer epitope and receptor data selected from the literature. The homogeneous sections that make up the handwriting images are analyzed to determine whether there are any horizontal, vertical, or diagonal segments present. A linear classifier and a nonlinear classifier use the total number of such segments detected in these regions as input. More sophisticated features have been recovered in several other investigations using CEDAR data recognition, producing results that are comparable to or somewhat better. It requires to examine the original image to confirm that working with continuous lines are mandatory. It also requires to check which points on the infinite line actually form a continuous line and that need to count to get the exact length of the line. Adaptive immune responses that identify cancer cells may be crucial to the way a disease develops clinically by helping to partially or totally stop tumor development. These reactions may arise spontaneously or be purposefully caused or enhanced by therapeutic vaccination. Creating query interfaces that are both user-friendly and enable granular query execution is a problem for biomedical databases. It is also possible to download all of the data kept in CEDAR in bulk (Liu et al., 2003).

5. IMPACT OF DATASETS IN COMPUTER VISION AND MACHINE LEARNING

The handwritten character datasets are the important document used for optical recognition of the letters and digits of different languages. The hardcopy of the writing will be converted to soft copy by using several machine learning and deep learning classification. The classification algorithms need to be developed and for that purposes existing data sets are very much essential. The popular language like Python is used to implement the deep learning algorithms and for the training and testing purposes. Prior to text recognition handwritten text recognition based on analytical word recognition segments words at the character level. Therefore, since word segmentation is not required, holistic word recognition techniques are frequently used to produce error-free results on documents including overlapped or mixed characters. The writers have also used a comprehensive approach in this work. For instance, the well-known CNN model was created from the ground up using multiple convolutional layers, including polling, flattening, and thick layers in the architecture. Moreover, an evaluation of the suggested CNN model's performance was carried out across various epoch counts and batch sizes, revealing that the model has demonstrated strong performance at each epoch. However, text recognition based on holistic word identification is a segmentation-free method that uses the word's contour or shape information to recognize the entire word at once. Given that missing data is a frequent occurrence in the optical character recognition dataset size is seen as a critical concern. Investigating how dataset quality affects supervised classification models' overall performance is the goal of this work (Bahlmann et al., 2002). One important factor that influences how well a machine learning model performs is the quantity of the dataset. Small datasets may cause over-fitting, while large datasets typically result in superior classification results. Sorting through the various categories is a difficult undertaking. When working with little datasets, it gets more difficult.

6. COMPARATIVE PERFORMANCES OF THE DATASETS

For NIST People can utilize these statistics with confidence and use them as the foundation for important decisions since they have been evaluated by specialists and are reliable. For the quick matching and identification of obtained tandem mass spectrometry, NIST provides the libraries which are extensive annotated sets of mass spectrum reference collections for a variety of species and proteins. The datasets in this database are included certified values that can be used to evaluate the precision of software for analysis of variance, linear regression, nonlinear regression and univariate statistics. One common dataset used in deep learning and computer vision applications is the MNIST handwritten digit classification issue. Despite the fact that the dataset is successfully addressed it can serve as a foundation for understanding and training convolutional deep learning neural networks for image categorization from scratch. This includes exploring ways to improve the model creating a reliable test harness for gauging the model's performance and saving and loading the model to make predictions on fresh data. The variations in the fonts sizes and styles utilized to create the images provide issues that the database attempts to solve. Additionally, attention is focused on low-resolution photos where noise is produced by anti-aliasing on recognizable characters (Shi et al., 2002). A dictionary of 113 psila 284 words ten Arabic fonts and ten letter sizes and four font styles are used to artificially create the database. To assess the score of the recognition rate of any particular Arabic OCR system there hasn't been a standard corpus that contains printed Arabic writing in a variety of fonts, styles and sizes. The Arab-Muslim world has highly developed calligraphy. Due to the Arabic calligraphic styles, the examination of the Arabic texts reveals the presence of a variety of writing fonts for printed paper. The Arabic printed text image dataset was chosen from El chourouk the official website of the Tunisian daily. Arabic printed text recognition and font identification may find use for this database. These text-block datasets produced Arabic image words with a vast vocabulary. For the big vocabulary Arabic printed text recognition competition the APTID/MF is set to host. For IAM Online Document Database the InkML standard is used for document storage in the database. The complexity of the InkML standard may make it difficult to access the document's content. The part of InkML needed to read the documents is implemented by a software library called libinkml which was released to address this problem. It carries out the following functions: rotate and mirror the documents, add and modify annotations, export as PDFs, pictures, and feature vectors, and display the documents (Teow & Loe, 2002). For Street View Text a list of potential words to be labeled with bounding boxes and an image. We solely label terms linked with businesses, which is different from the ICDAR Robust Reading data set. The SVHN dataset has an order of magnitude more labeled data more than 600,000-digit images but it has a taste that is comparable to MNIST. For example the images are of little, cropped digits. It originates from a much more difficult, open-ended real-world problem: identifying numbers and digits in photos of natural scenes. Original pictures including bounding boxes at the character level and 32-by-32 MNIST-like pictures with a single character at the center; several of the pictures do have some side distractions. These are the original color house number photos, with bounding boxes at the character level and varying resolution. In NEOCR metadata that includes brightness, contrast, inversion, texture, resolution, noise, blur, distortion, rotation, character arrangement, occlusion, typeface, and language information is added to the dataset. The annotation is given in XML and is based on the Label Me specification. For text detection The variety of the sentences in the dataset and the intricate background of the photos make it difficult to work with. The uses of automatic character recognition are expanding more than ever in light of today's ubiquitous cameras. For hars74K Data in limited scenarios the scanned document images with standard character fonts and a consistent background, this is essentially regarded as a solved problem for Latin script. Character recognition remains a daunting task when dealing with photos captured by popular cameras and handheld devices. Zoning often known as dividing the image into equal-sized sub blocks which is a technique that is comparable to JPEG compression. Every sub-block is regarded as an independent entity with a unique set of characteristics. The linearly organized features of each of the image's unique sub-blocks will comprise the image's overall features. Here observation is that the character's overall structure and complicated shape can be simplified by viewing it as composed of several tiny individual line strokes and their intersections. For CEDAR Database the Cancer Epitope Database and Analysis Resource (CEDAR) any epitope data related to cancer has been indexed. Epitopes are the precise chemical targets that activate B and T cell adaptive immunity. In a character image the Hough transform can be used to recognize characteristics of a specific shape such as circles, curves and straight lines. Several methods are used to determine which window contains the recognized lines. The first possibility we look into is where the windows are positioned. The identified lines fit into a variety of distinct angles. Depending on which class they most closely resemble, these can be categorized as horizontal, vertical, or one of two diagonals. The dataset is provided such that it their recognition by immune cells that fight cancer the epitopes produced from cancer antigens are extremely important. Thorough comprehension and planning of preventive and therapeutic applications, as well as precise characterization of naturally occurring immune responses and cancer immunopathology, require a detailed understanding of the molecular characteristics of cancer epitopes and the information associated with them (Cheung et al., 1998; Tsang et al., 1998).

7. DATASETS USED FOR RENEWABLE ENERGY

A thorough compilation of worldwide renewable energy data from 1965 to 2022 is available in the Renewable Energy World Wide dataset. The collection contains global data on geothermal, hydroelectric, solar, wind, and biofuel energy generation. Since the Industrial Revolution, fossil fuels have dominated the energy mix in the majority of the world's countries. Both the health of people and the environment are greatly affected by this. For our energy systems to become carbon-free in the next decades, renewable energy will be essential. How quickly is our renewable energy generation changing though different energy sources. All these phenomenon generates some data which is maintained by the repositories maintained by the popular datasets. The above datasets are the repositories of handwritten datasets and few of the datasets contains the renewable data also (Kim & Gorivindaraju, 1997). The NIST datasets contains data of several popular domains and renewable energy is also among these domains. The datasets covers domain like, renewable-share-energy data, renewable-share-energy data, renewable-share-energy data, total installed wind energy capacity measured in giga watts etc. All these applications have hand written documents for the data collection and maintenance. To convert the hand written documents of this activities into soft copy the OCR based handwritten character recognition system has huge impact (Downton et al., 1991).

8. CONCLUSION

In this paper a general analytical survey was done on publicly available OCR based hand written datasets. Starting from NIST and MNIST other popular available datasets are discussed which are important to develop the automated systems of optical character recognition. Handwritten characters are read properly and the computerized softcopy will be prepared based on that. These dataset are also important for feature extraction and classification of images of different languages like English, Devanagari, Bengali, Arabic etc. Future computing will heavily rely on handwritten character recognition. It's essential to have handwritten character recognition since it may be used to automate a variety of tasks, including bank receipts, which might be laborious. In order to achieve optimal efficiency, we can apply a spellchecker to the output and apply cleaning algorithms to the input photos to improve accuracy. Separate characters provide good accuracy, although differing writing styles have an impact on word recognition. Since the classifier must deal with a finite number of variations, good accuracy is seen in cases when the word scope is restricted to fixed numbers. In this research we effectively use unsupervised feature learning for a deployable application to identify numbers in natural settings. Although the primary benefits are demonstrated of learning representations over manually created ones, we have also discovered that there is still tremendous space for development of computer performance on our recently introduced tasks greatly trails human performance.

REFERENCES

Al-Yousefi, H., & Udpa, S. S. (1988). Recognition of handwritten Arabic characters. *Proc. SPIE 32nd Ann. Int. Tech. Symp. Opt. Optoelectric Applied Sci. Eng.*

Altuwaijri, M., & Bayoumi, M. A. (1994). Arabic Text Recognition Using Neural Network. *ISCAS 94. IEEE International Symposium on Circuits and systems, 6*(30). 10.1109/ISCAS.1994.409614

Badi, K., & Shimura, M. (1982). Machine recognition of Arabic cursive script. *Trans. Inst. Electron. Commun. Eng., E65*(2), 107–114.

Bahlmann, C., Haasdonk, B., & Burkhardt, H. (2002). Online Handwriting Recognition with Support Vector Machine – A Kernel Approach. *Proceeding of the 8th Int. Workshop in Handwriting Recognition (IWHFR)*, 49-54. 10.1109/IWFHR.2002.1030883

Beigi, H. S. M. (1993). An Overview of Handwriting Recognition. *Proceedings of the 1st Annual Conference on Technological Advancements in Developing Countries*, 30-46.

Cheung, K., Yeung, D., & Chin, R. T. (1998). A Bayesian framework for deformable pattern recognition with application to handwritten character recognition. *IEEE Transactions on Pattern Analysis and Machine Intelligence, 20*(12), 382–1388.

Dalal, N., & Triggs, B. (2005). Histograms of oriented gradients for human detection. *Proceedings of the IEEE Computer Society Conference on Computer Vision And Pattern Recognition. CVPR 2005*, 886–893. 10.1109/CVPR.2005.177

Downton, A. C., Tregidgo, R. W. S., & Kabir, E. (1991). Recognition and Verification of Handwritten and Hand- Printed British Postal Addresses. *International Journal of Pattern Recognition and Artificial Intelligence*, *5*(01n02), 265–292. doi:10.1142/S0218001491000168

Jung, J., Lee, S. H., Cho, M. S., & Kim, J. H. (2011). Touch TT: Scene Text Extractor Using Touch Screen Interface. *ETRI Journal*, *33*(1), 78–88. doi:10.4218/etrij.11.1510.0029

Khan, S. B., Ahmad, G., Ali, F., Faisal, F., Ahmed, I., & Elahi, S. (2018) Classification performance of linear binary pattern and histogram oriented features for Arabic characters images: A. Applied Physics of Condensed Matter (APCOM 2019), 5.

Kim, G., & Gorivindaraju, V. (1997). A Lexicon Dnven Approach to Handwritten Word Recognition for Real-Time Applications. *IEEE Transactions on Pattern Analysis and Machine Intelligence*, *19*(4), 366–379. doi:10.1109/34.588017

Lee, Cho, Jung, & Kim. (2010). *Scene Text Extraction with Edge Constraint and Text Collinearity Link. 20th International Conference on Pattern Recognition (ICPR)*, Istanbul, Turkey.

Liu, C. L., Nakashima, K., Sako, H., & Fujisawa, H. (2003). Benchmarking of state-of- the-art techniques. *Pattern Recognition*, *36*(10), 2271–2285. doi:10.1016/S0031-3203(03)00085-2

Nadal, C., Legault, R., & Suen, C. Y. (1990). Complementary Algorithms for Recognition of totally Unconstrained Handwritten Numerals. in Proc. 10th Int. Conf. *Pattern Recognition*, *1*, 434–449.

Nagy, R., Dicker, A., & Meyer-Wegener, K. (2011a). NEOCR: A Configurable Dataset for Natural Image Text Recognition. *CBDAR Workshop 2011 at ICDAR 2011*, 53-58.

Nagy, Dicker, & Meyer-Wegener. (2011b). *Definition and Evaluation of the NEOCR Dataset for Natural-Image Text Recognition.* University of Erlangen, Dept. of Computer Science, Technical Reports, CS-2011-07.

Nawaz, S., Calefati, A., Ahmed, N., & Gallo, I. (2018). Hand Written Characters Recognition via Deep Metric Learning. *Proceedings of the 2018 13th IAPR International Workshop On Document Analysis Systems (DAS),* 417–422. 10.1109/DAS.2018.18

Netzer, Wang, Coates, Bissacco, Wu, & Ng. (n.d.). *Reading Digits in Natural Images with Unsupervised Feature Learning NIPS Workshop on Deep Learning and Unsupervised Feature Learning*. Academic Press.

Newell, A. J., & Griffin, L. D. (2011). Multiscale histogram of oriented gradient descriptors for robust character recognition. *Proceedings of the 2011 International Conference On Document Analysis And Recognition (ICDAR),* 1085–1089. 10.1109/ICDAR.2011.219

Shi, M., Fujisawa, Y., Wakabayashi, T., & Kimura, F. (2002). Handwritten numeral recognition using gradient and curvature of gray scale image. *Pattern Recognition*, *35*(10), 2051–2059. doi:10.1016/S0031-3203(01)00203-5

Teow, L. N., & Loe, K. F. (2002). Robust vision-based features and classification schemes for off-line handwritten digit recognition. *Pattern Recognition*, *35*(11), 2355–2364. doi:10.1016/S0031-3203(01)00228-X

Tsang, I. J., Tsang, I. R., & Dyck, D. V. (1998). Handwritten character recognition based on moment features derived from image partition. *Int. Conf. Image Processing, 2*, 939–942.

Watanabe, T., Ito, S., & Yokoi, K. (2009). Co-occurrence histograms of oriented gradients for pedestrian detection. *Proceedings of the Pacific-Rim Symposium on Image and Video Technology,* 37–47. 10.1007/978-3-540-92957-4_4

Yao, C., Bai, X., Liu, W., Ma, Y., & Tu, Z. (2012). *Detecting Texts of Arbitrary Orientations in Natural Images*. CVPR.

Chapter 7
Anti-Poaching of Tree Monitoring and Forest Fire Detection System Using Renewable Energy

Palash Pal
Haldia Institute of Technology, India

Dilip Dey
Haldia Institute of Technology, India

Anushka Pandey
Haldia Institute of Technology, India

Monishita Chakraborty
Haldia Institute of Technology, India

Deepti Kumari
Haldia Institute of Technology, India

Komal Kumari
Haldia Institute of Technology, India

Priya Kumari
Haldia Institute of Technology, India

ABSTRACT

Wireless sensor networks (WSN) have emerged as an effective solution for monitoring and recording physical conditions of the environment. Forests, being an important part of the ecosystem, are prone to several challenges including forest fires and tree poaching. Uncontrolled anthropogenic activities, abnormal natural conditions, and tree smuggling are some of the factors that contribute to these problems. To address these issues, the authors propose a WSN-based system for preventing tree poaching and forest fires. The system will consist of spatially distributed sensors that will continuously monitor the physical conditions of the forest, including temperature, humidity, and presence of human activity. The data collected from these sensors will be processed in a central location to detect any abnormal changes in the environment. In case of a potential forest fire or tree poaching activity, the system will send alerts to the relevant authorities, allowing them to take timely action. By leveraging the power of WSN, the proposed system aims to prevent and monitor forest fires and tree poaching.

DOI: 10.4018/979-8-3693-2355-7.ch007

1. INTRODUCTION

Forests are a vital resource for human survival and social development, playing an essential role in maintaining the ecological balance of the earth. However, the occurrence of forest fires due to uncontrolled anthropogenic activities and abnormal natural conditions is frequent. Forest fires are considered among the most serious disasters, posing a threat to forest resources and the environment. Preventing and monitoring forest fires has become a global concern for forest fire prevention organizations (Ananthi J, et, al., 2022).

Additionally, incidents of tree cutting and smuggling, particularly of valuable tree woods like sandalwood, are increasing due to the significant amount of money involved. To curb such illegal activities and protect forests worldwide, preventive measures need to be adopted (Faroudja, A. B. I. D., et, al., 2020).

Current forest fire prevention methods primarily involve patrols, observation from watchtowers, satellite monitoring, and recently, wireless sensor networks (WSN) (V. Parthipan, et, al., 2019). However, the observation from watchtowers method has several limitations as it requires human supervision, which is not always possible in dense and remote forests (Eunna Jang, et, al., 2019).

WSN refers to a group of dedicated and spatially dispersed sensors that monitor and record physical environmental conditions and send the collected data to a central location (Sakr, G. E., et; al., 2010). These sensors measure temperature, sound, pollution levels, humidity, wind, etc., and rely on wireless connectivity for data transfer, similar to wireless ad hoc networks (Daxini Nidhi, et, al., 2015).

2. EXISTING SYSTEM AND CHALLENGES

The destruction of forests is a significant worldwide problem that endangers the ecological balance of the planet. The primary reason for this damage is the unlawful felling of trees, resulting in the loss of millions of acres of forest every year. Additionally, extreme weather events such as heavy rainfall can also damage trees and contribute to the destruction. It is crucial to halt the illegal trade of trees and implement strategies to safeguard forests from such activities.

The current system employs manual labor to combat forest fires, which can be hazardous and ineffective. In order to minimize the risk to human life and decrease reliance on manual labor, research is underway to explore wireless system-based monitoring (Venkatesan, Prof. Latha, et, al., 2013).

Current research on wireless networks for wildlife tracking has resulted in standardized solutions, which may not be entirely effective. This 'one size fits all' approach involves the design of a single tracking device for all types of wildlife (Jin, X. et, al., 2012). However, this approach may not be suitable for tracking trees and preventing forest fires, since each tree species may have distinct needs and susceptibilities. As a result, it is necessary to create more specialized wireless systems that can be customized for specific forest environments and tree species (Wark, T., et, al., 2007). To achieve this, cooperation among researchers, forest management organizations, and other stakeholders is necessary to establish effective and sustainable methods of safeguarding our valuable forest resources.

The challenges are stated below-

- Designing a wireless Sensor network and managing each is always a challenging task.
- Fixing of Sensors to wild animals is also big task.
- Sensors used in WSN should resist weather changes and environmental conditions.

- Detection of tree cutting in thick forest is also a risky task.

3. PROPOSED METHODOLOGY

This topology combines embedded technology and Zigbee wireless communication to create a system that monitors and tracks the health of animals by using a ZigBee transceiver and GPS. Forests are essential resources, and preventing and detecting forest fires has been a hot research topic in forest fire prevention departments worldwide. To address the shortcomings of conventional forest fire and conservation detection in terms of real-time monitoring accuracy, a wireless sensor network technique for forest fire and conservation detection was introduced, along with satellite monitoring. This enables related government departments to make timely and accurate decisions regarding fire extinguishing and conservation prevention.

This scheme proposes a wireless sensor network paradigm for real-time detection of forest fires and conservation efforts, which can detect and forecast forest fires, increases in carbon dioxide, decreases in soil moisture, and falling trees more efficiently. The paper mainly focuses on data collection and processing in wireless sensor networks for real-time forest fire and conservation detection, and applies a neural network method for in-network data processing. Additionally, a lightweight system is attached to the neck of animals to sense body temperature, which is sent to the microcontroller. The GPS modem receives a string from satellites, extracts latitude and longitude information from the string, and sends it to the Zigbee modem for transmission to the server room.

In here the design (DFD) and flowchart for the proposed topology is discussed. This is implemented as follows:

Case Diagram

A use case diagram is a graphical representation of a user's interaction with a system, showing in Figure 1, how the user is involved in different use cases and the relationships between them. It can also identify the types of users and their involvement in the system. Use case diagrams serve as a blueprint for the system and are often accompanied by other types of diagrams.

The flowchart is shown in Figure 2 and Data Flow Diagram (DFD) is shown in figure 3 provide a detailed explanation of the system. As depicted in the Use Case diagram, the DFD level diagram is used to retrieve information and send it to the microcontroller for implementation. The flowchart shows the process of reading data from sensors, conditioning the data from temperature, humidity, and tilt sensors, and then in putting it into the Blynk sensor.

Figure 1. Representation of a user's interaction

Figure 2. Flowchart of a user's interaction

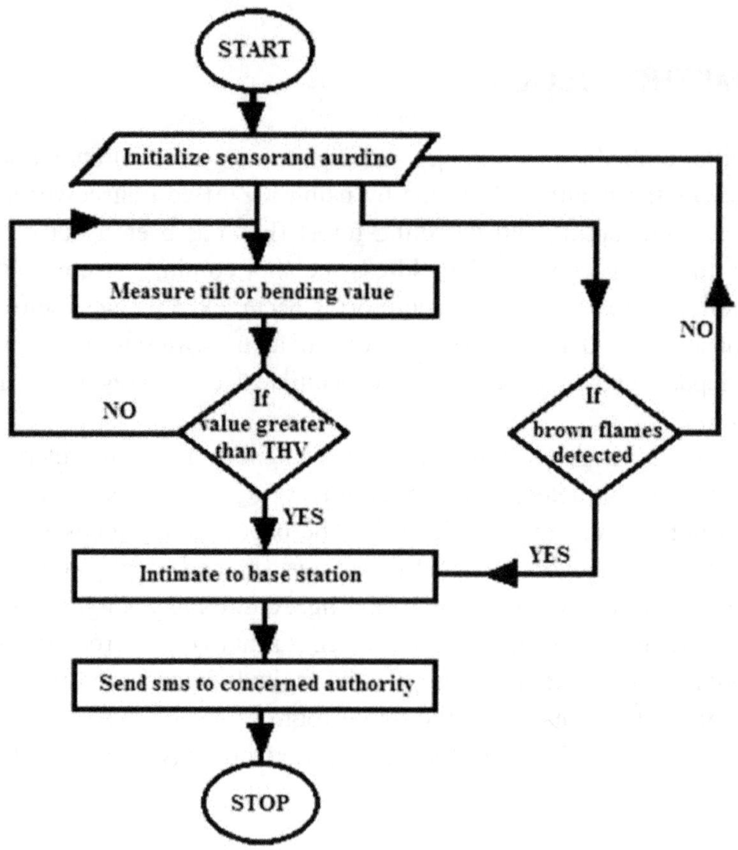

Class Diagram

A class diagram is a static structure diagram in the Unified Modelling Language (UML) used to depict the structure of a software system as shown in figure 4. It presents the system's classes, their attributes, operations (or methods), and the relationships between the classes, showing which class holds particular information.

4. SEQUENCE DIAGRAM

A sequence diagram is a type of diagram that illustrates the interactions between objects in a particular scenario in a time-based order. The diagram shows the objects and classes involved in the scenario and the messages they exchange to achieve the desired functionality. Sequence diagrams are usually related to the implementation of use cases in the system, and they may also be referred to as event diagrams or event scenarios.

Figure 3. Data flow diagram of a user's interaction

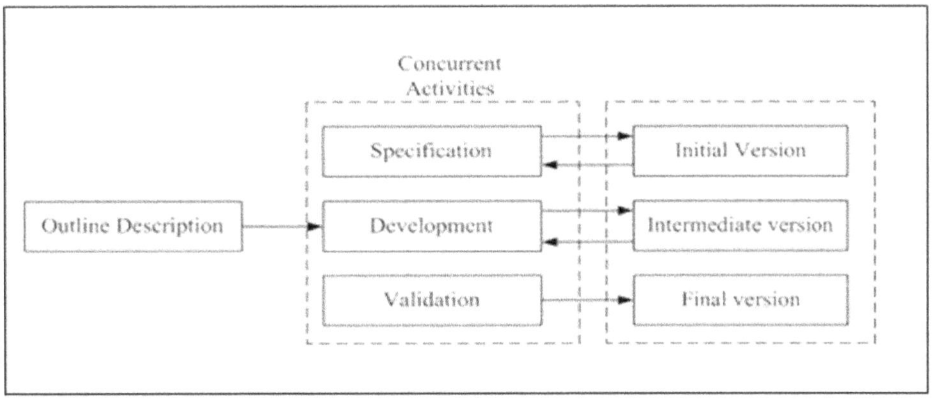

Figure 4. Class diagram

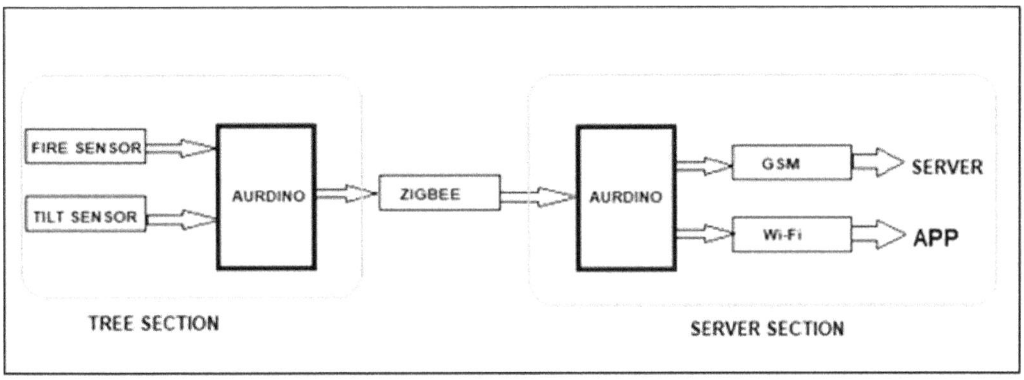

Activity Diagram

An Activity Diagram is a visual representation that depicts a sequence of actions or steps in a process as shown in figure 5. It includes various elements that support decision-making, repetition, and simultaneous execution. In the Unified Modeling Language (UML), Activity Diagrams are designed to model workflows and processes, including the flow of data associated with these activities. While the primary focus of Activity Diagrams is to illustrate the overall flow of control, it can also contain elements that show the movement of data between activities via one or more data stores.

5. INCREMENTAL DEVELOPMENT MODEL AND SYSTEM ARCHITECTURE

The incremental development approach involves creating an initial implementation, receiving feedback from users, and gradually improving the system through multiple iterations until it is complete. This approach involves a combination of specification, development, and validation activities, with each version building on the previous one by adding new features and functionalities as shown in Figure 6.

Figure 5. Activity diagram

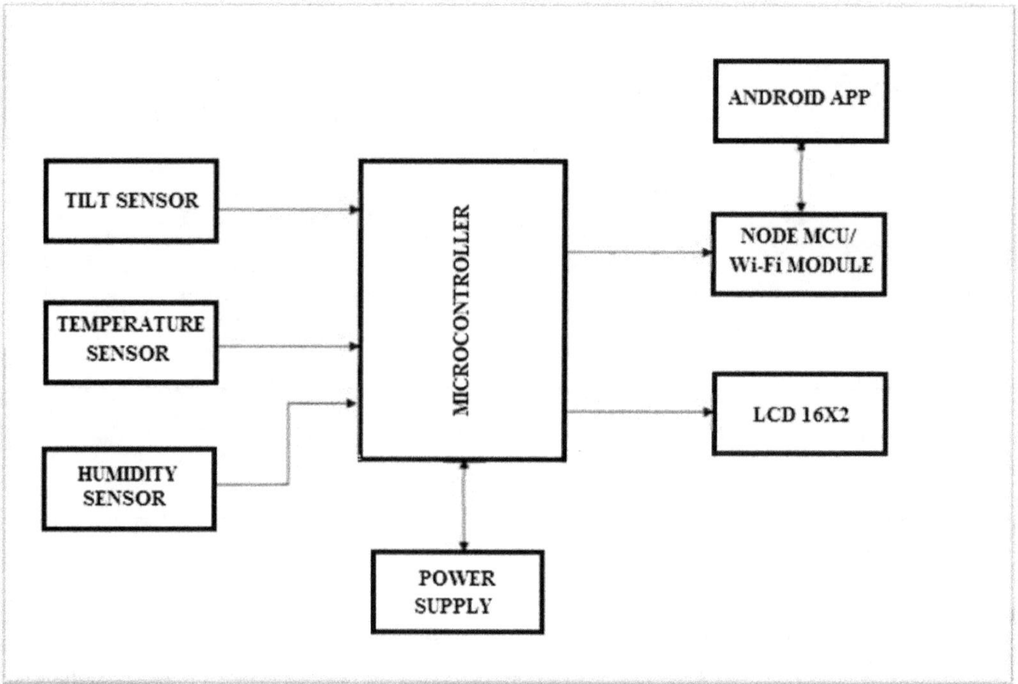

The proposed topology has adopted the agile software development model, which allows for flexibility in accommodating changing requirements and adding new features to improve overall functionality. However, the hardware aspect of the scheme has limited flexibility in design and implementation as shown in Figure 9. Therefore, it use conceptual diagrams and temporary breadboards to model the complete design, making it easy to integrate and disintegrate without having to produce and redesign end products repeatedly in response to changes or failures.

The proposed work combines the embedded technology with the Zigbee /wi-fi wireless communication technology as shown in Figure 7.

The limitations of traditional methods for detecting forest fires and conservation issues in real-time and their monitoring accuracy led to the introduction of wireless sensor network technology for these purposes. This technology is coupled with satellite monitoring to enable government departments to make timely and appropriate decisions regarding fire-extinguishing and conservation measures.

The scheme proposes a wireless sensor network paradigm for real-time detection of forest fires and conservation as shown in figure 8. The network is designed to detect and predict forest fires, changes in carbon-dioxide, decrease in soil moisture, and falling trees in a timely manner. The proposed system is based on a cluster-based wireless sensor network approach.

The paper focuses on detecting forest fires and promoting conservation through data collection and processing in wireless sensor networks in real-time using renewable energy. In-network data processing is carried out using a neural network approach. The GPS modem receives a string from satellites, which is then transmitted to the microcontroller. The microcontroller extracts the latitude and longitude information from the string and transmits it to the ZIGBEE modem, which, in turn, sends it to the server room.

Figure 6. Concurrent activities model

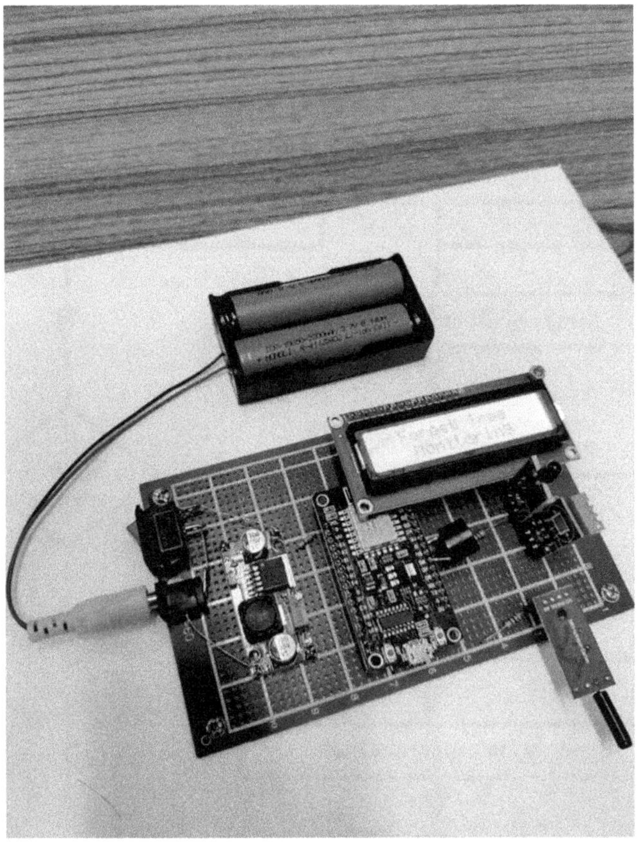

Figure 7. Embedded technology with the Zigbee/wi-fi wireless communication technology

6. SYSTEM TESTING

System Testing is a type of testing that assesses the overall system to determine its adherence to specified requirements. This testing involves testing the system's functionalities from end-to-end. In our specific system testing, there are two main test scenarios that verify the system's ability to search for available information as well as information that is not available.

Testing Methods

There are different methods that can be used for software testing are stated below,

Figure 8. Sensor network paradigm for real-time detection

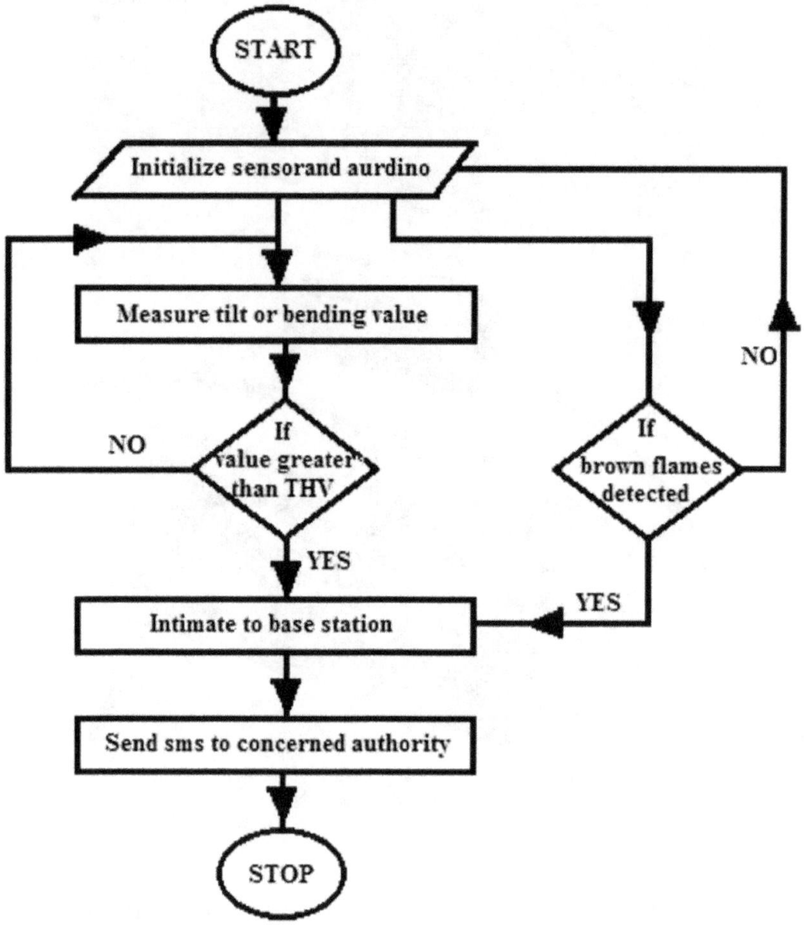

Blackbox Testing

The Black-box testing is a testing technique in which the tester does not have access to the internal workings or source code of the application. Instead, the tester interacts with the system's user interface by providing inputs and examining outputs, without any knowledge of how the system processes the inputs.

Whitebox Testing

White-box testing is a method of examining the internal structure and logic of software code. It is also known as open-box testing or glass-box testing. To conduct white-box testing, a tester should have knowledge of the application's internal operations. They need to examine the source code and identify which part of the code is not performing as expected.

Figure 9. Hardware test kit

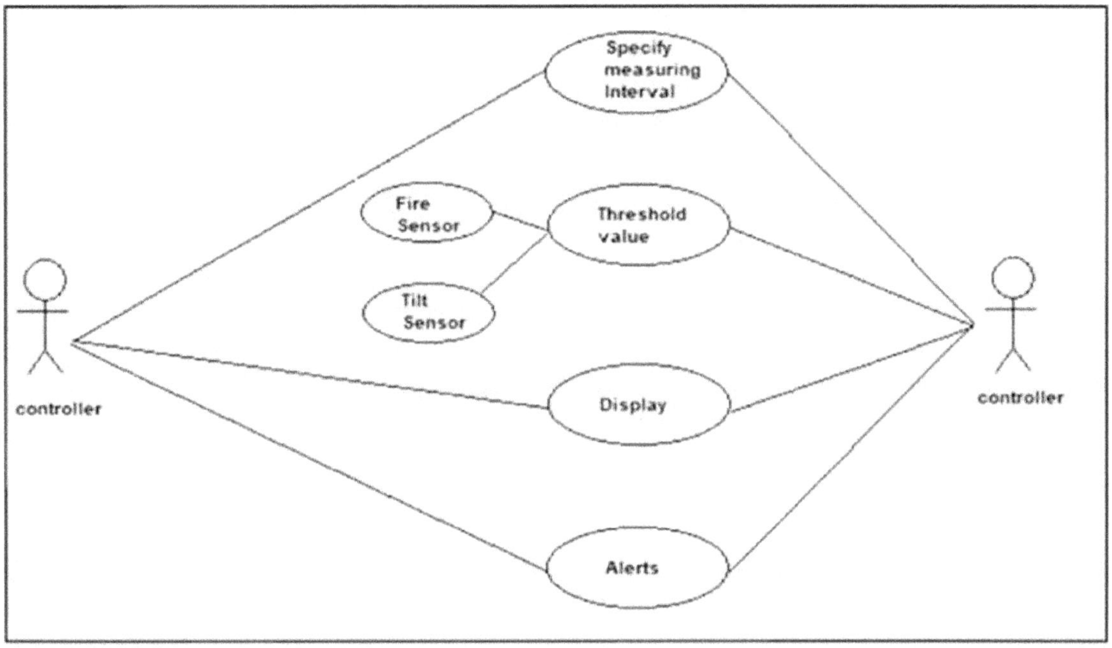

Levels of Testing

There are several stages of testing that are employed in software testing to ensure that the software functions correctly. These stages include different approaches or techniques that can be utilized. The primary stages of software testing are:

Functional Testing

Functional testing is a black-box testing method that verifies whether an application or software product operates according to its specified requirements. The process involves providing inputs to the software and analyzing the output to confirm if it performs as intended. The testing is conducted on the complete and integrated system to evaluate its compliance with the specified requirements. The functional testing of software includes five key steps.

1. Identify the intended functionality of the application to be tested.
2. Generate test data according to the application specifications.
3. Verify the output produced by the application with the expected results.
4. Create test scenarios and execute test cases.
5. Compare the actual test results with expected outcomes to ensure the software functions correctly.

Non-Functional Testing

This section pertains to the testing of an application's non-functional attributes. Non-functional testing involves assessing software against non-functional requirements, such as performance, security, user interface, etc. This type of testing can be conducted at various stages of the software development life cycle

Unit Testing

Unit testing is a crucial aspect of software development that involves testing the smallest functional units or modules of an application, referred to as "units," to ensure that they operate correctly and in line with requirements. Unit testing can be performed either manually or automatically. Its main objective is to isolate each component of the software and verify that it operates correctly and meets the expected functionality. Test cases and results are shown in Table 1.

Integrating Testing

Integration testing is a software testing level that involves testing individual units as a group to identify errors in their interaction. The main objective of this testing is to detect faults in the combined parts of an application. Test drivers and test stubs are used to support Integration Testing. It is performed after unit testing and before validation testing as shown in Table 2. There are two approaches to Integration Testing: Bottom-up integration testing and Top-down integration testing.

Bottom-Up Integration

In this type of testing, the testing process starts with unit testing, and then gradually moves towards testing of larger combinations of units, which are referred to as modules or builds.

Top-Down Integration

Top-down integration testing is a testing approach where the highest-level modules of an application are tested first and then the testing proceeds with the lower-level modules.

Table 1. Arduino test cases and results

SI # Test Case: -	UTC-1
Name of Test: -	Arduino Test
Items being tested: -	Power up the Arduino and boot
Sample Input: -	Turn on the Supply
Expected output: -	Arduino Should boot
Actual output: -	Power on test and upload successful
Remarks: -	Pass.

Table 2. Tree cutting and fire detection

Sl # Test Case: -	ITC-3
Name of Test: -	Tree Cutting and Fire Detection
Item being tested: -	Tilt Sensor variation and Update
Sample Input: -	Light up the fire and Change Tilt Values
Expected output: -	Fire and Tree Cutting Detection
Actual output: -	Detection of tree Cutting and fire
Remarks: -	Pass.

7. CONCLUSION

In conclusion, the Internet of Things (IoT) has enabled us to develop a system that monitors tree movement and forest fires, contributing to the overall preservation of the environment. The system consists of various technologies such as tilt sensors, temperature sensors, and buzzer notifications that detect the occurrence of tree logging and forest fires. Through our experiments, we have successfully demonstrated that the system is capable of sensing the sound generated due to axing of trees and immediately notifying the authorities via the Blynk application. Additionally, the system is capable of detecting any abnormal temperature changes in the surroundings and sending alerts to the forest officer if any event occurs, allowing them to take appropriate action.

The successful implementation of this system has highlighted the potential of IoT in preserving our environment by preventing illegal logging and forest fires. By leveraging various technologies and sensors, our system has demonstrated how IoT can be used to monitor and prevent forest-related issues effectively. With further advancements in technology, we believe that IoT can play a significant role in addressing environmental concerns, paving the way for a sustainable future.

REFERENCES

Ananthi, J., Sengottaiyan, N., Anbukaruppusamy, S., Upreti, K., & Dubey, A. K. (2022). Forest fire prediction using IoT and deep learning. *International Journal of Advanced Technology and Engineering Exploration*, *9*(87), 246–256. doi:10.19101/IJATEE.2021.87464

Daxini, N., Sharma, S., & Patel, R. (2015). Real Time Animal Detection System using HAAR Like Feature. *International Journal of Innovative Research in Computer and Communication Engineering*, *3*(6), 5177–5182. doi:10.15680/ijircce.2015.0306032

Faroudja & Izeboudjen. (2020) Decision tree based system on chip for forest fires prediction. In *International Conference on Electrical Engineering (ICEE)*. IEEE.

Jang, E., Kang, Y., Im, J., Lee, D.-W., Yoon, J., & Kim, S.-K. (2019). Detection and Monitoring of Forest Fires Using Himawari-8 Geostationary Satellite Data in South Korea. *Remote Sensing (Basel)*, *11*(3), 271. doi:10.3390/rs11030271

Jin, X., Sarkar, S., Ray, A., Gupta, S., & Damarla, T. (2012). Target Detection and Classification Using Seismic and PIR Sensors. *IEEE Sensors Journal*, *12*(6), 1709–1718. doi:10.1109/JSEN.2011.2177257

Parthipan, V., & Dhanasekaran, D. (2019). Preventing and Monitoring of Framework for Forest Fire Detection and Data Analysis Using Internet of Things. *International Journal of Engineering and Advanced Technology*, *8*(3).

Sakr, G. E., Elhajj, I. H., Mitri, G., & Wejinya, U. C. (2010). Artificial intelligence for forest fire prediction. *IEEE/ASME International Conference on Advanced Intelligent Mechatronics*, 1311-1316. 10.1109/AIM.2010.5695809

Taluja, C., & Guha, P. (2018). Detection of Forest Fires at Early Stages using Wireless Sensor Network and Graphical User interfacing. *4th International Conference on Computers and Management (ICCM),* 1-6.

Venkatesan, Farooq, Imraan, Kumar, & Kumar. (2013). Animals and Vehicle Collision Avoidance Using Wireless Sensor Actuator Network. *International Journal of Scientific & Engineering Research, 4*(5). https://www.ijser.org/researchpaper/Animals-and-Vehicle-Collision-Avoidance-Using-Wireless-Sensor-Actuator-Network.pdf

Wark, T. (2007). The Design and Evaluation of a Mobile Sensor/Actuator Network for Autonomous Animal Control. *6th International Symposium on Information Processing in Sensor Networks*, 206-215. 10.1109/IPSN.2007.4379680

Chapter 8
Applications of Renewable Energy Sources and Sustainable Innovations to Transform the Energy Sector

Neeta Nathani
Gyan Ganga Institute of Technology and Sciences, Jabalpur, India

Silky Pareyani
Gyan Ganga College of Technology, India

Jagdeesh Kumar Ahirwar
Gyan Ganga Institute of Technology and Sciences, Jabalpur, India

ABSTRACT

Green energy deployment in any country has the primary goal of assisting economic development, improving energy security, increasing energy access, and combating global climate change. Sustainable development can be achieved through utilizing renewable energy sources and ensuring inhabitants' access to affordable, reliable, and modern electricity. The term "green energy" refers to energy produced by natural resources such as sunshine, wind, and water. Environmentally hazardous mining and drilling operations are typically avoided by green providers. The advancement of technologies such as automation, additive manufacturing, artificial intelligence (AI), and others can facilitate the shift to a cleaner future by improving environmental quality and controlling and monitoring infrastructures through the replacement of outdated equipment with more advanced options.

IMPORTANCE OF GREEN ENERGY

Green energy is beneficial for the environment since it replaces the negative effects of fossil fuels with more environmentally friendly alternatives. Since green energy is frequently clean and renewable, it

DOI: 10.4018/979-8-3693-2355-7.ch008

produces little to no greenhouse gas emissions and is widely available. Its creation involves the usage of natural resources. Even when taking into consideration all of their life cycles, green energy sources emit far fewer greenhouse gases and relatively little air pollution than fossil fuels. Both the environment and the health of the animals and humans who breathe the air benefit from this.

Energy prices are more stable due to the fact that green energy is primarily produced locally, making it less vulnerable to supply chain interruptions, price surges, and geopolitical crises. One of the economic benefits is the creation of jobs through the building of structures that help the communities in which the workers are employed. 2018 saw the creation of 11 million new employments globally as a result of renewable energy.

The localized production of energy from sources like solar and wind power makes the energy infrastructure more flexible and less reliant on centralized sources that are subject to disruptions. It is thus less vulnerable to weather-related climate change. A cost-effective answer to many of the world's energy demands is also provided by green energy. Green energy will get easier to access as costs go down, especially in poorer nations.

Historically, the energy demands of humans have been met by commodities like coal, oil, and even kerosene. However, these fuels are non-renewable and detrimental to the biosphere and ecosystem. These resources will eventually run out as a result, raising the possibility of shortages and scarcity. Even worse, the damage they do to the ecology.

When traditional energy sources are consumed, they release greenhouse gases into the atmosphere. Toxic pollutants released into the atmosphere by coal and oil have a negative impact on general health, cause respiratory problems, and shorten life expectancy (National Center for Biotechnology, n.d.). The extraction of oil and coal alone has the potential to severely damage nearby businesses, people, and landscapes (National Resources Defense Council, n.d.). Green energy may be able to mitigate or even prevent some of these problems, thus the sooner we move to renewable energy sources, the better.

Potential of Global Primary Energy Consumption

Advancement in society requires access to an energy source. The past few decades have seen a rise in the demand for energy due to the rapid economic growth of developing nations. It is likely that this pattern will persist (Kaygusuz & Bilgen, 2008). To investigate appropriate environmental and economic measures, an anticipated future electricity demand is necessary (Chang et al., 2010). Likewise, forecasting future energy needs aids in planning future renewable energy expenditures. The growth of human society as well as global political and economic trends (Wang et al., 2014) have made energy supply and security critical issues. Therefore, past, present, and future power use can be ascertained with the aid of worldwide comparisons.

Global primary energy consumption is displayed in Table 1 according to the BP Energy Outlook 2018 publications. With an annual growth rate of 4.2 percent, India's total energy consumption was 724 million tonnes of oil equivalent (Mtoe) in 2016 and is predicted to reach 1921 Mtoe by 2040. Modern renewable energy sources and economically traded fuels are used to generate electricity in several major nations. India, following the United States, China, and the Organization for Economic Cooperation and Development (OECD) in Europe, was the fourth-largest energy consumer in the world in 2016 (EIA, 2019).

India is anticipated to have the highest energy consumption in the world by 2035–2040, and this trend is expected to continue (BP Energy, 2018). India's energy consumption is currently rising. India's growing energy consumption is expected to increase its share of the world's energy demand from 5% in

2016 to 11% by 2040. China, India, and Brazil are examples of emerging economies that have quickly industrialized, grown in importance within the world economy, and are currently exporting massive amounts of manufactured goods to developed nations. This shift in international economic activity has an impact on the nation's energy consumption (EIA, 2018).

Projected Primary Energy Consumption in India

The demand for energy is significantly influenced by a nation's population and rate of expansion. As of January 2019, India has 1.368 billion people living in it, making it the second most populous nation in the world (TWI-Global, n.d.). It makes up about 17.74 percent of the world's population and grows at a rate of 1.18 percent every year. By the end of 2020, 2030, 2040, and 2050, the population of the nation is projected to be 1.383 billion, 1.512 billion, 1.605 billion, and 1.658 billion, respectively. The world's largest immigration source each year is India; in fact, the population of several of its states exceeds that of several other countries.

By 2040, India will have the fastest-growing energy consumption of any large nation. The majority of this demand will be met by coal, with renewable energy coming in second. By 2020, renewable energy has become the second most significant domestic power generation source, surpassing both gas and then oil. Table 2 projects that India's consumption of renewable energy will rise from 17 Mtoe in 2016 to 256 Mtoe in 2040, an annual growth of 12 percent.

The main energy consumption of renewable in the BRIC nations from 2016 to 2040 is displayed in Table 3. (Brazil, Russia, India, and China). In 2016, India's use of renewable energy was estimated to be 17 Mtoe; by 2040, that number is expected to rise to 256 Mtoe. By 2040, it is predicted that India's energy consumption would increase at the highest rate among all major economies, with coal playing the largest role and renewables coming in second. By 2040, renewable energy is expected to account for 13% of total energy consumption, up from 2% in 2016.

The range of power generation capacity from renewable resources, which includes 7122 GW of solar PV and 5445 GW of wind power, is around 2000 GW to 16,000 GW, representing an eight-fold growth. The yearly capacity additions from the two would be three times and twice as much as those from 2017. There should be no new coal plants commissioned and a phase-out of 95% of the presently operating coal plants. An annual investment of USD 500 billion should be made in new renewable power capacity until

Table 1. Projected primary energy consumption of different nations from 1990 to 2040 (EIA, 2019)

Region	Consumption (Million Tones Oil Equivalent)											Change (Percentage per Annum)	
	1990	1995	2000	2005	2010	2016	2020	2025	2030	2035	2040	1990-2016	2016-2040
India	195	251	316	394	537	724	880	1118	1365	1624	1921	5.2%	4.2%
China	683	889	1008	1800	2491	3053	3387	3753	4017	4207	4319	5.9%	1.5%
US	1966	2119	2310	2349	2284	2273	2334	2344	2341	2325	2299	0.6%	0.0%
Brazil	126	158	188	211	268	298	330	378	419	451	477	3.4%	2.0%
Africa	222	244	274	327	389	440	509	603	710	840	1002	2.7%	3.5%
Middle East	264	351	423	565	734	895	980	1085	1189	1287	1382	4.8%	1.8%
EU	1672	1661	1732	1819	1754	1642	1667	1623	1570	1513	1460	-0.1%	-0.5%
Russia	865	662	620	647	673	674	711	720	723	722	716	-1.0%	0.3%

Table 2. India's estimated primary energy consumption from 2016 to 2040 (including renewable energy) (EIA, 2019)

	Level (Mtoe)												
	1990	1995	2000	2005	2010	2016	2020	2025	2030	2035	2040	1990-2016	2016-2040
Renewable	0	0	1	2	7	17	41	86	133	191	256	35.1%	12.0%
Coal	110	140	164	211	290	412	485	593	710	824	955	5.2%	3.6%
Nuclear	1	2	4	4	5	9	11	16	27	35	44	7.1%	7.0%
Hydro	15	17	17	22	25	29	36	43	47	50	52	2.6%	2.5%
Gas (Bcf/dl)	11	17	24	32	54	45	57	72	89	106	128	5.6%	4.5%
Oil (Mb/dl)	58	75	106	122	155	212	251	308	359	419	485	5.1%	3.5%
Total	195	251	315	394	537	724	880	1118	1365	1624	1921	5.2%	4.2%

Table 3. Renewable energy consumption-BRIC countries percentage: 2015-2035

Between 2016 - 2040	India		China		US		Russia		Brazil	
	2016	2040	2016	2040	2016	2040	2016	2040	2016	2040
Primary energy consumption renewable (Including bio fuels)	2%	13%	3%	18%	5%	17%	0%	2%	13%	24%

2050. Until 2050, about $1 trillion in yearly investments will be required to decarbonize the electricity sector. Fig. 4 displays a geographic split of the enhanced capacity for the production of renewable power. More than one-third comes from China, with the US, India, and EU trailing closely behind.

Environmental Health Strategy

Population growth has raised the need for fossil fuel energy sources, which has exacerbated pollution. Fossil fuels are known to be non-renewable energy sources, and using them excessively would cause them to disappear. Because fossil fuels contain a significant amount of carbon, using them releases greenhouse gases into the atmosphere, including CO2, CH4, N2O, 03, and H20 vapors. These gases have a negative impact on the environment because they cause global warming, which in turn causes climate change. Excessive reliance on fossil fuels and urbanization have also contributed to rising energy costs. As a result, the world economy looks for inexpensive, alternative energy sources that also enhance environmental health. Renewable energy sources include sun, air, rivers, oceans and geothermal energy create thermal and electrical energy and reduce greenhouse gasses (Jaiswal, 2022).

Solar energy is abundant and can be converted into electrical energy using sustainable, waste-free technology. Another consideration is the development of advanced solar power facilities, which require meticulous planning, financing, and implementation. For example, biomass energy turns plant, animal, and food waste into biofuels by using organic waste. The leftovers can be fed to animals or used as manure. Additionally, this minimizes and controls appropriate waste disposal and management procedures. Renewable energy should take the place of fossil fuels since it is better for the environment.

Figure 1. Regional distribution of additional capacity for renewable power production between 2018 and 2050

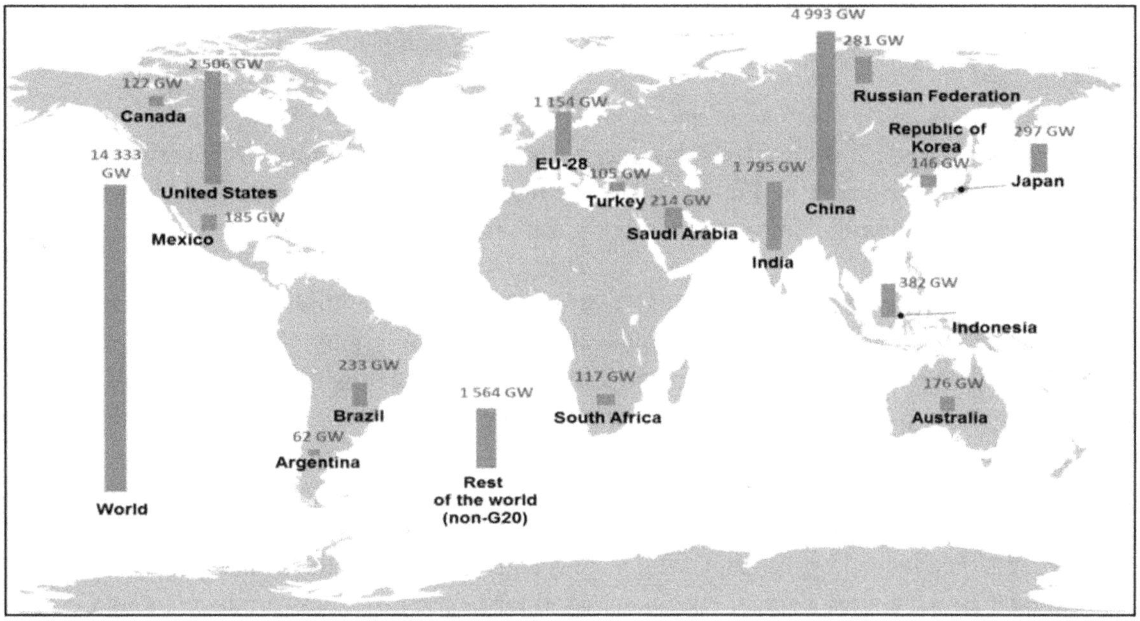

The Main Benefits of Green Energy

Investing in green energy has several benefits, including the ability to reduce carbon emissions, prevent further environmental harm, and provide employment. Everyone can help move that future closer by investing in green energy.

The sun produces more than enough energy each day to power the entire planet, but something is wrong. Since solar energy cannot be converted into electricity using current technologies, many people find that using other traditional sources is more economical. Being green involves increasing funding for renewable energy projects like solar, wind, and others, as well as developing technologies to better utilize these sources and make them more widely available and affordable.

Types of Green Energy

The main energy sources are solar, wind, and hydroelectric power, which also includes tidal energy, which harnesses the energy of the sea's tides. Large-scale production of solar and wind energy is possible in companies, as well as small-scale production in residences. The six most widely used forms of green energy are as follows:

1. Solar Power: Photovoltaic cells are a popular renewable energy source that is safe for the environment because they absorb sunlight and transform it into power. Not only does solar energy provide hot water and heat buildings, but it is also used for cooking and lighting. These days, solar energy is cheap enough to power whole cities on a wider scale as well as be used for home uses like lighting gardens.

Direct energy from the sun is solar energy, which is renewable and safe for the environment. Nuclear fusion, which is the process of fusing smaller atoms together under pressure and heat to form heavier atoms, produces a tremendous amount of energy in stars. Solar radiation carries this energy to us, which we can absorb and transform into energy that we can use.

The most popular method of harnessing solar energy is through solar panels. Photovoltaic cells are put in panels. When light from the sun enters these cells, the photoelectric effect generates an electrical current (Liu, 2013).

The photovoltaic cell's energy-generation mechanism is as follows: First, the photovoltaic solar panels attached to the dc links collect solar energy. Next, a dc to ac converter (inverter) converts the electricity into alternating current. Finally, a 3-phase line filter with transformer and circuit breaker are used in that order. Following that, you can either put it into the national grid mix or utilize it to power your home (Jaiswal, 2022).

2. Wind Power: Wind energy is best suited to offshore and higher altitude locations since it harnesses the power of the global airflow to drive turbines that produce electricity. Wind energy is another sort of solar-powered energy since winds are created by the uneven heating of the atmosphere. The shape, spin, and orbit of our planet all have an impact on this unevenness. The surface that winds travel over, whether it be land or water, also has an impact on them.

The most popular wind turbines are those with three blades. These operate on a simple concept, in contrast to solar energy. Wind drives the turbine's blades, which in turn rotate an internal rotor. The primary shaft is then moved by the rotor, which spins a generator to produce energy. Even though wind appears to be a strong energy source, enormous wind farms are needed to produce enough electricity to run the entire nation.

3. Hydropower: This form of renewable energy, commonly referred to as hydroelectric power, uses the flow of water in rivers, streams, dams, and other bodies of water to create electricity. Small-scale hydropower can be generated in the house by using rainfall, ocean tides, evaporation, and water flowing

Figure 2. Transformation of solar power into electrical power

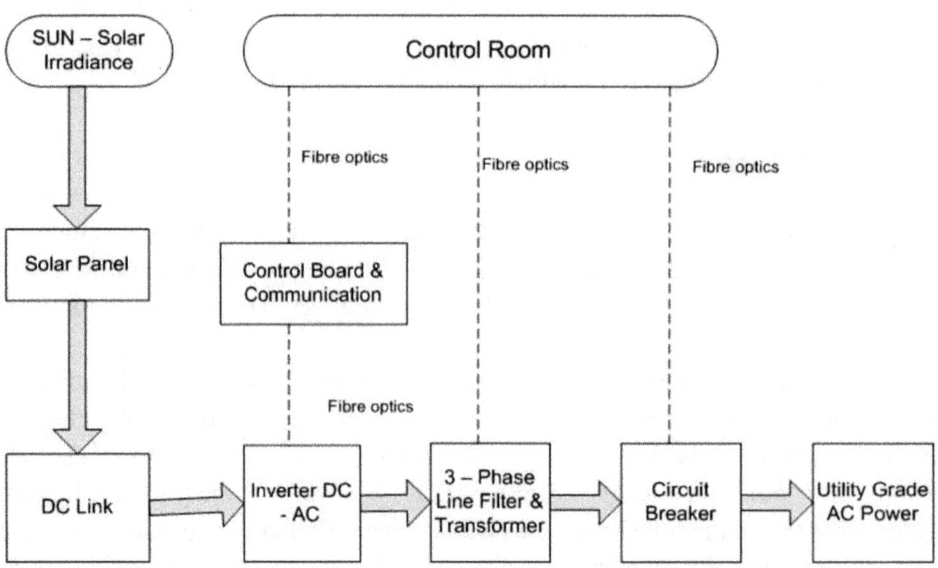

through pipes. The following three sources of green energy are classified as "green" to varied degrees depending on how they are produced.

By harnessing the energy that comes from flowing water, hydroelectric energy can be produced and used to power a machine. It is most effective to accomplish this by boosting the energy per square meter by forcing water down a small path. This is typically accomplished by holding water in a dam or reservoir and using an intake opening to purify it only.

Gravitational potential energy is stored when water is kept higher up in the dam, which quickens the flow into the intake. Release of the flow causes a turbine to spin, which in turn triggers a generator to start producing power.

Hydroelectricity is one of the most widely used renewable energy sources due to its efficiency. 4 billion tons less greenhouse gases were emitted into the environment in 2017 as a result of the usage of hydropower to produce electricity (Hydropower, 2018).

4. Geothermal Energy: Thermal energy that has been stored close below the surface of the earth is used in this form of green power. If utilized, this resource is enormous, even if extracting it requires drilling, which presents environmental issues. Although it has been utilized for thousands of years to heat up hot springs, geothermal energy may also be used to run turbines and generate electricity. There could be ten times as much energy underground in the United States as there is coal-based electricity produced today. Geothermal resources are easily accessible in certain nations, like Iceland, but their utilization depends on the geography and ongoing regulations are needed to ensure that drilling practices are entirely "green."

Geothermal energy can be utilized to heat buildings or create power by utilizing the natural heat that exists beneath the surface of the earth. Despite the fact that it harnesses a power that exists beneath our feet, geothermal energy is of small importance in the UK when compared to nations like Iceland, where geothermal heat is significantly more easily available.

5. Biomass: For this renewable resource to be called a "green energy" source, it must also be adequately managed. Biomass power plants use sawdust, wood debris, and combustible organic agricultural waste as fuel sources to generate electricity. These materials do produce greenhouse emissions, but they do so far less frequently than fuels made from petroleum.

Biomass is a sustainable green energy source since both plants and animals store solar energy as sugar or cellulose. A large portion of biomass is converted into liquid biofuels like ethanol and biodiesel and biogas when it comes to become useable energy. By themselves, other solid materials, like wood, can be burned to heat buildings and provide energy. According to the EPA, about 5% of the country's total energy consumption in 2017 (Energy Information Administration, n.d.) came from biomass fuels.

6. Biofuels: These organic materials can be processed into fuels like ethanol and biodiesel instead of being burned as biomass. It is projected that by 2050, biofuels will have surpassed petroleum as the primary fuel source, accounting for almost 25% of the world's transportation fuel use, up from just 2.7% in 2010.

The allure of biogas, presuming its existence, lies in its dual nature as an environmentally friendly and waste-derived energy source. Biogas is the byproduct of the breakdown of organic matter and is present in food waste, manure, sewage, and agricultural waste. Methane, carbon dioxide, and other gases are produced during fermentation and the holding of these components in oxygen-depleted containers. After that, the gas created can be sold and utilized to power cars, provide electricity, and heat houses. In addition, the waste that is digested in a biogas digester yields fertilizer that is rich in nutrients and can be used both at home and on farms.

Figure 3. Types of green energy

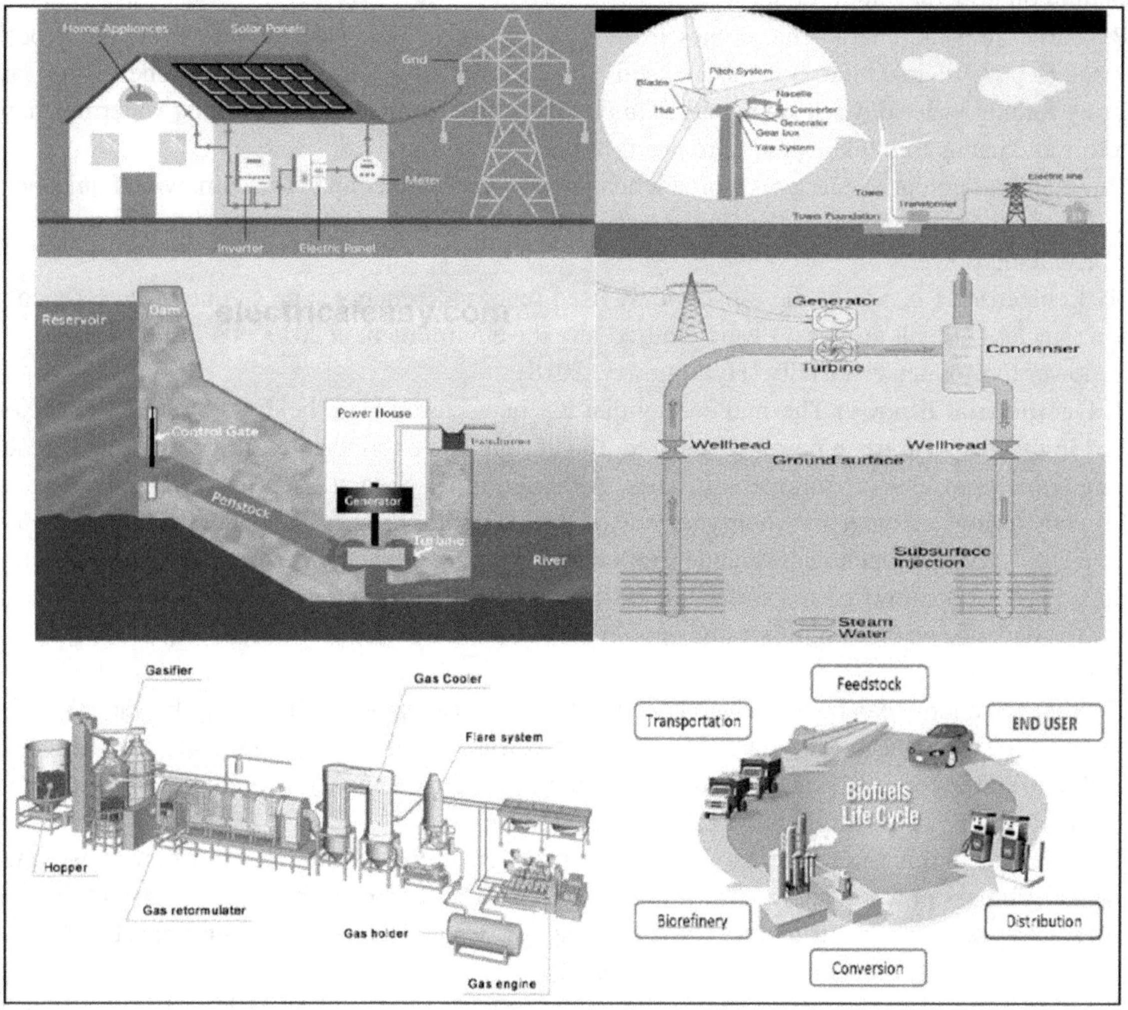

Green Energy: Can It Replace Fossil Fuels

Green energy may someday take the place of fossil fuels, however different production techniques would be required. For instance, geothermal energy works very well in places where it is easily accessible, whereas other regions may be better suited for solar or wind energy (TWI-Global, n.d.).

But given the variety of renewable energy sources we can use to meet our needs and the advancements in their development and production, it is reasonable to believe that fossil fuels may eventually be phased out. Though it will take some time to happen, it is obvious that reducing climate change, improving the environment, and making the shift to a more sustainable future are crucial.

Table 4. Different sources of biomass

Sources of Biomass	Feedstocks	Bioenergy and Biofuels	Applications
Agricultural by-products	Agricultural crop residues, Forestry residues, fruit pits, Corn cobs, grass trimmings, straw, leaves, lumber, bark, branches, algae,	Energy from light and heat sources, electrical generation, cellulosic ethanol, and compost	Natural fertilizers, soil conditioners, mulches for agriculture, and household heating and cooking
Wood by-Products	leftovers from wood processing, firewood, wood chips, timber, bark, wood pellets, and sawdust from furniture manufacturers	Energy from light and heat sources, electrical generation, cellulosic ethanol, and compost	Natural fertilizers, soil conditioners, mulches for agriculture, and household heating and cooking
Solid waste	rubbish from homes and businesses, Paper or cardboard, fruit or vegetable peels, and leftover or wasted food	Energy from light and heat sources, electrical generation, cellulosic ethanol, and compost	Natural fertilizers, soil conditioners, mulches for agriculture, and household heating and cooking
Non-crop foods plants, corn, sugarcane, agricultural by-products	Cellulosic components	methanol, ethanol-blended gasoline, cellulosic ethanol, and bio-alcohol fuels	Use of transportation vehicles
Landfills, municipal wastewater, manure, and industrial waste water	Biogas	electricity generation, liquid fuel production, methane and methanol production	Domestic cooking fuel and transportation

Economic Viability of Green Energy

Green energy needs to be contrasted with fossil fuels in order to be economically feasible. The price of fossil fuels will rise dramatically when easily accessible supplies become scarce. Greener energy sources become increasingly accessible at the same time that fossil fuels are more expensive. The ability to create reasonably priced localized energy solutions, like solar farms, is another aspect that supports green energy. As we deepen our understanding and build upon past successes, there is a decrease in costs associated with the interest in, investment in, and development of green energy solutions. As a result, green energy might emerge as the best choice in addition to being financially feasible.

Most Efficient Type of Green Energy

Since it is easy to construct a rapid and efficient energy solution if you have the correct conditions, like lots of sun, green energy efficiency is mostly location-dependent. To properly assess various energy kinds, however, a complete analysis of an energy source's entire life cycle is needed. This include figuring out how much energy was used to create the renewable energy source, determining the amount of energy that can be turned into electricity, and clearing any necessary land for the energy solution's production. Even if environmental harm would prohibit a source from being entirely "green," a "Levelised Energy Cost" (LEC) is determined after accounting for all of these considerations. Wind farms are now thought to be the most efficient kind of green energy because they require less processing and refining than, say, solar panels. Improvements in testing and composites technologies have made wind turbines last longer and have lower LEC. Solar panels, which are also experiencing significant developments, may be an example of this.

Green energy solutions also offer the benefit of minimum further energy expenditure after construction because they often employ a readily renewable energy source, like wind. In actuality, wind power generates 64% of its initial energy value from its usable energy efficiency, compared to only 29% for coal.

Green Energy: Helping the Environment

Green energy significantly improves the environment because it is generated from renewable resources like sunlight, and wind. These renewable energy sources are in direct opposition to the non-renewable, carbon-emitting fossil fuels that have sustained humans for more than a century.

A significant step in the direction of a better future is producing power with a low carbon impact. If we can use the environment to suit our demands for industry, transportation, and electricity, we can significantly minimize our impact on it.

Green Energy vs. Renewable Energy vs. Clean Energy

There is a distinction between clean, green, and renewable energy, as was previously mentioned. Although these terms are sometimes used synonymously, a resource might have more than one of these characteristics. For instance, it can be renewable but not clean or green (like certain types of biomass energy).

"Green energy" is energy that comes from the sun or other natural sources. Sustainable energy originates from regenerative sources, like solar electricity, wind power, and hydropower, and does not discharge any pollutants into the atmosphere.

Renewable energy and alternative energy are sometimes confused, while there is ongoing discussion on which is better. If a hydroelectric dam affects the ecosystem and changes streams, can it still be considered "green"? In contrast, wind power comes from an environmentally favorable, non-polluting source, making it clean, green, and renewable.

Working of Green Energy Products

It's crucial to remember that, whichever product you select, the energy you use will be a combination of conventional, renewable, and green energy. This is due to the fact that all energy sources in the electric grid are integrated upon joining the infrastructure for power transmission. In Canada and the US, a variety of regional systems provide power to homes and businesses.

Buying green energy only entails paying a little premium to offset the costs of adding more renewable energy to the grid, not buying green energy for your house. This is the greatest approach to lessen your carbon footprint related to your energy use if you want to go green at home but don't have the room or money for a solar panel installation. It's also the most economical approach to promote significant investments in renewable energy and increase the accessibility of green energy for households and companies.

Few Applications of Green Energy

Green energy is currently used in a variety of applications, including energy production, building thermal heating, off-highway mobility, and transportation. Here are a few examples of industries that are considering green alternatives:

Figure 4. Few green energy products

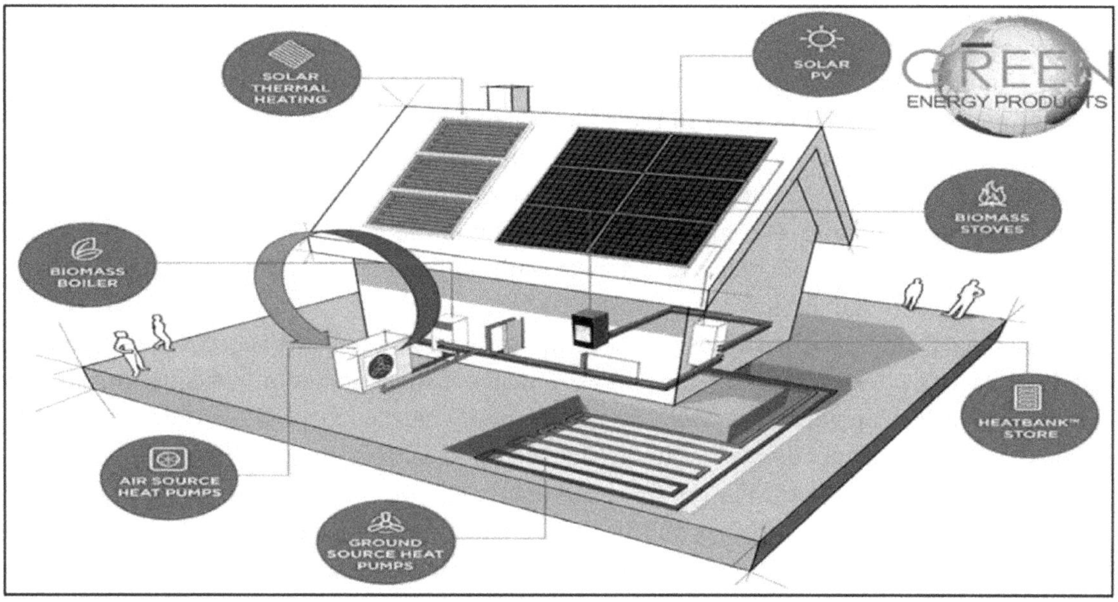

1. **Turbine Generator Systems for Heat Recovery:** The finest turbine generator for you will have heat recovery power applications such as geothermal, solar thermal, ocean thermal energy conversion (OTEC), waste heat from engines, and industrial operations.
2. **Wind Turbines and Compressors for Low Flow:** Many completely novel and extremely effective axial and radial impeller designs are now being development for application in both industrial and urban environments. As little as 5 mph of wind will be needed to start up and reach full efficiency with these little wind turbines. Turbine wind turbine technology is also being developed to lessen noise pollution and bird mortality. It is especially critical for offshore areas along migratory flight pathways to develop new blade forms that are safe for birds. A compressor is powered by a wind turbine in an additional wind-power cycle. There are occasions when compressed air is used to run a turbine generator (or another use).
3. **Solar Turbines**: More recent and advanced solar technologies promise to use solar energy more effectively, efficiently, and practically for thermal applications such as absorption chillers or for electrical generation by using concentrated solar heat (at 400°F or higher) to power a steam or gas turbine. Increasing turbine and pump efficiency to reduce installed cost per watt is the main turbomachinery problem across all solar technologies and applications.
4. **Hydroelectric Turbines:** Another fantastic potential is to develop innovative hydroelectric designs that can provide steady, predictable electricity from a range of low-head, low-power river flows that can vary greatly by location and season without the need for dams. Even sluggish moving streams can be used to power a highly efficient hydrokinetic turbine, which has an energy density 850 times that of wind. Low-flow underground streams and falling water are projected to benefit from the same turbo-green technologies. Underwater streams in ocean estuaries are a very reliable source of energy for driving a hydrazine turbine.

5. **Wave-Compressor and Tidal-Current Turbines:** A tidal-current turbine is one possible invention that gathers energy from various sea currents. As the tides come in and out, the Golay vertical-axis tidal-current turbine runs bidirectionally because of hydrofoil blades arranged helically around an axis. The operation of an underwater tidal turbine farm would be comparable to that of an offshore wind farm, with the added complexity of a relational environment. Another way for capturing and converting ocean-wave energy into compressed air, which is then utilized to power an air turbine generator, is to use an oscillating water column (OWC). The main component of the technology is a high-efficiency variable-pitch turbine with an electromechanical blade-pitch control that enables the turbine to spin in a single direction regardless of the OWC system's bidirectional airflow.
6. **Geothermal Pumps and Turbines:** These turbomachinery systems will need more efficient pumps and turbines if they are to see noticeable improvements. A small, incredibly effective turbo-expander pump is located deep within the wellbore of one gravity head energy system (GHES). Our innovative turbo-green concept delivers at least 20% and up to 30% more overall cycle efficiency.
7. **Biomass Steam Turbines:** For on-site CHP plants with access to a nearby supply of inexpensive biomass fuel, steam turbines have long been a natural green energy solution. Many utility companies are switching from coal to biomass wood chips since the latter is more cleaner. Advanced turbomachinery design expertise will soon enable more efficiency, lower emissions, and lower pricing.
8. **Energy Storage:** In order to stay up with the ever-accelerating development and acceptance of green energy technologies for distributed generation and smart grid supply, energy storage must be paired with renewable energy. All other forms of storage, with the exception of battery systems, depend on the effectiveness of different types of turbo machines to compress gases and pump fluids. Recently, there has been interest in a new generation of flywheels, with multiton carbon-fiber rotors spinning in a vacuum on magnetic bearings. Many megawatts of backup power can be stored in each flywheel, which will be used to stabilize electricity for fluctuating demands and absorb power fluctuations from intermittent energy sources like solar and wind. In one version, a green-energy motor pushes a flywheel to its maximum speed. Then, by reversing the process and converting the motor into a generator, the rotor retains its inertial energy "indefinitely" until it is released.
9. **Passive space heating by solar energy:** By using architectural features like wide south-facing windows and materials in the walls or flooring that absorb heat during the day and release it at night when it's most needed, passive solar room heating makes advantage of the sun's warmth. There are three types of passive solar design systems:
    a. Direct gain, the simplest technology, warms materials like tile or concrete by absorbing heat from the sun's direct beams into the building and slowly releasing it.
    b. Indirect gain, which is comparable to direct gain but involves the use of heat-storing, releasing, and holding materials between the sun and the living area (usually the wall).
    c. Isolated gain: This system gathers solar energy at a distance from the primary living area. For instance, warmer air from a sunroom that is attached to a house is collected and naturally circulated throughout the house.
10. **Electric vehicle with wind and solar energy:** Wind and solar energy may be one of the options for dealing with energy shortages and reducing greenhouse gas emissions. Electric vehicles can significantly improve air quality in cities. A significant method for guaranteeing that solar and

wind-generated electricity is available at all times is to store excess solar and wind energy and then release it on demand (Liu, 2013).

11. **Wind-powered sports such as windsurfing, kitesurfing:** Windsurfing is a sport that incorporates elements of both surfing and sailing. The kite is propelled along the water by the wind.
12. **Heating and cooling in buildings:** A wide range of structures, from massive corporate buildings to private homes, are utilizing green energy solutions. These include geothermal direct heat systems, biomass-fueled boilers, solar water heaters, and cooling systems that run on renewable energy.
13. **Industrial processes:** For industrial processes, biomass or renewable energy can be employed to produce renewable heat. Nowadays, hydrogen plays a significant role as a sustainable energy source for the chemical, iron, steel, and cement sectors.
14. **Transport:** In several industries, the use of sustainable biofuels and renewable energy for transportation is growing in popularity. Automobiles are an obvious example of how electrification is developing to replace fossil fuels, but electrification is also being researched in aircraft and construction.

Cost and Benefits of Energy Transition

The study demonstrates the cost-effectiveness to decarbonize the energy system. While the overall energy expenditure needed is substantial, by 2050 the additional expenses associated with the transition to a low-carbon energy sector will account for 0.4% of global GDP. Over the period of 2015–2050, the increased total investments would amount to 27 trillion USD, or an average of 0.77 trillion USD year during the same period (Gielen, 2019). More investments will be needed to decarbonize the energy sector, with a primary focus on end-use sectors (buildings and transportation) to increase energy efficiency. Infrastructure (such as energy storage, hydrogen and CO2 pipelines, transmission and distribution lines, and electric vehicle recharging stations) will also be included in these expenditures.

According to the E3ME econometric model, the REmap Case is expected to result in a 1% rise in global GDP by 2050 when compared to the Reference Case. The potential benefit from increasing GDP might be approximately USD 52 trillion between 2015 and 2050. Pro-growth policies, the use of carbon pricing and recycling of its proceeds to lower income taxes, and increased investment in energy efficiency and renewable energy sources are the main drivers of economic growth. According to the REmap Case, the global energy shift and faster overall economic growth might result in the creation of over 19 million additional direct and indirect jobs in the flexibility and grid augmentation, renewable energy, and energy efficiency sectors by 2050.

Innovation and R&D to Enable the Energy Transition

It will take a strong supporting infrastructure to combine technologies. These will include low-loss cross-border electricity interconnections, smart district heating systems, smart electric vehicle charging networks, smart high-voltage transmission lines (perhaps underground) to transport large amounts of power from areas with abundant solar or wind resources to demand centers, and biomass feedstock management and distribution strategies, as shown in figure 5. It would be hard for low-carbon technology to become commercially viable and widely adopted without this kind of infrastructure. The connection of different energy applications also enables the integration of clean technology. Two industries that use electric cars are transportation and power.

Figure 5. Innovations in power systems

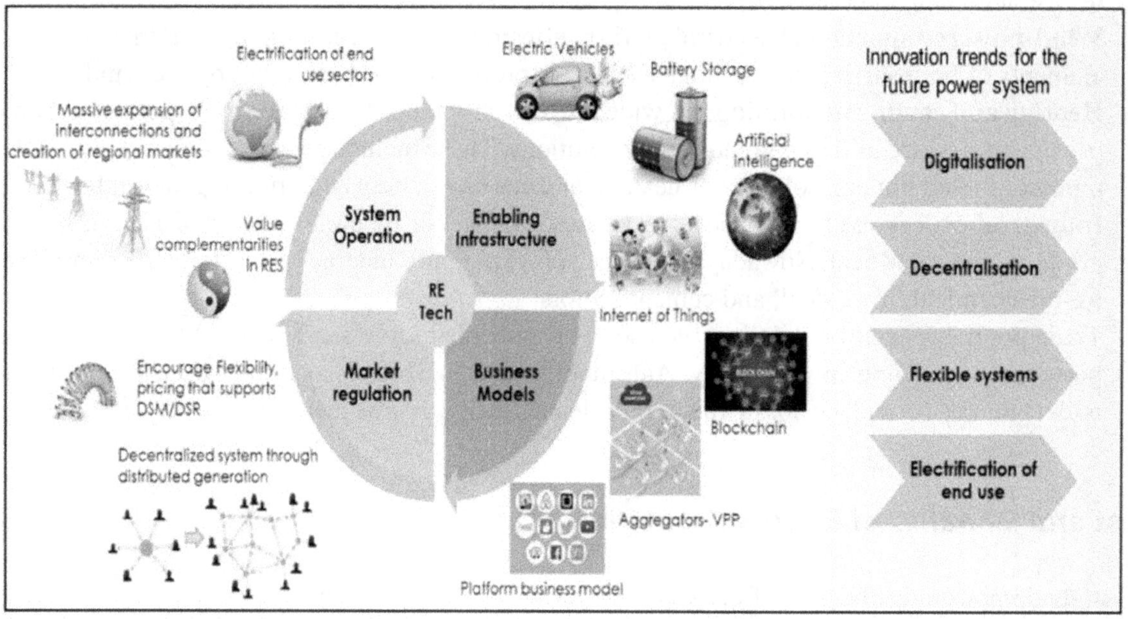

Collaboration across borders can foster innovation. The Clean Energy Ministerial, the Breakthrough Energy Coalition, and Mission Innovation—an international initiative unveiled at COP21 that aims to double government R&D investment in clean energy technologies—are a few recent examples of initiatives launched with the purpose of fostering R&D and innovation for clean energy technologies.

CONCLUSION

Green energy offers a more ecologically friendly substitute for many of the energy sources available now, suggesting that it is a part of the world's future. These readily replenishable energy sources are good for the environment, help create jobs, and seem to be economically viable in the long run. In actuality, fossil fuels are not a sustainable energy source and should be gradually phased out of our energy mix. We can build diverse green energy sources that don't harm the environment in which we all live, and thereby produce a truly sustainable energy supply in the future. It is imperative to use renewable energy for preserving, enhancing, and maintaining the state of the environment globally. Complete transition from fossil fuel-based, non-renewable energy to renewable energy is necessary. They come with a host of safety advantages that are good for the environment and the people who use renewable resources. That is only possible if certain problems are resolved. These challenges include the public's acceptability, government rules, political pressures, old infrastructure, expensive plant building costs, surplus capacity, and corporate dominance. Realizing the role that people play in maintaining environmental health, many countries have devised strategies to allow the corporate sector to profit from now-affordable renewable energy sources. Although the renewable energy market is becoming more favorable over time, more has to be done to promote the wider usage of these sources. Focusing on building the technology and infrastructure needed to capture renewable energy sources, such wind turbines, dams, and levees, is essential. It is also essential to keep the marine and aquatic environments clean and healthy if you want to get the most out of hydropower. The market for these technologies will expand and public acceptance of them will increase with the encouragement and promotion of solar panel installation in residential buildings. The development and utilization of renewable energy must be given top priority, particularly in light of the increasing awareness of the harmful environmental effects linked to the industrial process that uses fossil fuels. The need for sustainable energy development is growing rapidly on a worldwide scale. Widespread usage of renewable energy is necessary to attain sustainability in the energy sectors of both developed and developing countries.

REFERENCES

BP Energy. (2018). *Outlook country and regional insights-India*. https://www.bp.com/ content/dam/ bp/ en/corporate/ pdf/energy-economics/energy-outlook/bp-energy-outlook-2018-country-insight-india.pdf

Chang, Y., Lee, J., & Yoon, H. (2010). Alternative projection of the world energy consumption-in comparison with the 2010 international energy outlook. *Energy Policy*, *50*, 154–160. doi:10.1016/j.enpol.2012.07.059

EIA. (2018). *International energy outlook 2018 (IEO2018)*. https://www.eia.gov / pressroom/presentations/capuano_07242018.pdf

EIA. (2019). *Energy outlook 2019 with projections to 2050*. https://www.eia.gov/outlooks/aeo/pdf/aeo2019.pdf

Energy Information Administration. (n.d.). https://www.eia.gov/energyexplained/?page=biomass_home

Gielen, D. (2019). *The role of renewable energy in the global energy transformation*. Elsevier.

Hydropower. (2018). https://www.hydropower.org/publications/2018-hydropower-status-report

Jaiswal. (2022). *Renewable and sustainable clean energy development and impact on social, economic, and environmental health*. Elsevier.

Kaygusuz & Bilgen. (2008). Energy related environmental policies in Turkey. *Energy Sources Part B*, *3*, 396-410.

Liu, Q. (2013). *Electric car with solar and wind energy may change the environment and economy: A tool for utilizing the renewable energy resource*. AGU Earth's Future.

National Center for Biotechnology. (n.d.). https://www.ncbi.nlm.nih.gov/pmc/articles/PMC4740163

National Resources Defense Council. (n.d.). *The Gulf Spill*. https://www.nrdc.org/sites/default/files/gulfspill-impacts-summary-IP.pdf

TWI-Global. (n.d.). https://www.twi-global.com/technical-knowledge/faqs/what-is-green-energy

Wang, W., Zhang, M., & Li, P. (2014). Exploring temporal and spatial evolution of global energy production and consumption. *Renewable & Sustainable Energy Reviews*, *30*, 943–949. doi:10.1016/j.rser.2013.11.027

Chapter 9
Biomass-Based Biohydrogen Production Metabolic Pathway, Challenge, Future Economy, and AIML Implication

Shamba Chatterjee
Haldia Institute of Technology, India

Sucheta Das
Haldia Institute of Technology, India

ABSTRACT

The urgent need for a cleaner, more sustainable energy source has been sparked by the fast-dwindling supply of fossil fuels and the catastrophic consequences of climate change. The thriving market of hydrogen as a fruitful alternative is experiencing as an important environmentally friendly industrial feedstock. There are significant techno-financial and environmental obstacles to the commercialization of hydrogen as a fuel. Microalgal biohydrogen generation has drawn increasing attention as a means of overcoming these difficulties. Different metabolic pathways can be used to make microalgal biohydrogen, but a concise explanation of the economics and technical aspects of improving microalgae-driven biohydrogen generation is highly needed. Thus, assured prolonged sustainability can be achieved by improvising economical hydrogen generation technologies through the amendment of vital trial and error factors, possible gene-level, metabolic engineering modification and AIML employment.

INTRODUCTION

Recent climate change scenario has propelled the necessity to explore for possible fossil fuel substitute with an environment friendly source of energy. Researchers have pointed towards hydrogen (H_2) as potential contender following its superior level energy output and reasonable heating capacity (Wu *et al.*, 2012; Chandrasekhar *et al.*, 2020). The International Renewable Energy Agency (IRENA) portrayed

DOI: 10.4018/979-8-3693-2355-7.ch009

that most of the industrially manufactured hydrogen is attained from steam-methane reforming, which generates greenhouse gases, resulting in climate change. The same is applicable for oil and coal gasification (CG) process. Hydrogen production via electrolysis method is also not a carbon-neutral one as it demands a considerable amount of energy (IRENA, 2018). Whereas hydrogen attained via microbial routes (microalgal biohydrogen production) has a lot of advantages like elevated growth rate, less energy need, more environment-friendly nature and no pretreatment process as of lignin based material (Muhammad *et al.*, 2021; Ahmed *et al.*, 2022; Singh *et al.*, 2020; El-Dalatony *et al.*, 2020; Jiménez-Llanos *et al.*, 2020; Mona *et al.*, 2020; Salakkam *et al.*, 2021). Therefore, this review unveiled 1) the different metabolic routes of improved biohydrogen manufacture; 2) the foremost challenges and notable limitations; and 3) future insights, opportunities and economic considerations.

Biohydrogen Manufacturing Pathways From Biomass

Biohydrogen mass-production from microalgae based biomass is strong economical strategy in comparison to existing conventional energy approaches (Ge *et al.*, 2020). This can be achieved by either direct microalgae based production approach or by using that biomass as potential microbial fermentative substrate. The second one is having higher energy yield than the first one, but both of them can serve in combo as well (Oncel *et al.*, 2015). These areas are elaborated in the following sub-sections.

A. Photofermentation

Photofermentation utilizes sunlight in place of sugars as an energysource. This procedure can be portrayed by following equation:

$$16ATP + N_2 + 16H_2O + 10H^+ + 8e^- \rightarrow 16ADP + 2NH_4^+ + 16pi + H_2$$

Here, heterotrophic fermentation as well as the catabolic reaction of the in-house substrate supplies electron to green algae. Additionally, light energy along with oxidative carbon metabolism contributes for the electron extraction. Generated electrons successfully transferred to the plastoquinone pool coupled with chloroplast based nicotinamide adenine dinucleotide phosphate-plastoquinone oxidoreductase. Basically, the transferred electron to the [Fe]-hydrogenase and ferredoxin yields in elevated redox potential (Baeyens *et al.*, 2020). Induced [Fe]-hydrogenase aids in high amount of biohydrogen generation in a dark as well as anoxic atmosphere.

Photofermentation was termed to be the most efficacious process compared to dark-photo fermentation, if one consider the hydrogen yield [~141.42 ml $(g\ TS)^{-1}$] (Zhang *et al.*, 2020). In contrast, the aforementioned process is less efficient in daylight conversion and ultimate product yield, when compared to electrochemical as well as biophotolysis method (Zhang *et al.*, 2020).

B. Dark Fermentation

Carbohydrate-rich green microalgae routinely produce biohydrogen by a light-independent process popularly known as dark fermentation. The complete conversion of glucose molecule can be shown by following equation:

$$C_6H_{12}O_6 + 6H_2O = 12H_2 + 6CO_2$$

The dark fermentation then produces 4 mol of hydrogen along with possible by-products under organic acids group like acetic acid (Mona *et al.*, 2020). Two pathways contribute to the production of biohydrogen at some stage in the dark fermentation pathway. The hydrogenase pathway catalyses one of them through the reoxidation of nicotinamide adenine dinucleotide hydride, while the other one occurs through the catabolic transformation of formic acid (El-Dalatony *et al.*, 2020). Though this process has shown a higher reaction rate, but due to the low hydrogen concentration yield a final purification step is necessary (Mona*et al.*, 2020). On the other hand, an elevated outcome can be achieved by the optimal pre-treatment and after proper enzymolysis regarding *Alternantheraphiloxeroides* (alligator weed) (Song*et al.*, 2020).

C. Biophotolysis Process

Photosystem II in green algae and cyanobacteria possess the potential to employ energy from light to break water and produce biohydrogen. Here, electrons obtained from this process lead through the electron transport chain comprising of Photosystem I and II. The ferredoxin molecule is then reduced and adenosine triphosphate is produced. These participate in numerous metabolic processes that generate hydrogen. The term "biophotolysis" relates to a process driven by light that falls into two groups, viz., direct as well as indirect biophotolysis (Razu *et al.*, 2019).

In the photosynthesis process, *Chlamydomonas reinhardtii* (microalgae) and *Synechocystis* (cyanobacteria) promptly biophotolyze the water in their environment to produce hydrogen and oxygen. The microalgae can be easily cultivated and have negligible nutrient needs. They are also effective carbon dioxide fixers. Additionally, they grow in saltwater or wastewater in a similar fashion as they do in freshwater (Eroglu *et al.*, 2016; Chen *et al.*, 2016). Some of the principal hurdles include requirement of massive cultivation, inhabitation of ultimate product following accumulation of oxygen, excess light intensity need, and poor photochemical efficiency (Jiménez-Llanos *et al.*, 2020).

Microalgae under anoxic conditions perform indirect biophotolysis where oxygen evolution materializes in a different stage from hydrogen production. The initial phase portrays photosynthesis, where carbon dioxide and water are transformed to chemical energy and oxygen. Later on, the organic materials are further processed into hydrogen, carbon dioxide and additional metabolites. Major barriers in this case are reduced hydrogenate uptake, high operational costs, and generation of carbon dioxide by hydrogenase enzyme resulting in low yield of hydrogen (Razu *et al.*, 2019). Alternatively, biohydrogen can also be generated by oleaginous microalgae combined with the additional production of lipid (Sengmee *et al.*, 2017).

D. Electrochemical Procedure

The microbial electrolytic cell generates biohydrogen with the help of electro-chemically active microalgae capable of carbon dioxide (CO_2) and electron production. These electrons are conveyedvia the anode directing towards the cathode and finally coalesce with hydrogen (H_2) atoms that release H^+. An outer voltage feed is essential to obtain biohydrogen at the cathode as the process is not spontaneous. Microbial electrolytic cells are well efficient in manufacturing 12 mol H_2/mol substrate, which is pretty low in comparison to the traditional one as it consumes residual organic ingredients (Park *et al.*, 2021).

Recently, organic waste based biohydrogen production has drawn immense interest in the form of microbial photoelectrochemical cell. This front-line technology employs membrane separated semiconductors as electrodes in an electrolyte, where water splits into hydrogen and oxygen in existence of sunlight. Photoelectrochemical cell is extremely environmentally prospective method for biohydrogen manufacture (El-Dalatony *et al.*, 2020). Current report unveiled that use of hematite nanowire based photoanode improved output power at visible light (Zhu *et al.*, 2017).

Challenges With Biohydrogen Production

Biohydrogen has been identified as an environmental-friendly and sustainable replacement in comparison to hydrocarbon deposit based energy resources till date. Challenges are still there not only to properly scale up the available processes, but also to enhance the ultimate yield. Additionally, economic feasibility is a crucial factor for its unanimous commercial level production in a sustainable manner.

Biophotolysis method has dealt with the problem regarding the process sensitivity to oxygen. One has to either bring down the level by employing oxygen-binding proteins or by holding back the oxygen production through inert gas sparging at headspace of the reactor. This may add additional operation cost to the entirety of the process. One way to tackle this problem is to negate certain nutrients (e.g. sulfur, nitrogen) from the algal cultivation medium that in turn will raise biohydrogen yield. It has already been observed that metabolically engineered cyanobacteria can reduce hydrogenase based hydrogen consumption. Alternatively, a report has unveiled improved light conversion efficiency by modified green algae and cyanobacteria through genetic engineering. Some other major concerns are safety issues with product level explosive mishmash, necessity of synthetic illumination for closed systems, light-capturing as well as carbon dioxide fixation efficiency of the involved microalgae species (Nagarajan *et al.*, 2017; Kosourov *et al.*, 2017; Kumar *et al.*, 2019; Khetkorn *et al.*, 2014).

Replacing conventional substrate with alternative choice like environmental wastes can potentially lower the operational cost in case of fermentative biohydrogen production. Selection of suitable microorganism is of utmost concern as it substantially influences the hydrogen yield. Besides this, sometimes proper pretreatment is essential to reduce the complex sugars into simpler ones for better fermentation. A separation process has to be incorporated as carbon dioxide can drastically hamper the carbon neutrality of the surrounding environment as obvious byproduct. Reports also suggested the presence of possible metabolite by-products, methane and minute quantity of hydrogen sulfide gas that in turn affects hydrogen yield (Sarangi *et al.*, 2020; Łukajtis *et al.*, 2018).

Microbial electrolysis cell based process still has a lacuna in pointing out the actual mechanism behind electron transfer in between a microbial species and the electrode. Moreover, methanogenic activity near the anode and cathode contest with the species of interest for organic substrates as well as they utilize generated hydrogen (Kadier *et al.*, 2017). Though chemical inhibitors have been found to be useful in combating the methanogens but operational costs become higher (Chae *et al.*, 2010). Finally, one has to search for low cost materials with high conductivity for electrodes. Biocathodes or microbial cathodes are good alternative as they are cheaper in this respect (Croese *et al.*, 2011).

Future Insights and Opportunities

Biohydrogen technologies are still at its nascent stage as it needs low-cost design materials for reactor as well as inexpensive substrates. Even some prospective approaches such as amalgamation of biohydrogen production with waste water treatment and implication of genetically engineered microorganisms have not truly alleviated its unanimous usage in comparison with existing hydrogen production approaches. Researchers have reported that hydrogen production from energy sources like coal can provide high hydrogen yield if compared to natural gas and other notable sources viz., solar, wind and nuclear power (Bartels *et al.*, 2010). But with recent advances in renewable sectors these data may have altered in a substantial manner. Interestingly, one can effectively raise the hydrogen yield by down-regulating targeted metabolic pathways followed by limiting hydrogen consuming reactions. Besides this, competently designed bioreactors can potentially escalate light penetration that upturn ultimate yield. In certain cases, where carbon dioxide is a byproduct along with hydrogen, implementation of gas separation technique can effectively ensure the purity of the product. Moreover, implementation of hybrid hydrogen production methods employs combined culture of green algae and photosynthetic species which utilizes the different light spectrum regions to improve ultimate gain (Melis *et al.*, 2006). Other studies have stated the ability to increase product yield by assimilating both dark fermentation and photofermentation (Ding *et al.*, 2015; Chookaew *et al.*, 2015; Ghimire *et al.*, 2015; Morsy *et al.*, 2015; Rai *et al.*, 2015; Xie *et al.*, 2016, Salem *et al.*, 2018). In addition to this, if dark fermentation can be combined with microbial electrolysis cell, then the yield is getting significantly improved (Nguyen *et al.*, 2020).

Japan is efficiently growing towards the concept of building up a fully driven hydrogen society. Harumi Flag is making that dream into reality right in the middle of Tokyo. This flagship project has 23 residential constructions with 5,632 units and retail assets. The entire energy requirement will be fulfilled by hydrogen and solar power. Japan is also soundly implementing Fuel Cell Vehicles, where Toyota and Honda's contribution have shown substantial future prospects (Toyota, 2020). Furthermore, Yamanashi Prefecture is a perfect example of public-private partnership model which potentially aspire towards decarbonization efforts (Japan Science and Technology Agency, 2021). The Californian government has already promoted fuel cell electric vehicles since 2015 (Sinha *et al.*, 2021).

South Korea aimed to establish six cities under the roof of the 'Hydrogen City Creation Project' whose daily life will be powered by hydrogen (Hydrogeninsight, 2023). The government is insistently endorsing fuel cell electric vehicles by providing subsidy on purchase. Hyundai is leading the market in this aspect.

China has also improved its decarbonization efforts by elevating the figure of fuel cell electric vehicles as well as refueling stations in the past few years (Factsheet-China-FC-EN; now-gmbh.de, 2020).

In Europe, Germany and its associates (e.g. the Netherlands, Austria, as well as Norway) have spent substantial amount of capital to compete with rest of the world by encouraging hydrogen production from bioresources(IRENA, 2019).

With futuristic policies, the current popular method of hydrogen production via electrolysis can be slowly replaced by biohydrogen as per the IRENA report (IRENA, 2019).

Economic Considerations

Hydrogen gas acquired from biological procedures came out as a thriving area for bioenergy manufacture. Still, several issues have to be considered for the unanimous application of the hydrogen economy. Presently, researchers and technicians have inclined their notion towards cost-effectiveness of generated hydrogen (~\$0.30 kg−1) to face substantial challenge by gasoline driven market [Goswami *et al.*, 2021].

Biophotolysis system based on microalgae cultivation functioning at 90% attainment, intends to manufacture 1200 TJ of product per year (El-Dalatony *et al.*, 2020; Mona *et al.*, 2020). A U.S.-based venture has proposed a two-stage biohydrogen production design that in turn reduced the photobioreactor cost, but later on faced problem regarding sulphur deprivation (Nagarajan *et al.*, 2021). Besides this, economically feasible open pond based microalgae cultivation has shown a considerable amount of biohydrogen production (1,239 kg/ h) with suitable market price ($10 kg-1) (Gholkar *et al.*, 2021). One has to keep an eye on storage cost by lowering the production cost in case of direct biophotolysis. This can be further managed by controlling the conversion efficiency and biomass outcome (Jiménez-Llanos *et al.*, 2020).

Design of certain bioreactors demands specific manufacturing materials as well as nutrients, which usually cover majority of expenditure. This may lead into high biohydrogen production cost (Show *et al.*, 2018).

Microalgae biomass based biohydrogen production cost through supercritical and thermal gasification is still unfavorable as the cost seems to be fourfold higher than the already existing natural gas driven hydrogen production (Kumar *et al.*, 2019).

High production costs are also caused by problems including differential proton gradients in thylakoid, disturbances in light capture, and inefficient photosynthetic processes (Eroglu *et al.*, 2016).

If one effectively compare the capital cost for direct biophotolysis with photofermentation and dark fermentation, then it seems quite clear that the later one is economically unfavorable (Dincer *et al.*, 2015; Bolatkhan *et al.*, 2019; Peng *et al.*, 2019; Limongi *et al.*, 2021; Wang *et al.*, 2021).

However, synthesis of biohydrogen characteristically costs more than other fuel alternates. Forthcoming research should concentrate on detailed assessment of amenity and operation perspective, budget of raw ingredients and operation process, light, meteorological conditions, and site-conditions, mixing of biomass, and product solidarity. Recycling waste metabolites and biomass can increase economic viability.

Artificial Intelligence and Machine Learning-Assisted Hydrogen-Based Systems

Artificial intelligence (AI) has drawn peer attention over the last few years and successfully been applied in industry arena together with other phases. Traditional researches regarding AI comprises of data representation, structured planning, methodical learning, natural language processing, portraying perception, and aid for robotics. In principal, the sole aim is to develop general intelligence that can potentially complete any duty as human can perform. Present generation has adapted and amalgamated a vast range of problem solving mathematical optimization techniques along with formal logic and different network modelling systems depending on statistics as well as operation research. On the other hand, Machine learning (ML) emphases on the practice of data and algorithms to replicate the humanized mode of learning in an improved and accurate manner.

The extensive usage of AI in current industrial sectors has potentially changed the scenario in relation to high production cost and performance accuracy. A major improvement has been observed specifically in hydrogen refining and filling system which has radically changed the modus operandi regarding prediction, planning and control. AI based systems thus well divided as swarm-biological, non-swarm biological and physico-chemical heuristic algorithms for sound implication in real time situation. Whereas machine learning algorithms has improvised artificial neural network following its simplification capability (Xiao *et al.*, 2020; Xiao *et al.*, 2021; Ahmad *et al.*, 2021; Rangel-Martinez *et al.*, 2021).

In 2021, a hydrogen manufacturingsetup prediction prototype buttressed by photovoltaic technology has been established with the help of agnostic deep learning (DL) and artificial neural network

(ANN) models based on insolation value dataset (Mert *et al.*, 2021). Proper planning can steadily help us to minimize the hydrogen storage systems by implementing DL algorithm (Shams *et al.*, 2021). Present situation of mother earth positively supports to build environmentally-friendly and sustainable strategies for autonomous energy systems as was shown in a DL algorithms based supply-demand forecasting model system to mimic an independent hydrogen blended renewable grid (Hwangbo *et al.*, 2019).

AI methods like non-dominated sorting genetic algorithm (NSGA) has also become popular for system optimization in case of short term hydrogen energy storage to fulfil the load demand (Beshr *et al.*, 2018). However, the particle swarm optimization algorithm(PSO) is basically a simple and efficient swarm intelligence based technique which has commendably unveil the optimal proportion and energy designing of hydrogen storage system (HassanzadehFard *et al.*, 2020). Improved ones can optimize the hydrogen fuel cell system for techno-economic, environmental and socio-political goals (Maleki *et al.*, 2016; Eriksson *et al.*, 2019). Here, artificial swarm optimization (ABSO) algorithm has proven its efficacy to optimize life span value and loss of power-supply probability in hydrogen based systems. Besides this, simulated annealing (SA) can be implemented to achieve an improved outcome in comparison to the outdated response surface methodology (RSM) system(Maleki *et al.*, 2018). Cost optimization can be attained by employing a discrete harmony search algorithm (DHS) for successful size amelioration of fuel cell storage devices (Maleki *et al.*, 2015). Additionally, a hybrid algorithms for optimization (HRES) has been applied to achieve minimum lifecycle cost for storage system and it portrayed acceptable robustness with lower index value in comparison to other methods (Zhang *et al.*, 2018).

Currently, green hydrogen production (GHP) systems from water manufactories has focused on machine learning (ML) dependentforecasting and optimization of scaling up issues. They applied different relevant models to forecast and optimize the proton exchange membrane and dark fermentation methodologies to unveil the scaling up issues confirming a sustainable hydrogen economy (Kabir *et al.*, 2023). Moreover, researchers from the KEPCO International Nuclear Graduate School have used a hydrogen economic evaluation programme to compute the manufacturing cost employing two types of small modular reactors evolved by the Korea Atomic Energy Research Institute and NuScale Power, LLC. They have unveiled the critical parameters like heat consumption and electricity rating that potentially influence production cost from a ML-based approach (Kim *et al.*, 2022).

CONCLUSION

Hydrogen is a environment-friendly transportation and vehicular fuel as it forms only water in the time of combustion. The State of California acknowledged significant effects on environment that has resulted into the severe wildfire in 2020. The state then realized the value of immediate implementation strategy of zero-carbon electric system that can potentially be achieved by 2030. Right now electrolysis process is the leading contributor of clean hydrogen as biohydrogen based system is still not feasible for mass-scale level. The cost and output of the current biohydrogen technologies, however, is still lagging behind those of the traditional methods of manufacturing hydrogen. Additional research initiatives are of utmost importance that can potentially elevate the process optimization, product yield and economic stand up. Genetic engineering can also help us to enhance microbial properties. Besides this, specifically designed bioreactors can greatly improve the same. Also, the lacuna in latest economic assessment remains as majority of information were carried out few years back. Therefore, systematic economic studies should be conducted in order to minimize the cost regarding biohydrogen storage and transportation.Finally, Artificial intelligence plays a crucial role in simulating renewable energy systems. Artificial Intelligence techniques have the potential to enhance energy generation and supply chain variable prediction by enabling more intelligent planning, management, and control. The renewable energy systems can be optimized to have a more appropriate size or configuration by using a variety of Artificial Intelligence optimization algorithms.Important factors that influence the cost of producing green hydrogen can be accurately anticipated using machine learning-based methods as well.

ACKNOWLEDGMENT

The authors would like to express their sincere appreciation regarding the support given by the home institution for providing the space and resources.

REFERENCES

Ahmad, T., Zhang, D., Huang, C., Zhang, H., Dai, N., Song, Y., & Chen, H. (2021). Artificial intelligence in sustainable energy industry: Status Quo, challenges and opportunities. *Journal of Cleaner Production*, *289*, 125834. doi:10.1016/j.jclepro.2021.125834

Ahmed, S. F., Mofijur, M., Parisa, T. A., Islam, N., Kusumo, F., Inayat, A., Lee, V. G., Badruddin, I. A., Khan, T. M. Y., & Ong, H. C. (2022). Progress and Challenges of Contaminate Removal from Wastewater Using Microalgae Biomass. *Chemosphere*, *286*, 131656. doi:10.1016/j.chemosphere.2021.131656 PMID:34325255

Baeyens, J., Zhang, H., Nie, J., Appels, L., Dewil, R., Ansart, R., & Deng, Y. (2020). Reviewing the Potential of Bio-Hydrogen Production by Fermentation. *Renewable & Sustainable Energy Reviews*, *131*, 110023. doi:10.1016/j.rser.2020.110023

Bartels, J. R., Pate, M. B., & Olson, N. K. (2010). An economic survey of hydrogen production from conventional and alternative energy sources. *International Journal of Hydrogen Energy*, *35*(16), 8371–8384. doi:10.1016/j.ijhydene.2010.04.035

Beshr, E. H., Abdelghany, H., & Eteiba, M. (2018). Novel optimization technique of isolated microgrid with hydrogen energy storage. *PLoS One*, *13*(2), e0193224. doi:10.1371/journal.pone.0193224 PMID:29466433

Bolatkhan, K., Kossalbayev, B. D., Zayadan, B. K., Tomo, T., Veziroglu, T. N., & Allakhverdiev, S. I. (2019). Hydrogen Production from Phototrophic Microorganisms: Reality and Perspectives. *International Journal of Hydrogen Energy*, *44*(12), 5799–5811. doi:10.1016/j.ijhydene.2019.01.092

Chae, K. J., Choi, M. J., Kim, K. Y., Ajayi, F. F., Chang, I. S., & Kim, I. S. (2010). Selective inhibition of methanogens for the improvement of biohydrogen production in microbial electrolysis cells. *International Journal of Hydrogen Energy*, *35*(24), 13379–13386. doi:10.1016/j.ijhydene.2009.11.114

Chandrasekhar, K., Kumar, S., Lee, B. D., & Kim, S. H. (2020). Waste based hydrogen production for circular bioeconomy: Current status and future directions. *Bioresource Technology*, *302*, 122920. doi:10.1016/j.biortech.2020.122920 PMID:32029301

Chen, C., Chang, H., & Chang, J. (2016). Producing carbohydrate-rich microalgal biomass grown under mixotrophic conditions as feedstock for biohydrogen production. *International Journal of Hydrogen Energy*, *41*(7), 4413–4420. doi:10.1016/j.ijhydene.2015.05.163

Chookaew, T., O-Thong, S., & Prasertsan, P. (2015). Biohydrogen production from crude glycerol by two stage of dark and photo fermentation. *International Journal of Hydrogen Energy*, *40*(24), 7433–7438. doi:10.1016/j.ijhydene.2015.02.133

Croese, E., Pereira, M. A., Euverink, G. J. W., Stams, A. J. M., & Geelhoed, J. S. (2011). Analysis of the microbial community of the biocathode of a hydrogen-producing microbial electrolysis cell. *Applied Microbiology and Biotechnology*, *92*(5), 1083–1093. doi:10.1007/s00253-011-3583-x PMID:21983651

Dincer, I., & Acar, C. (2015). Review and Evaluation of Hydrogen Production Methods for Better Sustainability. *International Journal of Hydrogen Energy*, *40*(34), 11094–11111. doi:10.1016/j.ijhydene.2014.12.035

Ding, L., Xia, A., Lin, R., Li, Y., Zhou, J., & Cen, K. (2015). Hydrogen production using amino acids obtained by protein degradation in waste biomass by combined dark- and photo-fermentation. *Bioresource Technology*, *179*, 13–19. doi:10.1016/j.biortech.2014.11.109 PMID:25514397

El-Dalatony, M. M., Zheng, Y., Ji, M.-K., Li, X., & Salama, E.-S. (2020). Metabolic Pathways for MicroalgalBiohydrogen Production: Current Progress and Future Prospectives. *Bioresource Technology*, *318*, 124253. doi:10.1016/j.biortech.2020.124253 PMID:33129070

Eriksson, E. L. V., & Gray, E. M. (2019). Optimization of renewable hybrid energy systems – A multi-objective approach. *Renewable Energy*, *133*, 971–999. doi:10.1016/j.renene.2018.10.053

Eroglu, E., & Melis, A. (2016). Microalgal Hydrogen Production Research. *International Journal of Hydrogen Energy*, *41*(30), 12772–12798. doi:10.1016/j.ijhydene.2016.05.115

Factsheet-China-FC-EN. (n.d.). https://www.now-gmbh.de/wp-content/uploads/2020/09/Factsheet-China-FC-EN.pdf

Ge, S., Foong, S. Y., Ma, N. L., Liew, R. K., Wan Mahari, W. A., Xia, C., Yek, P. N. Y., Peng, W., Nam, W. L., Lim, X. Y., Liew, C. M., Chong, C. C., Sonne, C., & Lam, S. S. (2020). Vacuum Pyrolysis Incorporating Microwave Heating and Base Mixture Modification: An Integrated Approach to Transform Biowaste into Eco-Friendly Bioenergy Products. *Renewable & Sustainable Energy Reviews*, *127*, 109871. doi:10.1016/j.rser.2020.109871

Ghimire, A., Valentino, S., Frunzo, L., Trably, E., Escudié, R., Pirozzi, F., Lens, P. N. L., & Esposito, G. (2015). Biohydrogen production from food waste by coupling semi-continuous dark-photofermentation and residue post-treatment to anaerobic digestion: A synergy for energy recovery. *International Journal of Hydrogen Energy*, *40*(46), 16045–16055. doi:10.1016/j.ijhydene.2015.09.117

Gholkar, P., Shastri, Y., & Tanksale, A. (2021). Renewable Hydrogen and Methane Production from Microalgae: A Techno-Economic and Life Cycle Assessment Study. *Journal of Cleaner Production, 279*, 123726. doi:10.1016/j.jclepro.2020.123726

Goswami, R. K., Mehariya, S., Obulisamy, P. K., & Verma, P. (2021). Advanced Microalgae-Based Renewable Biohydrogen Production Systems: A Review. *Bioresource Technology, 320*, 124301. doi:10.1016/j.biortech.2020.124301 PMID:33152683

HassanzadehFard, H., Tooryan, F., Collins, E. R., Jin, S., & Ramezani, B. (2020). Design and optimum energy management of a hybrid renewable energy system based on efficient various hydrogen production. *International Journal of Hydrogen Energy, 45*(55), 30113–30128. doi:10.1016/j.ijhydene.2020.08.040

Hwangbo, S., Nam, K., Heo, S., & Yoo, C. (2019). Hydrogen-based self-sustaining integrated renewable electricity network (HySIREN) using a supply-demand forecasting model and deep-learning algorithms. *Energy Conversion and Management, 185*, 353–367. doi:10.1016/j.enconman.2019.02.017

Hydrogeninsight. (2023). https://www.hydrogeninsight.com/policy/south-korea-to-create-six-hydrogen-cities-that-would-use-h2-in-buildings-and-transport-as-part-of-daily-life/2-1-1385821

IRENA. (2018). Hydrogen from renewable power. IRENA.

IRENA. (2019). Hydrogen: a renewable energy perspective. IRENA.

Japan Science and Technology Agency. (2021). https://sj.jst.go.jp/stories/2021/s0728-01a.html

Jiménez-Llanos, J., Ramírez-Carmona, M., Rendón-Castrillón, L., & Ocampo-López, C. (2020). Sustainable Biohydrogen Production by Chlorella Sp. Microalgae: A Review. *International Journal of Hydrogen Energy, 45*(15), 8310–8328. doi:10.1016/j.ijhydene.2020.01.059

Kabir, M. M., Roy, S. K., Alam, F., Nam, S. Y., Im, K. S., Tijing, L., & Shon, H. K. (2023). Machine learning-based prediction and optimization of green hydrogen production technologies from water industries for a circular economy. *Desalination, 567*, 116992. doi:10.1016/j.desal.2023.116992

Kadier, A., Jiang, Y., Lai, B., Rai, P. K., Chandrasekhar, K., Mohamed, A., & Kalil, M. S. (2017). Biohydrogen production in microbial electrolysis cells from renewable resources. Bioenergy and Biofuels. Taylor & Francis. doi:10.1201/9781351228138-12

Khetkorn, W., Khanna, N., Incharoensakdi, A., & Lindblad, P. (2014). Metabolic and genetic engineering of cyanobacteria for enhanced hydrogen production. *Biofuels, 4*(5), 535–561. doi:10.4155/bfs.13.41

Kim, J., Rweyemamu, M., &Purevsuren, B. (2022). Machine Learning-Based Approach for Hydrogen Economic Evaluation of Small Modular Reactors. *Science and Technology of Nuclear Installations*. doi:10.1155/2022/9297122

Kosourov, S., Murukesan, G., Seibert, M., & Allahverdiyeva, Y. (2017). Evaluation of light energy to H2 energy conversion efficiency in thin films of cyanobacteria and green alga under photoautotrophic conditions. *Algal Research, 28*, 253–263. doi:10.1016/j.algal.2017.09.027

Kumar, M., Oyedun, A. O., & Kumar, A. (2019). A Comparative Analysis of Hydrogen Production from the Thermochemical Conversion of Algal Biomass. *International Journal of Hydrogen Energy*, *44*(21), 10384–10397. doi:10.1016/j.ijhydene.2019.02.220

Kumar, S., Sharma, S., Thakur, S., Mishra, T., Negi, P., Mishra, S., Hesham, A. E-L., Rastegari, A. A., Yadav, N., & Yadav, A. N. (2019). Bioprospecting of Microbes for Biohydrogen Production: Current Status and Future Challenges. Bioprocessing for Biomolecules Production. John Wiley & Sons Ltd. doi:10.1002/9781119434436.ch22

Limongi, A. R., Viviano, E., De Luca, M., Radice, R. P., Bianco, G., & Martelli, G. (2021). Biohydrogen from Microalgae: Production and Applications. *Applied Sciences (Basel, Switzerland)*, *11*(4), 1616. doi:10.3390/app11041616

Łukajtis, R., Hołowacz, I., Kucharska, K., Glinka, M., Rybarczyk, P., Przyjazny, A., & Kamiński, M. (2018). Hydrogen production from biomass using dark fermentation. *Renewable & Sustainable Energy Reviews*, *91*, 665–694. doi:10.1016/j.rser.2018.04.043

Maleki, A. (2018). Modeling and optimum design of an off-grid PV/WT/FC/diesel hybrid system considering different fuel prices. *The International Journal of Low Carbon Technologies*, *13*(2), 140–147. doi:10.1093/ijlct/cty006

Maleki, A., & Pourfayaz, F. (2015). Sizing of stand-alone photovoltaic/wind/diesel system with battery and fuel cell storage devices by harmony search algorithm. *Journal of Energy Storage*, *2*, 30–42. doi:10.1016/j.est.2015.05.006

Maleki, A., Pourfayaz, F., & Ahmadi, M. H. (2016). Design of a cost-effective wind/photovoltaic/hydrogen energy system for supplying a desalination unit by a heuristic approach. *Solar Energy*, *139*, 666–675. doi:10.1016/j.solener.2016.09.028

Melis, A., & Melnicki, M. R. (2006). Integrated biological hydrogen production. *International Journal of Hydrogen Energy*, *31*(11), 1563–1573. doi:10.1016/j.ijhydene.2006.06.038

Mert, İ. (2021). Agnostic deep neural network approach to the estimation of hydrogen production for solar-powered systems. *International Journal of Hydrogen Energy*, *46*(9), 6272–6285. doi:10.1016/j.ijhydene.2020.11.161

Mona, S., Kumar, S. S., Kumar, V., Parveen, K., Saini, N., Deepak, B., & Pugazhendhi, A. (2020). Green Technology for Sustainable Biohydrogen Production (Waste to Energy): A Review. *The Science of the Total Environment*, *728*, 138481. doi:10.1016/j.scitotenv.2020.138481 PMID:32361358

Morsy, F. M. (2015). CO_2-free biohydrogen production by mixed dark and photofermentation bacteria from sorghum starch using a modified simple purification and collection system. *Energy*, *87*, 594–604. doi:10.1016/j.energy.2015.05.044

Muhammad, G., Alam, M. A., Mofijur, M., Jahirul, M. I., Lv, Y., Xiong, W., Ong, H. C., & Xu, J. (2021). Modern Developmental Aspects in the Field of Economical Harvesting and Biodiesel Production from Microalgae Biomass. *Renewable & Sustainable Energy Reviews*, *135*, 110209. doi:10.1016/j.rser.2020.110209

Nagarajan, D., Dong, C.-D., Chen, C.-Y., Lee, D.-J., & Chang, J.-S. (2021). Biohydrogen production from microalgae-Major bottlenecks and future research perspectives. *Biotechnology Journal, 16*(5), 2000124. doi:10.1002/biot.202000124 PMID:33249754

Nagarajan, D., Lee, D. J., Kondo, A., & Chang, J. S. (2017). Recent insights into biohydrogen production by microalgae – From biophotolysis to dark fermentation. *Bioresource Technology, 227*, 373–387. doi:10.1016/j.biortech.2016.12.104 PMID:28089136

Nguyen, P. K. T., Das, G., Kim, J., & Yoon, H. H. (2020). Hydrogen production from macroalgae by simultaneous dark fermentation and microbial electrolysis cell. *Bioresource Technology, 315*, 123795. doi:10.1016/j.biortech.2020.123795 PMID:32659424

Oncel, S. S. (2015). Biohydrogen from Microalgae, Uniting Energy, Life, and green Future. Handbook of Marine Microalgae. Elsevier. doi:10.1016/B978-0-12-800776-1.00011-X

Park, J. H., Chandrasekhar, K., Jeon, B. H., Jang, M., Liu, Y. & Kim, S. H. (2021). State-of-the-art Technologies for Continuous High-Rate Biohydrogen Production. *Bioresource Technology*, 320(Part A), 124304. doi:10.1016/j.biortech.2020.124304

Peng, L., Fu, D., Chu, H., Wang, Z., & Qi, H. (2019). Biofuel Production from Microalgae: A Review. *Environmental Chemistry Letters, 18*(2), 285–297. doi:10.1007/s10311-019-00939-0

Rai, P. K., Singh, S. P., Asthana, R. K., & Singh, S. (2014). Biohydrogen production from sugarcane bagasse by integrating dark- and photo-fermentation. *Bioresource Technology, 152*, 140–146. doi:10.1016/j.biortech.2013.10.117 PMID:24291314

Rangel-Martinez, D., Nigam, K. D. P., & Ricardez-Sandoval, L. A. (2021). Machine learning on sustainable energy: A review and outlook on renewable energy systems, catalysis, smart grid and energy storage. *Chemical Engineering Research & Design, 174*, 414–441. doi:10.1016/j.cherd.2021.08.013

Razu, M. H., Hossain, F., & Khan, M. (2019). Advancement of Bio-Hydrogen Production from Microalgae, Microalgae Biotechnology for Development of Biofuel and Wastewater Treatment. Springer, 423–462. doi:10.1007/978-981-13-2264-8_17

Salakkam, A., Sittijunda, S., Mamimin, C., Phanduang, O., & Reungsang, A. (2021). Valorization of Microalgal Biomass for Biohydrogen Generation: A Review. *Bioresource Technology, 322*, 124533. doi:10.1016/j.biortech.2020.124533 PMID:33348113

Salem, A. H., Brunstermann, R., Mietzel, T., & Widmann, R. (2018). Effect of pre-treatment and hydraulic retention time on biohydrogen production from organic wastes. *International Journal of Hydrogen Energy, 43*(10), 4856–4865. doi:10.1016/j.ijhydene.2018.01.114

Sarangi, P. K., & Nanda, S. (2020). Biohydrogen production through dark fermentation. *Chemical Engineering & Technology, 43*(4), 601–612. doi:10.1002/ceat.201900452

Sengmee, D., Cheirsilp, B., Suksaroge, T. T., & Prasertsan, P. (2017). Biophotolysis-based Hydrogen and Lipid Production by Oleaginous Microalgae Using Crude Glycerol as Exogenous Carbon Source. *International Journal of Hydrogen Energy, 42*(4), 1970–1976. doi:10.1016/j.ijhydene.2016.10.089

Shams, M. H., Niaz, H., Na, J., Anvari-Moghaddam, A., & Liu, J. J. (2021). Machine Learning-Based Utilization of Renewable Power Curtailments under Un-certainty by Planning of Hydrogen Systems and Battery Storages. *Journal of Energy Storage*, *41*, 1–13. doi:10.1016/j.est.2021.103010

Show, K.-Y., Yan, Y., Ling, M., Ye, G., Li, T., & Lee, D.-J. (2018). Hydrogen Production from Algal Biomass - Advances, Challenges and Prospects. *Bioresource Technology*, *257*, 290–300. doi:10.1016/j.biortech.2018.02.105 PMID:29506887

Singh, H., & Das, D. (2020). Biohydrogen from Microalgae, Handbook of Microalgae-Based Processes and Products. Elsevier.

Sinha, P., & Brophy, B. (2021). Life cycle assessment of renewable hydrogen for fuel cell passenger vehicles in California. *Sustainable Energy Technologies and Assessments*, *45*(21), 101188. doi:10.1016/j.seta.2021.101188

Song, W., Ding, L., Liu, M., Cheng, J., Zhou, J., & Li, Y.-Y. (2020). Improving Biohydrogen Production through Dark Fermentation of Steam-Heated Acid Pretreated *Alternanthera Philoxeroides* by Mutant *Enterobacter Aerogenes* ZJU1. *The Science of the Total Environment*, *716*, 134695. doi:10.1016/j.scitotenv.2019.134695 PMID:31837880

Toyota. (2020). New Mirai press information 2020. Author.

Wang, K., Khoo, K. S., Chew, K. W., Selvarajoo, A., Chen, W.-H., Chang, J.-S., & Show, P. L. (2021). Microalgae: The Future Supply House of Biohydrogen and Biogas. *Frontiers in Energy Research*, *9*, 1–9. doi:10.3389/fenrg.2021.660399

Wu, T. Y., Hay, J. X. W., Kong, L. B., Juan, J. C., & Jahim, J. M. (2012). Recent advances in reuse of waste material as substrate to produce biohydrogen by purple non-sulfur (PNS) bacteria. *Renewable & Sustainable Energy Reviews*, *16*(5), 3117–3122. doi:10.1016/j.rser.2012.02.002

Xiao, J., Bi, C., Bénard, P., Chahine, R., Zong, Y., Luo, M., & Yang, T. (2021). Neural network based optimization for cascade filling process of on-board hydrogen tank. *International Journal of Hydrogen Energy*, *46*(3), 2936–2951. doi:10.1016/j.ijhydene.2020.10.245

Xiao, J., Li, C., Fang, L., Böwer, P., Wark, M., Bénard, P., & Chahine, R. (2020). Machine learning–based optimization for hydrogen purification performance of layered bed pressure swing adsorption. *International Journal of Energy Research*, *44*(6), 4475–4492. doi:10.1002/er.5225

Xie, G. J., Liu, B. F., Wang, Q., Ding, J., & Ren, N. Q. (2016). Ultrasonic waste activated sludge disintegration for recovering multiple nutrients for biofuel production. *Water Research*, *93*, 56–64. doi:10.1016/j.watres.2016.02.012 PMID:26896823

Zhang, T., Jiang, D., Zhang, H., Jing, Y., Tahir, N., Zhang, Y., & Zhang, Q. (2020). Comparative Study on Bio-Hydrogen Production from Corn stover: Photo-Fermentation, Dark-Fermentation and Dark-Photo Co-fermentation. *International Journal of Hydrogen Energy*, *45*(6), 3807–3814. doi:10.1016/j.ijhydene.2019.04.170

Zhang, W., Maleki, A., Rosen, M. A., & Liu, J. (2018). Optimization with a simulated annealing algorithm of a hybrid system for renewable energy including battery and hydrogen storage. *Energy*, *163*, 191–207. doi:10.1016/j.energy.2018.08.112

Zhu, G., Yang, Y., Liu, J., Liu, F., Lu, A., & He, W. (2017). Enhanced Photocurrent Production by the Synergy of Hematite Nanowire-Arrayed Photoanode and Bioengineered *Shewanella oneidensis* MR-1. *Biosensors & Bioelectronics*, *94*, 227–234. doi:10.1016/j.bios.2017.03.006 PMID:28285200

Chapter 10
Blue Wind Energy for Sustainable Urbanization:
Ushering Green Energy Integrating Artificial Intelligence and IoT for Smart Energy Management

Bhupinder Singh
https://orcid.org/0009-0006-4779-2553
Sharda University, India

Ruchi Lal
https://orcid.org/0000-0003-4582-8488
Sharda University, India

Komal Vig
https://orcid.org/0009-0008-8771-0481
Sharda University, India

Bhupendra Kumar Gautam
https://orcid.org/0000-0002-2255-937X
Sharda University, India

Christian Kaunert
https://orcid.org/0000-0002-4493-2235
Dublin City University, Ireland

ABSTRACT

Blue wind energy, a pioneering concept, amalgamates offshore wind energy generation with cutting-edge technologies. There is a dire need to address the growing energy demands of urbanization while mitigating the environmental impact of conventional energy sources. Urban areas are substantial energy consumers, and traditional methods are both unsustainable and environmentally damaging. Blue wind energy, alongside AI and IoT integration, presents a compelling solution, but the challenge is in effectively implementing and optimizing these technologies to ensure sustainable, efficient, and eco-friendly energy management in cities. This chapter introduces a set of essential performance indicators to assess the advantages of diverse AI-accelerated workflows in blue-wind energy and the latest developments in applications of computer vision and employing AI in the enhancement of energy harvesting like as photovoltaics, storage including batteries, conversion, and management in smart grids.

DOI: 10.4018/979-8-3693-2355-7.ch010

1. INTRODUCTION

Blue-wind energy is to comprehend Smart Energy Management, its start by getting acquainted with the fundamentals of energy management and then explore how it seamlessly fits into the framework of smart cities (Moroke & Makatjane, 2022). The green energy solution essentially, is the science that encompasses the planning, directing and controlling of energy supply and consumption (Makatjaneet al., 2021). The goal is to maximize productivity and comfort while minimizing both energy costs and pollution through mindful, strategic, and efficient energy usage. In simpler terms, energy management is all about saving energy (Makatjane, 2022). When it talks about energy savings, it involves the ongoing process of monitoring, controlling and conserving energy in any given scenario (Kasinathan et al. 2022).

The Internet of Things (IoT) is an important data source for data science technologies, providing benefits such as quick discovery of trends and patterns, enhanced automation, continuous development, effective handling of multidimensional data, and cheap computing cost (Elrahmani et al. 2021). Predicting energy consumption is critical for promoting sustainable cities and urban planning, given that buildings are the world's top energy users, driven by variables such as population increase, development, and economic structural shifts. This study looked at the use of deep learning algorithms to forecast energy usage in smart residential buildings. It was discovered that the optimal window size has a substantial impact on prediction performance, determining the best N window size, and calculating model uncertainty (Li et al. 2019).

Artificial intelligence, big data, Internet of Things (IoT) devices, and blockchain are examples of technologies that have experienced widespread research and use in a variety of technical sectors. While previous review papers frequently focus on specific elements and deployments in the energy sector, these technologies, which are based on communication, information, and data analysis, have intrinsic coherence (ElFar et al. 2021). This encompasses optimizing energy utilization, managing energy resources effectively, and actively promoting energy efficiency. As cities rapidly expand, dealing with the challenge of energy consumption becomes a pivotal goal. As existing cities evolve into smart cities, the role of blue-wind energy management becomes indispensable in the urban transformation process (Singh, 2023). The smart cities in essence strive to be a sustainable and efficient hub, aiming to enhance the quality of life for its residents by optimizing available resources. The vision for smart cities includes achieving greater autonomy and improved efficiency in managing their energy consumption, all while considering local resources and the diverse needs of their communities. In this context, Smart Energy Management emerges as a fundamental building block for realizing the potential of smart cities (Eljack & Kazi, 2021).

1.1 Background of Study

The surge in socio-economic development and the inclination towards a more comfortable lifestyle have led to a substantial increase in energy demand in recent years. So, recognizing that energy resources are finite and should be conserved for the well-being of future generations, it becomes crucial to optimize our energy systems towards sustainability (Bogdanov et al., 2021). However, navigating the intricacies of integrated energy systems poses a considerable challenge. The computer-aided decision-making is critical in establishing the optimal supply-demand balance and providing transformational solutions. The objective of developing artificial intelligence (AI) and the Internet of Things (IoT) in the twentieth century was to give computers the ability to duplicate human learning and decision-making. Data mining and artificial intelligence have developed as critical topics in various academic disciplines over time, altering our approach to complex problems (Dent, 2012). Advanced digital technologies such as IoT and blockchain can help to improve smart energy management even further (Devabhaktuni et al. 2013). An IoT platform, which includes edge, fog, and cloud layers, makes it possible to integrate artificial intelligence with a variety of hardware and software devices and systems. Furthermore, an IoT platform improves data transmission, storage, and accessibility, enhancing stakeholders' data mining capabilities.

The blue-wind energy as renewable energy is taking on an increasingly vital role across sectors like lighting, transportation, and electricity. To harness the full potential of renewable energy, a range of intelligent Internet of Things (IoT) devices are deployed (Singh, 2024). However, within the realm of energy management, the efficiency of renewable energy faces a significant challenge due to a two-way mismatch between demand and supply. Furthermore, the security risks associated with energy data and potential privacy breaches for users pose obstacles to the continued advancement of smart IoT devices (Arent et al., 2022). Therefore, addressing the issues of achieving harmony between demand and renewable energy supply, along with ensuring the security and privacy of smart IoT devices, emerges as crucial factors in establishing an energy-efficient and secure smart environment.

1.2 Significance of the Chapter

There are both predominant global models of urban development in which smart cities and sustainable cities are dealing with unparalleled challenges brought about by the surge in urbanization, rapid urban expansion, environmental deterioration, and the impacts of climate change. The swift pace of urbanization emerges as a particularly perplexing issue for major cities globally, notably the sprawling "megacities," presenting a profound connection to some of the most intricate and pressing challenges of our era. Blue-Wind energy stands out as a renewable and eco-friendly power source, gaining substantial popularity in recent times and showing promise in replacing traditional energy sources like coal and oil (Doost Mohammadian & Rezaie, 2020).

Wind turbines efficiently convert wind power into electricity, with offshore wind farms proving particularly beneficial. They boast minimal impact on landscapes and ecosystems and harness the higher wind speeds found at sea. Despite these advantages, many wind energy systems face challenges, often being challenging to monitor and burdened with high maintenance costs (Singh, 2023). To ensure optimal performance and avert potential disasters, it becomes crucial to implement condition monitoring and fault detection systems. A typical framework for monitoring wind energy systems relies on signal processing methods, necessitating the installation of sensors and real-time collection of system signals. This chapter is relevant and important to add innovative knowledge and solutions for futuristic smart energy management and its solutions (Abualigah et al., 2022).

1.3 Objectives of the Chapter

The chapter has the broad objectives to cover the wide range of applications of AI and IoT in smart and sustainable management energy solutions as to:

- novel computer vision, Artificial Intelligence and the Internet of Things play pivotal roles in optimizing every facet of energy production, distribution, and consumption.

Figure 1. Synergy between blue wind energy, sustainable urbanization and comport vision for green energy

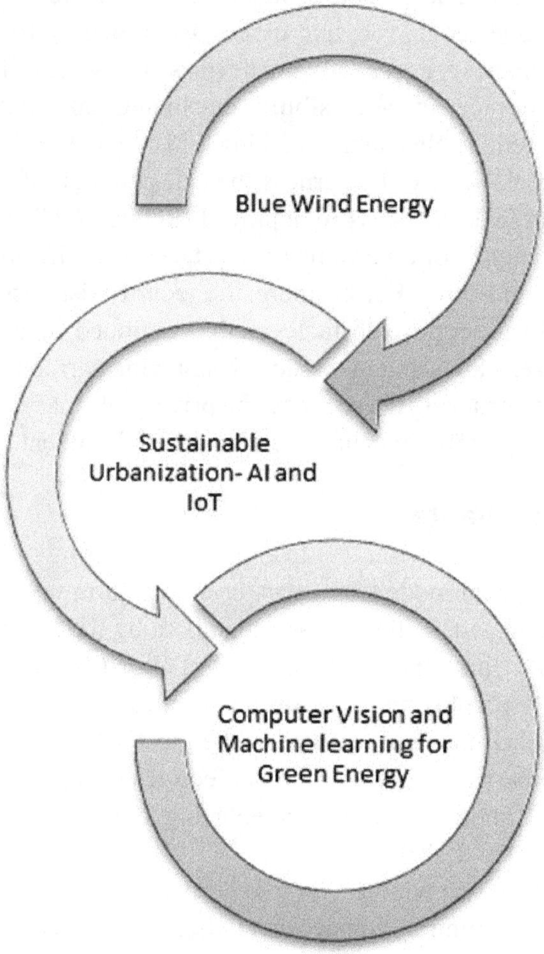

- AI algorithms facilitate predictive maintenance, allowing for more efficient use of wind turbines and reduced downtime. IoT devices offer real-time monitoring and control, enhancing grid stability and demand-side management.
- illuminates the myriad applications and benefits of integrating AI and IoT into the energy grid. However, the true power of this synergy comes to life when integrated into urban settings.
- agility of Blue Wind Energy, AI-driven predictive analytics, and the responsiveness of IoT-connected devices combine to create a dynamic, sustainable energy ecosystem. The result is reduced energy wastage, peak load shaving, and a significantly diminished carbon footprint.

1.4 Structure of the Chapter

This chapter comprehensively explores the diverse dimensions of blue wind energy for the sustainable urbanization with the assimilation of artificial intelligence, IoT and novel computer vision. Section 2

Figure 2. Objectives of the chapter from diverse aspects

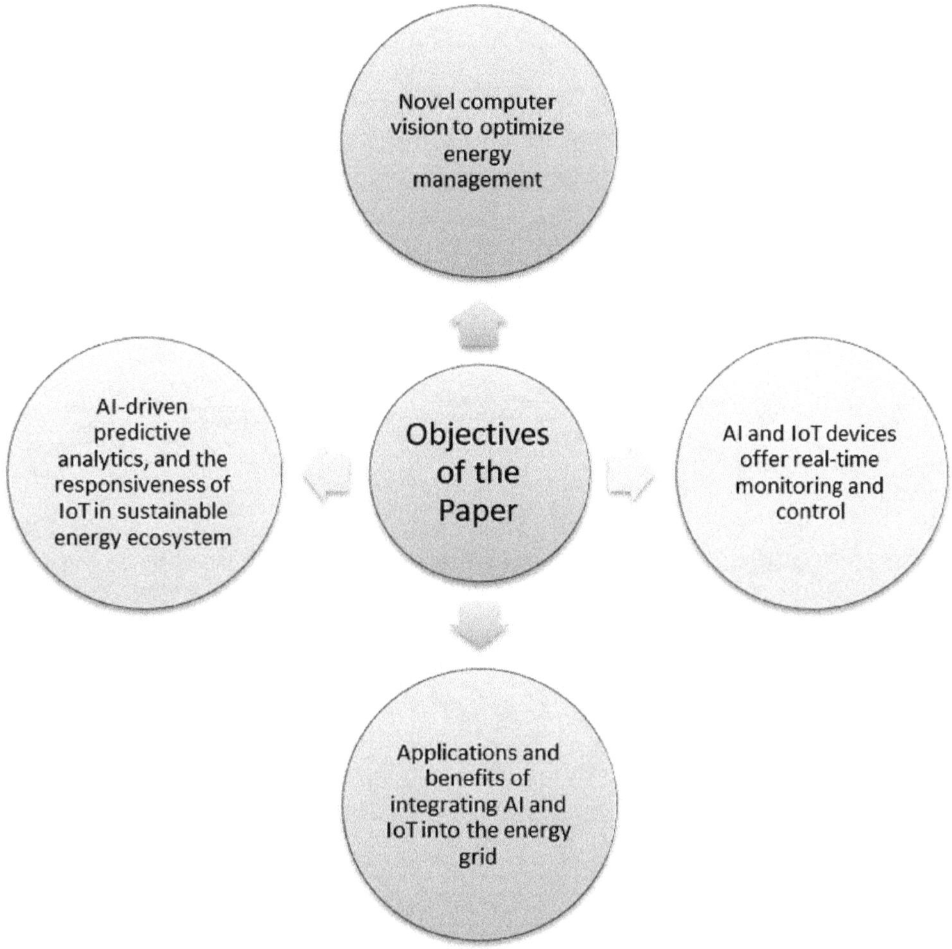

elaborates the Blue Wind Energy: An Overview. Section 3 explores the Computer Vision and Machine Learning for Renewable Energy- Novel Concepts. Section 4 lays down Sustainable Urbanization and Energy Demand. Section 5 examines the Integration of AI and IoT in Energy Management. Section 6 travels the Smart Grids and Energy Storage. Section 7 discusses the Environmental and Economic Considerations. Section 8 examines the Policy and Regulatory Framework- Government Incentives and Policies Promoting Blue Wind Energy; International cooperation and agreements. And, finally Section 9 deals with Conclusion and Future Scope.

2. BLUE WIND ENERGY: OVERVIEW, DEFINITION, AND PRINCIPLES OF BLUE WIND ENERGY

Artificial Intelligence (AI) and IoT stand out as a groundbreaking technology with immense potential across various fields, and energy management is no exception (Tabor et al., 2018) As the demand for effective energy use rises and energy systems become more intricate, AI has the power to transform how

Figure 3. Flow/organization of the chapter

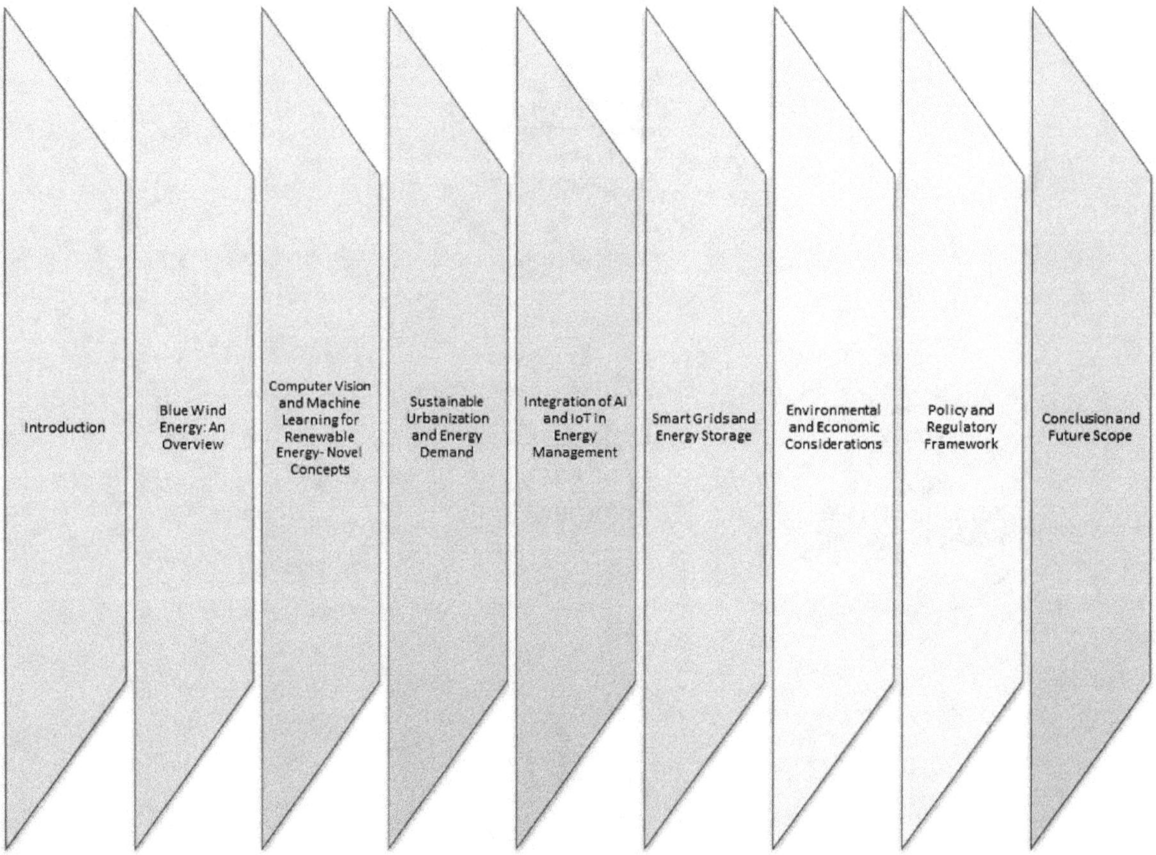

it handle and enhance our energy resources. This paper delves into the impact of AI on smart energy management, delving into its advantages, the hurdles it faces, and the promising possibilities that lie ahead (Szeliski, 2022).

The wind turbine is a device designed to transform the energy present in the wind into electricity. This differs from a 'windmill,' which converts wind power into mechanical power. Wind turbines, serving as electricity generators, are integrated into various electrical networks (Rao et al., 2022). These networks encompass battery-charging systems, power setups for residential use, networks for isolated or island locations, and expansive utility grids. In terms of quantity, the most common wind turbines are relatively small, typically around 10 KW or less. But, when considering total generating capacity, the majority of capacity comes from larger turbines, generally ranging from 1.5 to 5 MW. These larger turbines are primarily utilized in extensive utility grids, initially prevalent in Europe and the United States and more recently in China and India (Saini et al., 2023). The typical modern wind turbine, part of a wind farm configuration connected to a utility network. The showcased turbine is a General Electric 1.5 MW, and as of the writing of this text, this model from the manufacturer has seen over 10,000 units delivered.

2.1 Offshore Wind Energy

Offshore wind power also known as offshore wind energy, involves generating electricity through wind farms located in bodies of water, typically at sea. The advantage of offshore locations lies in the higher wind speeds experienced compared to onshore areas, resulting in a greater electricity yield per unit of installed capacity (Tomar et al., 2022). Moreover, offshore wind farms tend to be less contentious than their onshore counterparts since they have a smaller impact on both people and the landscape. It's worth noting that, in the context of offshore wind power, "offshore" extends beyond open seas and includes inshore water areas like lakes, fjords, and sheltered coastal regions, as well as deeper-water areas. While most offshore wind farms currently use fixed-foundation wind turbines situated in relatively shallow water, the exploration of floating wind turbines for deeper waters is in the early stages of development and deployment (Del Ser et al., 2022).

2.2 Benefits and Challenges of Blue Wind Energy

Wind energy's fundamental strength is its capacity to generate power without damaging the air or water and without adding to greenhouse gas emissions (Aslam et al., 2021). In addition, wind energy has a number of significant environmental and socio-economic benefits.

Environmental Benefits: Wind energy is not worsening climate change. Despite the energy costs connected with making and shipping turbines, studies on their life cycle shows that their carbon impact is covered by CO2 reductions after only six months of operation. In 2019, wind power avoided producing an amount of CO2 comparable to that produced by 43 million vehicles.

Source of Renewable Energy: Wind energy is a renewable and sustainable energy source since it is based on an infinite and unrestricted source is wind. Wind energy, unlike traditional energy sources, does not require extraction from the ground and transit via rail or truck, which adds extra energy, pollutants, and costs to fossil fuel burning facilities. Furthermore, continual improvements in wind turbine models lead to increased efficiency.

Emission-Free: A wind turbine or wind farm generates no wastewater or pollutants once constructed. Scrubbing of smoke stacks is no longer required, and the removal or processing of harmful pollutants is no longer required because they are not formed in the first place.

Economic Benefits- Minimum Operating Expenses: Wind turbines have cheap running expenses after installation and there are no source costs as because wind is a free and limitless resource, the cost of the energy source is basically nothing. Wind energy is a more cost-effective choice than coal-burning facilities, which are the most common power source.

Affordable: Wind energy may need less subsidies to remain affordable and all major energy distribution systems, including coal plants and wind farms, get government subsidies. However, depending on the exact criteria considered, the fossil fuel industry may receive far more substantial grants and tax advantages than renewable sources.

Mitigate Health and Pollution Hazards: Wind turbines do not add to air pollution or the health hazards linked with it. Coal-fired power plants, on the other hand, have been demonstrated to have negative impacts on human health, resulting in related medical costs. These health-related expenditures are sometimes overlooked when calculating the overall "cost" of generating power from coal. It is critical to recognize these factors as either expenses or cost-savings related with wind power, whether seen as an economic effect, a public health risk or both.

Flexibility: Wind energy is flexible and promotes energy independence. Wind power, unlike fossil fuels, which normally require centralized power facilities for optimal energy generation, is size and location adjustable. Even oil-powered domestic generators, which are frequently utilized during crises, are inefficient and contribute to neighborhood air pollution

Although some challenges are also associated with wind energy as-

Wind Dependability is Variable: While turbines may generate energy 90% of the time, their production may not achieve 100% capacity on a constant basis, averaging approximately 35% capacity. Because of this unpredictability, low or no wind can cause a wind turbine to shut down, as can overly high winds to protect the gear. Maintaining a regular electricity supply at such times will entail either using stored wind energy from batteries or switching to another power source.

Disturbances in Sound and Vision: Wind turbines have the capacity to emit noise in the 40-60 decibel range, which is comparable to the sound generated by a medium-sized window air conditioning unit. This noise can be bothersome, especially for people who live near even smaller turbines. However, reliable studies on the health effects of wind turbine noise are lacking.

Wildlife Effects: Wind turbine noise may also have an impact on wildlife, notably birds and bats, as well as other organisms that communicate through vocalizations.

Shadow Flickers: This phenomenon occurs when the whirling blades of a wind turbine collide with a low-lying sun, resulting in a shifting shadow that appears as a flicker as the blades rotate. While this

Figure 4. Multifarious benefits of blue wind energy

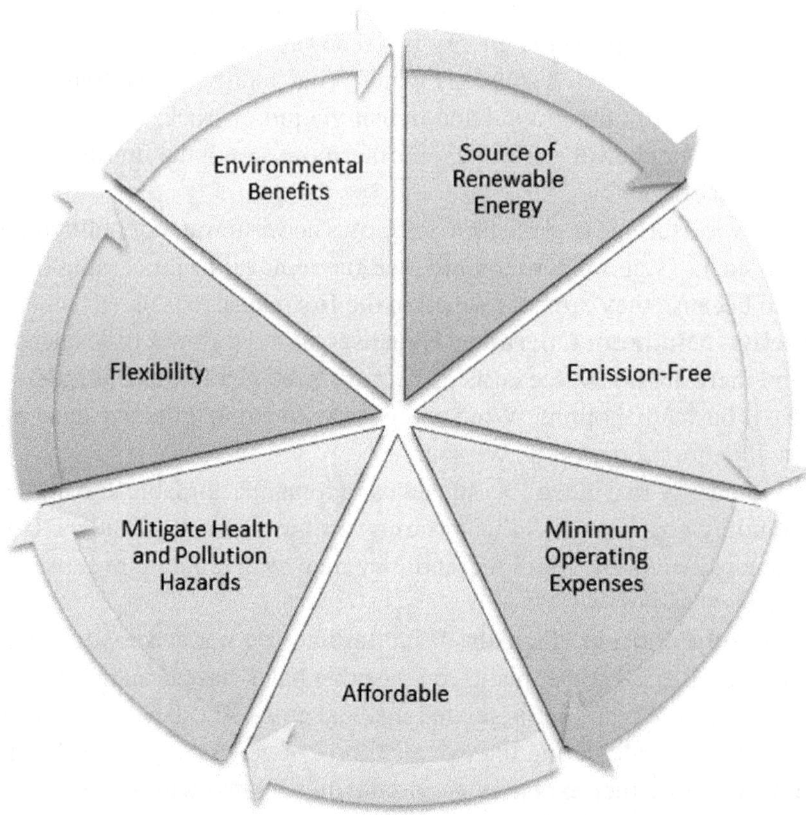

effect can be disconcerting and frightening for individuals who live near the turbine, it only occurs under particular, time-limited situations. Calculations and steps to reduce the influence of shadow flicker can be carried out. Due to their lesser height, smaller turbines often have fewer concerns with shadow flicker, making this matter more significant for bigger turbines.

2.3 Blue Wind Energy for Urbanization: Potential of Blue Wind Energy for Urban Areas

Energy is necessary for our everyday existence, since it is used to power our houses, electronics, companies, and other activities. Rising energy demand, driven by industrialization, urbanization, and increased public awareness of environmental concerns associated to energy consumption, highlights the need for continuous improvements in energy-generation technology (Watson, 2022). This initiative seeks to transition from traditional fossil fuels to renewable energy sources, with the goal of producing plentiful, clean, and environmentally beneficial energy naturally. Among the different renewable energy sources, wind power is fast gaining prominence (Cruz-Rojas et al., 2023).

Wind turbines are used to generate energy for local power systems, with an emphasis on spinning reserve dependability and environmental consciousness. Wind energy systems, in general, work best on a small to medium scale, making them well-suited for use in rural and suburban locations (Ozcanli et al., 2020). The previous research efforts concentrated on power density studies for micro-wind turbines erected in suburban areas. Recent researches have revealed an untapped potential for urban wind energy, as wind speed in metropolitan locations provides an extra aspect (Sharma & Singh, 2022). Buildings in metropolitan areas contribute to increased turbulence intensity, needing specific attention for successful use of this energy source. Rapid building constructions, widespread urban growth, rising technology, and the desire for comfortable, high-quality living have increased energy use and sparked environmental concerns (Rosiello & Maleki, 2021). So, given that buildings and urban environments contribute considerably to overall energy consumption, it is critical to prioritize energy-efficient solutions in building and urban design (Fischer & Preonas, 2010). Wind power has emerged as one of the most efficient techniques of ensuring energy efficiency in buildings. Wind, which is frequently connected with renewable energy sources, has not been widely used in construction applications until recently (Grubb et al., 2021).

3. COMPUTER VISION AND MACHINE LEARNING FOR RENEWABLE ENERGY: NOVEL CONCEPTS

The burning of fossil fuels, which currently satisfies a significant portion of the world's energy requirements, stands as the primary contributor to the escalating levels of greenhouse gas emissions and the

Figure 5. Challenges associated with wind energy

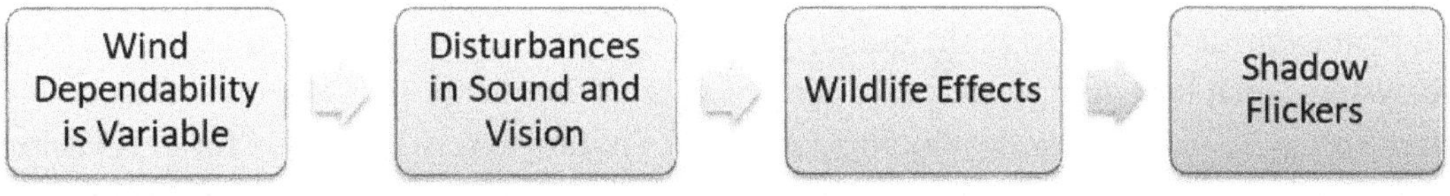

subsequent global temperature rise (Raihan et al., 2022). Addressing this challenge, the increased utilization of sustainable energy sources, particularly solar and wind power, presents a financially viable trajectory in alignment with the climate objectives outlined in the Paris Agreement. Nevertheless, the pace at which renewable energy adoption has progressed has struggled to keep pace with the continually expanding energy demands, resulting in a limited share of total energy derived from renewable sources (Singh, 2023). Consequently, expediting the shift to sustainable energy sources becomes imperative. Facilitating this transition necessitates the development of energy technologies, infrastructure, and policies that support and encourage the capture, storage, conversion, and effective management of renewable energy (Dai et al., 2016).

The careful selection of acceptable materials, particularly those used in photovoltaic applications, is critical in the pursuit of sustainable energy solutions. Following that, these selected materials must be synthesized with high yield and quality to match the standards for devices such as solar panels. The typical timescale for the materials discovery process is 15-20 years, implying plenty of room for improvement. Furthermore, the devices must be refined for durability and repeatability in order to fit easily into energy systems such as solar farms. To maintain economic viability, this integration demands effective control of energy use and generation patterns (Elavarasan et al., 2020).

4. SUSTAINABLE URBANIZATION AND ENERGY DEMAND-IMPORTANCE OF TRANSITIONING TO GREEN ENERGY SOURCES

As the severity of global warming and climate change increases, as evidenced by frequent and drastic changes in the natural environment affecting humans and other living beings around the world, international organizations such as the United Nations Organization, the G20, the G8, and individual nations become more aware of the environmental consequences of extensive fossil fuel consumption. This understanding emphasizes the economic and societal implications of climate change, particularly global warming (Zwane et al., 2022). The United Nations, notably through its climate change-focused environmental program, has continually worked to encourage greater international agreement. This agreement focuses on the need to meet an overall global carbon reduction target within the period given by the Intergovernmental Panel on Climate Change (IPCC). The goal is to gain international consensus that each country should have some responsibility for attaining the carbon emission reduction objective (Mohammadian, 2022).

Renewable energy and energy efficiency stand out as critical components of the transition to a low-consumption, low-carbon society. To make this transition easier, government policies should prioritize achieving region-specific objectives for renewable energy generation and energy efficiency. However, these objectives range between metropolitan regions, and there is a lack of comprehensive understanding of the link between energy transition and urbanization levels (Amran et al., 2020). The emphasis on growing and safeguarding energy in the current global context is around increasing energy producing capacity and, ideally, adding cleaner and more environmentally friendly energy sources into the growth process. Through a complex system, advancement and planning procedures frequently cause changes in the landscape and have a human influence on the surroundings (Hainsch et al., 2022). Resource extraction and land management operations in the building of energy infrastructure, as well as human monitoring of these systems, can have long-term and often unexpected implications. Despite increased interest in alternative energy sources, many developing countries continue to rely largely on fossil fuels for energy generation (Atalay et al., 2016). So, energy production has the capacity to influence land productivity, land cover, human migration, and other aspects of the energy production system, resulting in a complex interaction of various variables. The interdependence of land use, energy choices, infrastructure development, and the people for whom these facilities are developed necessitates careful thought. It is critical to examine and comprehend the interactions between these components (Singh, 2022).

5. INTEGRATION OF AI AND IOT IN ENERGY MANAGEMENT

The substantial portion of blue-wind as renewable energy sources and energy storage technologies should be unified in micro-grids to the inside electrical grids for the sake of to create a society with zero $CO2$ emissions. This is because of integration enables energy exchange so that the reducing demands on the transmission system (Xiao et al., 2019). However, managing energy inside and across these micro-grids is difficult due to a variety of uncertainties and there are erroneous data on energy output and consumption.

These uncertainties may make the energy optimization in this setting more difficult (Scarlat et al., 2015) and advancement of IoT platforms, in combination with artificial intelligence (AI), big data analytics and cloud computing, opens up new operational possibilities for smart grids. Even so, the centralized data processing on the cloud results in significant communication delays and overhead (Connolly et al., 2010). The edge computing also important for decentralizes computational activities at the network's edge, and to putting them closer to the user. These activities like which include data collecting, pre-processing and forecasting simulation, improve energy management decision-making processes (Salleh et al., 2020). As the result of which edge-cloud orchestration appears as a potential and effective architecture for improving energy trading flexibility, scalability, and performance. These AI-powered solutions for blue-wind have the potential to drastically cut energy prices and carbon footprints by actively managing use patterns in both residential and commercial settings (Singh, 2023). So, the AI plays a role in predictive maintenance of energy infrastructure, which is crucial for averting failures and prolonging asset lifespan, resulting in a more stable and dependable power supply.

5.1 Role of AI in Optimizing Energy Production and Consumption

Artificial intelligence (AI) is revolutionizing the management of energy systems and its capacity to handle large information, forecast trends and promote real-time decision-making makes it a valuable tool for navigating the complexities of modern energy networks (Bujang et al., 2016). The Artificial Intelligence has the potential to transform the energy sector in a variety of ways in which they including optimizing output, anticipating demand, enhancing grid efficiency and lowering operating costs. Also, the predictive skills of AI are critical in the field of renewable energy for forecasting weather patterns and maximizing the production of solar and wind energy (Lawrenz et al., 2018). The advances in enormous data processing, significant computer power, information technology, and the improvement of machine learning (ML) and deep learning (DL) algorithms are ascribed to the increase in AI technologies.

The Artificial Intelligence technologies essential the potential to help the energy business which capitalize on the expanding opportunities connected with the Internet of Things (IoT) and the incorporation of renewable energy sources. The supercomputers, power electronics, cyber technologies, information systems connectivity between the control center and equipment are all part of the smart energy sector's contemporary infrastructure (Westman et al., 2019). Despite these difficulties, the global economy is largely reliant on energy production, distribution planning, and financial sustainability (Cohen et al., 2021).

6. SMART GRIDS AND ENERGY STORAGE: ADVANCED ENERGY STORAGE SOLUTIONS

These energy storage technologies are critical for integrating renewable energy as the sources into current power systems and smart grids. They improve the system's balance and flexibility, allowing for the smooth integration of intermittent RES into the smart grid design (Yang et al., 2021). These technologies are important an for offering ancillary services in power generation sources inside the smart grid framework. Electric energy storage technologies have the potential to greatly improve the present electric power grid's operating capabilities, decreasing the requirement for massive expenditures in big transmission power lines (White et al., 2018). The capacity of energy storage devices to retain power generated at a specified moment for later usage distinguishes them, successfully balancing current supply.

The energy storage technologies have the potential to improve economic dispatch, efficiency and power quality in the distribution system by addressing modern difficulties in the electric power industry like as such as increased unpredictability and uncertainty (Nastasi et al., 2022). Energy storage technologies is expected to play a critical role in the construction of clean energy communities focused on renewable energy sources in the near future (Danish et al., 2023). As they handle both energy demand and supply in smart towns and cities, smart grids provide a big challenge to the energy business. Grid-connected energy storage devices are viewed as feasible alternatives for delivering ancillary services to electrical networks, and they will play an important part in the growth of smart grids (Basheer et al., 2022).

6.1 Role of Smart Grids in Urban Energy Distribution

Smart grids are improved electric power grid infrastructure that are intended to improve efficiency and reliability. They use automated control, high-power converters, advanced communication infrastructure, sensing and metering technologies, and modern energy management approaches to optimize demand, energy, and network availability (Al-tabatabaie et al., 2022).

Thus energy storage system integration in smart distribution networks has grown significantly, adding to the overall efficiency and efficacy of smart grids (Aldieri et al., 2021).The enhancement of different power quality and reliability parameters is a critical application of ESS technology. Because power quality concerns include precise parameters such as voltage variation amplitude, frequency, and duration, addressing and improving these metrics necessitates specialized technological considerations. When evaluating system dependability, technical indices that indicate the length and frequency of interruptions can be used to characterize the system's performance (Nakata, 2004).

6.2 AI and IoT Contribution to Efficient Energy Storage and Distribution

The city gains smart status by combining Artificial Intelligence (AI) with the Internet of Things (IoT) in blue wind energy management (Singh, 2022). This connection makes it easier to collect data from persons, devices, and buildings, which is then analyzed to optimize control over infrastructure, transportation, energy, and other factors. A smart city is essentially a complete structure that includes Information and Communication Technologies (ICT) and the Cloud, allowing its components to interact seamlessly (Winfield & Dolter, 2014). It digs into smart energy infrastructure, with an emphasis on monitoring energy use in the city to reduce costs and carbon emissions. The recent focus in energy utilization has switched toward renewable energy sources with low carbon emissions, emphasizing the need of continued efforts to preserve the environment and human health (Damman et al., 2021).

The energy consumption patterns tend to expand in a way that is strongly related to economic development as economies improve. This relationship stems from the strong and positive relationship between energy use and economic advancement. Energy is a critical input for societal development, and beyond energy, global geopolitical situations, new technical breakthroughs, and the availability of natural resources all impact socioeconomic progress (Pascaris, 2021). These influences of these many elements on socioeconomic growth, forecasting associated uncertainties becomes difficult, potentially influencing energy supply and demand patterns. The negative shocks to the energy industry might have a negative impact on socioeconomic growth. As a result, the quest of adequate, dependable, resilient, equitable, secure, and economical energy supply is prioritized (Budiman, 2018).

The Internet and other intelligent systems' importance in societal advancement is clear and projected to expand in the near future. Concerns regarding environmental safety have also been expressed, with advocates asking for future energy supply to adhere to low-carbon and other clean energy requirements (Reedman et al., 2021). As a result, smart and sustainable energy supplies are being advocated for in order to ensure socioeconomic progress. As a result, the effects of using renewable energy resources as an alternative to fossil fuel producing technologies on socioeconomic growth become increasingly obvious (Xie et al., 2023).

7. ENVIRONMENTAL AND ECONOMIC CONSIDERATIONS

The predicted increase in offshore wind farm projects to meet global decarbonization objectives has generated concern in the scientific community about the marine environment's biological well-being. It investigates the environmental consequences and changes related with offshore wind farms, with an emphasis on their compliance with the Marine Strategy Framework Directive's Good Environmental Status. The Environmental Impact Assessment (EIA) is an important tool for decision-makers to examine the environmental consequences of planned operations (Vinichenko, 2018). Regulatory bodies frequently use EIA to ensure that projects meet environmental protection goals. Furthermore, EIA acts as a critical link between scientific research and regulatory procedures, guaranteeing the appropriate use of current information while also supporting the production and dissemination of new scientific discoveries (Arbabzadeh et al., 2019).

The rapid growth of renewable energy efforts, particularly offshore wind farms, brings a plethora of artificial structures in maritime ecosystems that will ultimately require decommissioning (Ilyas et al., 2022). The Environmental Impact Assessment (EIA) is carried out prior to decommissioning activities in order to assess potential environmental implications. EIA provides a chance to study appropriate environmental decommissioning options when used early in the planning phase during the design of wind farms. EIA should be used to help make decisions by assessing effects, developing mitigation plans, and tracking the asset's life cycle.

7.1 Environmental Impact Assessment of Blue Wind Energy

The renewable energy has grown significantly globally in recent decades, owing to growing worries about climate change and a transition away from reliance on fossil fuels. Offshore wind energy has emerged as a prominent participant in the renewable energy market. Nonetheless, because wind farms are temporary facilities, they must be decommissioned at the end of their operating lives (Garrido et al., 2020).

8. POLICY AND REGULATORY FRAMEWORK

The conventional energy grid designs frequently underestimate the relevance of information and energy storage, despite the fact that a genuinely dynamic power system requires both. The authors urge for energy storage to be recognized as a unique asset class within the electric grid system, supported by appropriate regulatory and financial policies for its development and deployment within a storage-based smart grid framework in which storage plays a vital role (Steen & Weaver, 2017). This method would optimize load and market operations by allowing storage technologies to provide the full range and value of services. The energy storage technologies have the potential to significantly improve grid efficiency and performance (Hoffmann et al., 2020). Their capacity to supply energy services targeted to specific applications across diverse grid components uniquely positions them to respond to signals inside smart grid systems in a timely and effective manner. As a result, classifying energy storage as a separate asset class will increase the value of storage investments while also contributing to the overall efficiency of the smart grid.

With the advancement in smart grid technologies and designs, energy system planning must abandon the current view of energy storage in favor of a more effective and sustainable framework. Although energy storage systems have been incorporated into the energy grid design, their use has been limited due to the high prices and inefficiencies associated with many storage systems, reducing their total value and usefulness (Manoharan et al., 2021). The common perception of storage as only a tool for supporting generators in load balancing over time fails to recognize the vast potential of energy storage technology (Ghai et al., 2022). Smart grid design and storage technology improvements provide considerable opportunity to disrupt the current paradigm. To achieve a completely dynamic and efficient grid, especially while moving to a low-carbon energy system, planners must fully embrace the benefits of renewable electricity generation and post-generation energy storage (Pimitha et al., 2023).

9. CONCLUSION AND FUTURE SCOPE

The AI and Internet of Things (IoT) has evolved into a popular and scalable information and communication infrastructure. It is becoming increasingly important in smart cities because to its support for interoperability platforms, communication across heterogeneous devices, and ubiquitous access. A smart city's primary goal is to provide eco-friendly communication access to devices and users by harvesting natural energy, giving nature-inspired support, and contributing to a pollution-free environment. Smart cities have a dual role in climate change as they are large producers to global greenhouse gas emissions while also being particularly vulnerable to the effects of climate change and extreme weather events. The chapter focuses on practical techniques for constructing sustainable energy systems, with a major focus on lowering energy use, particularly in the construction and transportation sectors, and generating resilient, decentralized, and renewable energy sources. Power density advancements are critical for properly integrating renewable energy inside urban landscapes to fulfill the growing energy requirements of developing metropolitan regions. Addressing several challenges economic, technological, behavioral and political is critical for generating innovation that might improve urban sustainability. Renewable energy sources such as heat, light, and wind are infinite, but their power must be conserved for long-term usage. Smart grids, which are managed by IoT infrastructures, are used in smart city environments for optimal power management and distribution. As a result, the integration of grid management with IoT uses information and communication technologies to improve power management operational efficiency in smart cities. Carbon dioxide (CO2) emissions must be reduced in order to achieve sustainable development goals, protect the environment, address climate change, and preserve biodiversity. The adopting of comprehensive plans that promote energy efficiency, renewable energy and sustainable urbanization is critical for creating a cleaner, more resilient and bright future.

REFERENCES

Abualigah, L., Zitar, R. A., Almotairi, K. H., Hussein, A. M., Abd Elaziz, M., Nikoo, M. R., & Gandomi, A. H. (2022). Wind, solar, and photovoltaic renewable energy systems with and without energy storage optimization: A survey of advanced machine learning and deep learning techniques. *Energies*, *15*(2), 578. doi:10.3390/en15020578

Ahmad, T., Madonski, R., Zhang, D., Huang, C., & Mujeeb, A. (2022). Data-driven probabilistic machine learning in sustainable smart energy/smart energy systems: Key developments, challenges, and future research opportunities in the context of smart grid paradigm. *Renewable & Sustainable Energy Reviews*, *160*, 112128. doi:10.1016/j.rser.2022.112128

Al-tabatabaie, K. F., Hossain, M. B., Islam, M. K., Awual, M. R., TowfiqulIslam, A. R. M., Hossain, M. A., Esraz-Ul-Zannat, M., & Islam, A. (2022). Taking strides towards decarbonization: The viewpoint of Bangladesh. *Energy Strategy Reviews*, *44*, 100948. doi:10.1016/j.esr.2022.100948

Aldieri, L., Grafström, J., & Vinci, C. P. (2021). The effect of Marshallian and Jacobian knowledge spillovers on jobs in the solar, wind and energy efficiency sector. *Energies*, *14*(14), 4269. doi:10.3390/en14144269

Amran, Y. A., Amran, Y. M., Alyousef, R., & Alabduljabbar, H. (2020). Renewable and sustainable energy production in Saudi Arabia according to Saudi Vision 2030; Current status and future prospects. *Journal of Cleaner Production*, *247*, 119602. doi:10.1016/j.jclepro.2019.119602

Arbabzadeh, M., Gençer, E., Morris, J. F., Paltsev, S., & Armstrong, R. C. (2019). *Plausible Energy Futures: A Framework for Evaluating Options*. Impacts, and National Energy Choices.

Arent, D. J., Green, P., Abdullah, Z., Barnes, T., Bauer, S., Bernstein, A., & Turchi, C. (2022). Challenges and opportunities in decarbonizing the US energy system. *Renewable & Sustainable Energy Reviews*, *169*, 112939. doi:10.1016/j.rser.2022.112939

Aslam, S., Herodotou, H., Mohsin, S. M., Javaid, N., Ashraf, N., & Aslam, S. (2021). A survey on deep learning methods for power load and renewable energy forecasting in smart microgrids. *Renewable & Sustainable Energy Reviews*, *144*, 110992. doi:10.1016/j.rser.2021.110992

Atalay, Y., Biermann, F., & Kalfagianni, A. (2016). Adoption of renewable energy technologies in oil-rich countries: Explaining policy variation in the Gulf Cooperation Council states. *Renewable Energy*, *85*, 206–214. doi:10.1016/j.renene.2015.06.045

Basheer, M., Nechifor, V., Calzadilla, A., Ringler, C., Hulme, D., & Harou, J. J. (2022). Balancing national economic policy outcomes for sustainable development. *Nature Communications*, *13*(1), 5041. doi:10.1038/s41467-022-32415-9 PMID:36028486

Bogdanov, D., Ram, M., Aghahosseini, A., Gulagi, A., Oyewo, A. S., Child, M., & Breyer, C. (2021). Low-cost renewable electricity as the key driver of the global energy transition towards sustainability. *Energy*, *227*, 120467. doi:10.1016/j.energy.2021.120467

Budiman, I. (2018). Enabling community participation for social innovation in the energy sector. *Indonesian Journal of Energy*, *1*(2), 21–31. doi:10.33116/ije.v1i2.23

Bujang, A. S., Bern, C. J., & Brumm, T. J. (2016). Summary of energy demand and renewable energy policies in Malaysia. *Renewable & Sustainable Energy Reviews*, *53*, 1459–1467. doi:10.1016/j.rser.2015.09.047

Cohen, S. M., Iyer, G. C., Brown, M., Macknick, J., Wise, M., Binsted, M., & Hejazi, M. (2021). How structural differences influence cross-model consistency: An electric sector case study. *Renewable & Sustainable Energy Reviews*, *144*, 111009. doi:10.1016/j.rser.2021.111009

Connolly, D., Lund, H., Mathiesen, B. V., & Leahy, M. (2010). A review of computer tools for analysing the integration of renewable energy into various energy systems. *Applied Energy*, *87*(4), 1059–1082. doi:10.1016/j.apenergy.2009.09.026

Cruz-Rojas, T., Franco, J. A., Hernandez-Escobedo, Q., Ruiz-Robles, D., & Juarez-Lopez, J. M. (2023). A novel comparison of image semantic segmentation techniques for detecting dust in photovoltaic panels using machine learning and deep learning. *Renewable Energy*, *217*, 119126. doi:10.1016/j.renene.2023.119126

Dai, H., Xie, X., Xie, Y., Liu, J., & Masui, T. (2016). Green growth: The economic impacts of large-scale renewable energy development in China. *Applied Energy*, *162*, 435–449. doi:10.1016/j.apenergy.2015.10.049

Damman, S., Sandberg, E., Rosenberg, E., Pisciella, P., & Graabak, I. (2021). A hybrid perspective on energy transition pathways: Is hydrogen the key for Norway? *Energy Research & Social Science*, *78*, 102116. doi:10.1016/j.erss.2021.102116

Danish, M. S. S., & Senjyu, T. (2023). AI-Enabled Energy Policy for a Sustainable Future. *Sustainability (Basel)*, *15*(9), 7643. doi:10.3390/su15097643

Del Ser, J., Casillas-Perez, D., Cornejo-Bueno, L., Prieto-Godino, L., Sanz-Justo, J., Casanova-Mateo, C., & Salcedo-Sanz, S. (2022). Randomization-based machine learning in renewable energy prediction problems: Critical literature review, new results and perspectives. *Applied Soft Computing*, *118*, 108526. doi:10.1016/j.asoc.2022.108526

Dent, C. M. (2012). Renewable energy and East Asia's new developmentalism: Towards a low carbon future? *The Pacific Review*, *25*(5), 561–587. doi:10.1080/09512748.2012.728240

Devabhaktuni, V., Alam, M., Depuru, S. S. S. R., Green, R. C. II, Nims, D., & Near, C. (2013). Solar energy: Trends and enabling technologies. *Renewable & Sustainable Energy Reviews*, *19*, 555–564. doi:10.1016/j.rser.2012.11.024

Doost Mohammadian, H., & Rezaie, F. (2020). Blue-green smart mobility technologies as readiness for facing tomorrow's urban shock toward the world as a better place for living (Case studies: Songdo and Copenhagen). *Technologies*, *8*(3), 39. doi:10.3390/technologies8030039

Elavarasan, R. M., Shafiullah, G. M., Padmanaban, S., Kumar, N. M., Annam, A., Vetrichelvan, A. M., & Holm-Nielsen, J. B. (2020). A comprehensive review on renewable energy development, challenges, and policies of leading Indian states with an international perspective. *IEEE Access : Practical Innovations, Open Solutions*, *8*, 74432–74457. doi:10.1109/ACCESS.2020.2988011

ElFar, O. A., Chang, C. K., Leong, H. Y., Peter, A. P., Chew, K. W., & Show, P. L. (2021). Prospects of Industry 5.0 in algae: Customization of production and new advance technology for clean bioenergy generation. *Energy Conversion and Management: X, 10*, 100048.

Eljack, F., & Kazi, M. K. (2021). Prospects and challenges of green hydrogen economy via multi-sector global symbiosis in Qatar. *Frontiers in Sustainability*, *1*, 612762. doi:10.3389/frsus.2020.612762

Elrahmani, A., Hannun, J., Eljack, F., & Kazi, M. K. (2021). Status of renewable energy in the GCC region and future opportunities. *Current Opinion in Chemical Engineering*, *31*, 100664. doi:10.1016/j.coche.2020.100664

Fischer, C., & Preonas, L. (2010). Combining policies for renewable energy: Is the whole less than the sum of its parts? *Resource for the future discussion paper*, (10-19).

Garrido, S., Sequeira, T., & Santos, M. (2020). Renewable energy and sustainability from the supply side: A critical review and analysis. *Applied Sciences (Basel, Switzerland)*, *10*(17), 5755. doi:10.3390/app10175755

Ghai, D., Tripathi, S. L., Saxena, S., Chanda, M., & Alazab, M. (Eds.). (2022). *Machine Learning Algorithms for Signal and Image Processing*. John Wiley & Sons. doi:10.1002/9781119861850

Grubb, M., Drummond, P., Poncia, A., McDowall, W., Popp, D., Samadi, S., & Pavan, G. (2021). Induced innovation in energy technologies and systems: A review of evidence and potential implications for CO2 mitigation. *Environmental Research Letters*, *16*(4), 043007. doi:10.1088/1748-9326/abde07

Hainsch, K., Löffler, K., Burandt, T., Auer, H., del Granado, P. C., Pisciella, P., & Zwickl-Bernhard, S. (2022). Energy transition scenarios: What policies, societal attitudes, and technology developments will realize the EU Green Deal? *Energy*, *239*, 122067. doi:10.1016/j.energy.2021.122067

Hoffmann, M. W., Wildermuth, S., Gitzel, R., Boyaci, A., Gebhardt, J., Kaul, H., & Tornede, T. (2020). Integration of novel sensors and machine learning for predictive maintenance in medium voltage switchgear to enable the energy and mobility revolutions. *Sensors (Basel)*, *20*(7), 2099. doi:10.3390/s20072099 PMID:32276442

Ilyas, A., Parkinson, S., Vinca, A., Byers, E., Manzoor, T., Riahi, K., & Muhammad, A. (2022). Balancing smart irrigation and hydropower investments for sustainable water conservation in the Indus basin. *Environmental Science & Policy*, *135*, 147–161. doi:10.1016/j.envsci.2022.04.012

Kasinathan, P., Pugazhendhi, R., Elavarasan, R. M., Ramachandaramurthy, V. K., Ramanathan, V., Subramanian, S., & Alsharif, M. H. (2022). Realization of Sustainable Development Goals with Disruptive Technologies by Integrating Industry 5.0, Society 5.0, Smart Cities and Villages. *Sustainability (Basel)*, *14*(22), 15258. doi:10.3390/su142215258

Lawrenz, L., Xiong, B., Lorenz, L., Krumm, A., Hosenfeld, H., Burandt, T., & Von Hirschhausen, C. (2018). Exploring energy pathways for the low-carbon transformation in India—A model-based analysis. *Energies*, *11*(11), 3001. doi:10.3390/en11113001

Li, W., Lu, C., & Zhang, Y. W. (2019). Prospective exploration of future renewable portfolio standard schemes in China via a multi-sector CGE model. *Energy Policy*, *128*, 45–56. doi:10.1016/j.enpol.2018.12.054

Makatjane, K. (2022). Deep Learning for Sentiment Analysis to Predict the Probability of Bank Loan Default. *American Journal of Data Mining and Knowledge Discovery*, *7*(2), 5–12.

Makatjane, K., Moroke, N., & Ncube, B. (2021). Detecting Financial Fraud in South Africa: A Comparison of Logistic Model Tree and Gradient Boosting Decision Tree. Academic Press.

Manoharan, K. G., Nehru, J. A., & Balasubramanian, S. (2021). *Artificial Intelligence and IoT*. Springer. doi:10.1007/978-981-33-6400-4

Mohammadian, H. D. (2022). Mapping the Future Global SMEs Growth via Hybrid SMEs/SME 5.0/ Tomorrow's SMEs Concept through the 5th Wave, i-Sustainability Plus and DCT Theories. In *Management and Information Technology in the Digital Era* (Vol. 29, pp. 173–206). Emerald Publishing Limited. doi:10.1108/S1877-636120220000029011

Moroke, N. D., & Makatjane, K. (2022). Predictive Modelling for Financial Fraud Detection Using Data Analytics: A Gradient-Boosting Decision Tree. In Applications of Machine Learning and Deep Learning for Privacy and Cybersecurity (pp. 25-45). IGI Global. doi:10.4018/978-1-7998-9430-8.ch002

Nakata, T. (2004). Energy-economic models and the environment. *Progress in Energy and Combustion Science*, *30*(4), 417–475. doi:10.1016/j.pecs.2004.03.001

Nastasi, B., Markovska, N., Puksec, T., Duić, N., & Foley, A. (2022). Renewable and sustainable energy challenges to face for the achievement of Sustainable Development Goals. *Renewable & Sustainable Energy Reviews*, *157*, 112071. doi:10.1016/j.rser.2022.112071

Ozcanli, A. K., Yaprakdal, F., & Baysal, M. (2020). Deep learning methods and applications for electrical power systems: A comprehensive review. *International Journal of Energy Research*, *44*(9), 7136–7157. doi:10.1002/er.5331

Pascaris, A. S. (2021). Examining existing policy to inform a comprehensive legal framework for agrivoltaics in the US. *Energy Policy*, *159*, 112620. doi:10.1016/j.enpol.2021.112620

Pimitha, K., Anbarasi, S., & Balasubramanian, T. (2023). Machine Learning Techniques for Renewable Energy Resources. In IoT, Machine Learning and Blockchain Technologies for Renewable Energy and Modern Hybrid Power Systems (pp. 121-148). River Publishers.

Raihan, A., Begum, R. A., Said, M. N. M., & Pereira, J. J. (2022). Relationship between economic growth, renewable energy use, technological innovation, and carbon emission toward achieving Malaysia's Paris agreement. *Environment Systems & Decisions*, *42*(4), 586–607. doi:10.1007/s10669-022-09848-0

Rao, S., Katoch, S., Narayanaswamy, V., Muniraju, G., Tepedelenlioglu, C., & Spanias, A. (2022). *Machine learning for solar array monitoring, optimization, and control*. Springer Nature.

. Reedman, L., Chew, M. S., Gordon, J., Sue, W., Brinsmead, T., Hayward, J., & Havas, L. (2021). Multi-sector energy modelling.

Rosiello, A., & Maleki, A. (2021). A dynamic multi-sector analysis of technological catch-up: The impact of technology cycle times, knowledge base complexity and variety. *Research Policy*, *50*(3), 104194. doi:10.1016/j.respol.2020.104194

Saini, G., Kannan, R., Benini, E., & Kumar, K. (Eds.). (2023). *Enabling Methodologies for Renewable and Sustainable Energy*. CRC Press. doi:10.1201/9781003272717

Salleh, S. F., Mohd Roslan, M. E., Abd Rahman, A., Shamsuddin, A. H., Tuan Abdullah, T. A. R., & Sovacool, B. K. (2020). Transitioning to a sustainable development framework for bioenergy in Malaysia: Policy suggestions to catalyse the utilisation of palm oil mill residues. *Energy, Sustainability and Society*, *10*(1), 1–20. doi:10.1186/s13705-020-00269-y

Scarlat, N., Dallemand, J. F., Monforti-Ferrario, F., & Nita, V. (2015). The role of biomass and bioenergy in a future bioeconomy: Policies and facts. *Environmental Development*, *15*, 3–34. doi:10.1016/j.envdev.2015.03.006

Sharma, A., & Singh, B. (2022). Measuring Impact of E-commerce on Small Scale Business: A Systematic Review. *Journal of Corporate Governance and International Business Law*, *5*(1).

Singh, B. (2022). Relevance of Agriculture-Nutrition Linkage for Human Healthcare: A Conceptual Legal Framework of Implication and Pathways. *Justice and Law Bulletin*, *1*(1), 44–49.

Singh, B. (2023). Tele-Health Monitoring Lensing Deep Neural Learning Structure: Ambient Patient Wellness via Wearable Devices for Real-Time Alerts and Interventions. *Indian Journal of Health and Medical Law*, *6*(2), 12–16.

Singh, B. (2023). Federated Learning for Envision Future Trajectory Smart Transport System for Climate Preservation and Smart Green Planet: Insights into Global Governance and SDG-9 (Industry, Innovation and Infrastructure). *National Journal of Environmental Law*, *6*(2), 6–17.

Singh, B. (2023). Blockchain Technology in Renovating Healthcare: Legal and Future Perspectives. In Revolutionizing Healthcare Through Artificial Intelligence and Internet of Things Applications (pp. 177-186). IGI Global.

Singh, B. (2023). Federated Learning for Envision Future Trajectory Smart Transport System for Climate Preservation and Smart Green Planet: Insights into Global Governance and SDG-9 (Industry, Innovation and Infrastructure). *National Journal of Environmental Law*, *6*(2), 6–17.

Singh, B. (2024). Legal Dynamics Lensing Metaverse Crafted for Videogame Industry and E-Sports: Phenomenological Exploration Catalyst Complexity and Future. *Journal of Intellectual Property Rights Law*, *7*(1), 8–14.

Steen, M., & Weaver, T. (2017). Incumbents' diversification and cross-sectorial energy industry dynamics. *Research Policy*, *46*(6), 1071–1086. doi:10.1016/j.respol.2017.04.001

Szeliski, R. (2022). *Computer vision: algorithms and applications*. Springer Nature. doi:10.1007/978-3-030-34372-9

Tabor, D. P., Roch, L. M., Saikin, S. K., Kreisbeck, C., Sheberla, D., Montoya, J. H., Dwaraknath, S., Aykol, M., Ortiz, C., Tribukait, H., Amador-Bedolla, C., Brabec, C. J., Maruyama, B., Persson, K. A., & Aspuru-Guzik, A. (2018). Accelerating the discovery of materials for clean energy in the era of smart automation. *Nature Reviews. Materials*, *3*(5), 5–20. doi:10.1038/s41578-018-0005-z

Tomar, A., Malik, H., Kumr, P., & Iqbal, A. (2022, September). Machine Learning, Advances in Computing, Renewable Energy and Communication (MARC). In *Proceedings of 3rd International Conference on Machine Learning, Advances in Computing, Renewable Energy and Communication: MARC 2021* (pp. 1-19). Singapore: Springer Nature Singapore. 10.1007/978-981-19-2828-4_1

Vinichenko, V. (2018). Mechanisms of energy transitions: National cases and the worldwide uptake of wind and solar power. *PhD diss., Central European University*.

Watson, R. (2022). The virtual economy of the metaverse: Computer vision and deep learning algorithms, customer engagement tools, and behavioral predictive analytics. *Linguistic and Philosophical Investigations*, (21), 41–56.

Westman, L. K., Broto, V. C., & Huang, P. (2019). Revisiting multi-level governance theory: Politics and innovation in the urban climate transition in Rizhao, China. *Political Geography*, *70*, 14–23. doi:10.1016/j.polgeo.2019.01.002

. White, D., Winchester, N., Atkins, M., Ballingall, J., Coates, S., de Miguel Mercader, F., & Suomalainen, K. (2018). Energy-and multi-sector modelling of climate change mitigation in New Zealand: current practice and future needs. Academic Press.

Winfield, M., & Dolter, B. (2014). Energy, economic and environmental discourses and their policy impact: The case of Ontario's Green Energy and Green Economy Act. *Energy Policy*, *68*, 423–435. doi:10.1016/j.enpol.2014.01.039

Xiao, M., Simon, S., & Pregger, T. (2019). Scenario analysis of energy system transition-A case study of two coastal metropolitan regions, eastern China. *Energy Strategy Reviews*, *26*, 100423. doi:10.1016/j.esr.2019.100423

Xie, Y., Wu, X., Hou, Z., Li, Z., Luo, J., Lüddeke, C. T., & Liao, J. (2023). Gleaning insights from German energy transition and large-scale underground energy storage for China's carbon neutrality. *International Journal of Mining Science and Technology*, *33*(5), 529–553. doi:10.1016/j.ijmst.2023.04.001

Yang, D., Liu, D., Huang, A., Lin, J., & Xu, L. (2021). Critical transformation pathways and socio-environmental benefits of energy substitution using a LEAP scenario modeling. *Renewable & Sustainable Energy Reviews*, *135*, 110116. doi:10.1016/j.rser.2020.110116

Zwane, N., Tazvinga, H., Botai, C., Murambadoro, M., Botai, J., De Wit, J., & Mabhaudhi, T. (2022). A bibliometric analysis of solar energy forecasting studies in Africa. *Energies*, *15*(15), 5520. doi:10.3390/en15155520

Chapter 11
Detecting Aberrations in Renewable Energy With the One-Class Support Vector Machine Model

Ntebogang Moroke
https://orcid.org/0000-0001-8545-1860
North West University, South Africa

Katleho Makatjane
https://orcid.org/0000-0002-3687-4098
University of Botswana, Botswana

ABSTRACT

The traditional energy time series forecasting methods use accurate input data for prediction and then make single-step or multi-step predictions based on the established regression model. But because of the complexity of financial markets, the traditional forecasting methods are less reliable. Unsupervised anomaly detection techniques operate directly on an unseen dataset, under the assumption that outliers are sparsely present in it. A one-class support vector machine (SVM) is an extension of support vector machines for unlabeled data and can be used for anomaly detection. Even though outliers are accounted for in one-class SVMs, they greatly influence the learned model. Hence, one modification to one-class SVMs is presented here: eta one-class SVMs. For evaluation, eta one support vector machine-based algorithm is compared using four kernels. The results affirmed the proficiency of the RBF kernel, with the eta one-class SVM performing best. In regards to subspace division, the preliminary results showed a decreasing nonlinear trend.

INTRODUCTION

South Africa is one of the global leaders in green energy transition because of its abundant natural resources. The diverse topography of the entire country offers plenty of opportunities to generate

DOI: 10.4018/979-8-3693-2355-7.ch011

electricity from wind, solar, hydroelectric, and other energies, including wind-whipped coasts and sun-kissed landscapes. South Africa has taken the lead in investing heavily in green energy efforts, seeing the enormous potential that exists within its borders. Tall windmills collect the unrelenting sea winds along the coast, while vast solar farms absorb the sun's unending beams across parched landscapes. In addition, waterways across the country are crossed by flowing rivers, which provide the kinetic energy that hydroelectric facilities placed strategically can harness (Intergovernmental Panel on Climate Change report, 2020). This chapter's value extends beyond only enhancing the operational efficacy of renewable energy plants. By presenting a novel framework customized to the unique features of South Africa's rapidly changing energy sector, it plays a crucial role in promoting international attempts to tackle climate change (Uhunamure and Shale, 2021). This chapter develops a rigorous methodology for renewable energy anomaly identification, contributing to the regional context and serving as a model for similar global endeavors. This holistic approach functions as a lighthouse, directing initiatives in the direction toward establishing a more robust and sustainable world energy grid.

This shift is not devoid of its complexities, though. Renewable energy integration into current systems is a complex process. Unlike traditional power sources, which provide a regular and controlled output, the generated supply of energy from renewable sources is inherently unpredictable and affected by factors such as climate, seasonal fluctuations, and cyclical cycles. To ensure a steady and dependable energy supply, this unpredictability adds another level of complexity (Shahzad et al., 2022). Unexpected variations in energy output or consumption are referred to as anomalies, and they pose a significant threat. These abnormalities could upset the electricity grid's equilibrium and cause instability and possible outages if they are not quickly found and fixed (Jithish et al., 2023). The successful integration of renewable energy sources is reliant on the proficient handling of these irregularities. Anomalies can be caused by a variety of factors, including unexpected changes in wind speed or cloud cover, equipment failures, or irregularities in trends of energy utilisation. They can also take multiple forms. Strong anomaly detection systems that can quickly find and fix these deviations are essential for ensuring the stability and dependability of the electricity grid (Madabhushi and Dewri, 2023).

With an emphasis on the changing energy landscape of South Africa, this chapter attempts to create a customised framework for identifying aberrations in renewable energy infrastructure. The goal is to effectively manage abnormalities in energy generation and usage patterns to ensure the successful integration of renewable energy. These anomalies may arise from factors such as changing meteorological patterns, equipment malfunctions, and challenges in grid integration. Traditional anomaly detection techniques, effective in conventional power systems, may not readily apply to the dynamic nature of renewable energy (Mohammadpourfard et al., 2018). Hence our primary objective is to establish a comprehensive framework for identifying and mitigating anomalies in green energy systems, addressing unexpected variations in energy patterns that hinder the seamless integration of renewable sources into existing systems (Chahla et al., 2020). Leveraging the efficacy of the One-Class Support Vector Machine (One-Class SVM) model and advanced data analytics, the chapter aims to surpass conventional methods in anomaly detection tasks. The study utilises One-Class SVM because it is effective in high-dimensional feature spaces. In real-world scenarios, data is represented in a high-dimensional space, and traditional distance-based methods are plagued by several dimensions. One-Class SVM is less affected by this issue.

Moreover, the One-Class SVM algorithm is less prone to abnormalities present in the training data compared to other anomaly detection methods. Using kernel functions, it can simulate non-linear decision boundaries and concentrate on finding the area in the domain of features that contains most of the data points (normal examples), treating the remaining data points as anomalies. This gives it the ability to recognize complex trends in the data, which increases its adaptability to different anomalies. In addition, this chapter leverages the state-of-the-art powers of sophisticated data analytics to decrypt complex patterns from the extensive dataset of live renewable energy installations. To extract meaningful insights, we meticulously prepare the data, including cleaning, standardization, and extraction of features. Then, using cutting-edge analytical tools, we evaluate the data seeking any concealed discrepancies and developing a greater comprehension of the complex links between various factors that affect the production of renewable energy.

LITERATURE REVIEW

This section offers insights from several research on the use of One-Class SVM in renewable energy systems for the detection of anomalies. They highlight how well One-Class SVM works to find abnormalities in data from several renewable energy sources, including solar, wind, and energy usage. In the current study, we seek to increase the dependability and efficacy of renewable energy techniques by utilising the powers of One-Class SVM to explicitly identify and mitigate abnormalities within hydropower green energy data from South Africa gathered over a specified period.

METHODS AND MATERIALS

This section provides a discussion of cutting-edge statistical techniques and software applications used in this study.

Statistical Software and Data Pre-Processing

Sophisticated statistical approaches were employed in the methodology to examine the intricate patterns and trends present in the information. For example, *pandas, numpy, scipy*, and other pertinent libraries, and statistical software like R or Python were used to perform data in the preprocessing phase, in data exploration, and statistical modelling. These software programs are perfect for managing complex and complicated datasets that are frequently found in renewable energy research since they include a broad range of functions for data management, visualisation, and statistical analysis. Furthermore, for tasks involving anomaly detection and predictive modelling, machine learning techniques written in R or Python, like ensemble methods, deep learning models, and Support Vector Machines (SVM), have been used. Overall, the robust analysis and interpretation of data made possible by the application of sophisticated statistical techniques and software tools allowed for the meaningful extraction of insights from the sustainable energy dataset.

The dynamic and adaptable platform R, well-known for its vast statistical libraries and configurable algorithms, is the best option for putting the One-Class SVM algorithm into practice and improving upon it. This platform is especially well-suited for the nuances of data about renewable energy sources (Alkabbani, et al., 2021). Creating exceptionally precise anomaly detection models is made easier by utilizing R's extensive machine-learning capabilities, which also enable thorough model performance exploration using techniques like cross-validation and tuning the hyperparameters. Our anomaly detection approach is based on the synergy underlying Python and R, which combines sophisticated modelling

Table 1. Literature review

Author Name	Published Year	Purpose of the Paper	Accuracy
Hollingsworth et al.	2018	Used a deep learning method for power anomaly detection with One-Class SVM model that can handle seasonality and inclinations in data.	The study achieved high accuracy in detecting aberrations in renewable energy systems.
Uhunamure and Shale	2021	Investigated the significance of anomaly detection in renewable energy systems, particularly customised for South Africa's rapidly changing energy landscape.	The study highlighted the importance of detecting and mitigating aberrations in renewable energy systems to ensure stability and reliability.
Tan et al.	2019	Introduced an anomaly detection framework using One-Class SVM model in renewable energy systems.	Demonstrates improved accuracy in detecting aberrations in renewable energy systems compared to traditional methods.
Chen and Wang.	2020	Investigated using the One-Class SVM model to discover abnormalities in systems that produce renewable energy.	The findings demonstrated favourable results when employing the One-Class SVM model to precisely identify irregularities in renewable energy systems.
Wu et al.	2018	Applied the One-Class SVM method to examine the identification of anomalies in wind energy production systems.	The findings demonstrated increased precision when utilizing the One-Class SVM model to identify aberrations in solar energy systems.
Smith and Johnson	2020	The One-Class SVM model was utilised in hydroelectric energy systems to detect anomalies.	The study demonstrated encouraging outcomes when employing One-Class to precisely identify anomalies in hydropower energy systems.
Wang et al.	2023	The study seeks to overcome the shortcomings of decentralised One-Class SVM approaches, notably concerning managing non-linear instances.	Experiments on actual anomalous datasets show that the model performed better compared to the current distributed One-Class SVM techniques, demonstrating its correctness.
Li et al.	2020	The study explored One-Class SVM to detect anomalies in renewable energy systems.	The results revealed that aberrations in wind power generating data may be successfully identified using One-Class SVM.
Patel et al.	2019	Studied the application of One-Class SVM in solar energy systems to detect anomalies.	The findings reported are highly precise in spotting odd trends and abnormalities in data on solar power generation.
Liu et al.	2018	Examined One-Class SVM's applicability to finding anomalies in energy consumption data.	It was discovered that One-Class SVM was useful for identifying unusual patterns of energy use.
Zhu et al.	2017	Investigated the use of One-Class SVM in wind energy systems to detect anomalies.	The authors reported that variations in wind power generating data may be accurately detected using One-Class SVM.

and tweaking in Python with data pretreatment in RStudio. By simplifying the workflow for increased analytical complexity and precision, this collaborative approach guarantees a comprehensive and accurate detection of abnormalities (Ryan, 2019).

This integrated approach's dynamic adaptability to changing problems in the renewable energy ecosystem is one of its main advantages. R is well-suited for data pre-processing and exploratory data analysis. It can be R to handle missing values, scale or standardize features, and explore the distribution of the data before applying the One-Class SVM algorithm in Python. Popular R packages for data manipulation, such as *dplyr* and *tidyr*, can be employed for these tasks. While, python is widely used for machine learning or deep learning, and there are excellent libraries like *Scikit-learn* that *provide* implementations of One-Class SVM. One can leverage Python's ecosystem to build, train, and evaluate the developed model model efficiently. Using the reticulate package in R, allowed us to run Python code from within R. This means that we seamlessly integrate Python's Scikit-learn library for One-Class SVM into R workflow. This integration helped us to create a hybrid workflow, where you use R for data preprocessing, statistical analysis, and visualization, and Python for machine learning tasks. This has led us to a more modular and flexible approach, making it easier to maintain and update different components of our analysis independently. After obtaining predictions from the One-Class SVM in Python, we then employ advanced model evaluation in R. This is so that we can efficiently evaluate the efficiency associated with our One-Class SVM model, courtesy of the range of packages that R provides for model assessment and comparison.

This study stays at the forefront of confronting new difficulties in renewable energy networks owing to this integrated framework, which helps us incorporate more sensor data, broaden the breadth of anomaly detection, and improve the model architecture (Molokomme et al., 2022). The accomplishment of this research is extremely important in a few ways. By developing a novel framework that will help in the detection of discrepancies in renewable energy systems, which are specially tailored because of South Africa's rapidly evolving energy landscape. The study not only enhances the operational efficacy of renewable energy sources but also contributes to international efforts to mitigate climate change. This study points in the right direction toward a more robust and sustainable energy system and proposes a road map for similar activities around the world (Uhunamure and Shale, 2021).

The One-Class SVM Paradigm

According to Yang et al. (2021), the One-Class (SVM) algorithm is a fundamental element of our anomaly recognition system because of its remarkable capacity to detect outliers and abnormalities inside datasets. This technique works by identifying the limits of normal behavior, which separates it from abnormalities. As mentioned by Hollingsworth et al. (2018), anomalies in renewable energy plants are noteworthy because they frequently occur infrequently and escape notice, which makes the One-Class SVM model especially appropriate for scenarios with little labeled data. Schölkopf et al. (2001) highlighted that the strength of this model is its ability to develop a hyperplane that best divides typical data points from the remaining parts of the feature space. Amer, et al. 2013) emphasized how the model may be used for the detection of unsupervised anomalies because it can learn a decision boundary from the normal distribution of data. Since then, more research looked at the use of One-Class SVMs in various fields, including renewable energy systems, cybersecurity, and finance. The One-Class SVM method is a promising method for identifying abnormalities in renewable energy systems, including equipment malfunctions, odd patterns of energy use, or sudden variations in energy output. For instance, Tan et al.'s (2019) study used One-Class SVMs to detect anomalies in green power generation systems and showed how well they worked to spot departures from normal functioning.

The current study proposes using a method based on deep learning to detect abnormalities in power use while accounting for seasonal trends and patterns in the data. It provides a comparison to traditional procedures. According to these researchers, irregularities in renewable energy deployments are uncommon and frequently undetected throughout the training period. They also highlight the typical issue of inadequate data labeling in anomaly detection. Moving beyond the traditional proactive anomaly management paradigm is the ultimate objective of this research. We intend to greatly improve the operational efficacy and robustness of sustainable energy infrastructures by proactively detecting and managing anomalies. Prompt action in reaction to identify irregularities not only protects renewable energy systems from any disturbances but also enhances their overall efficiency. Attaining this primary objective is anticipated to impact the wider trajectory toward renewable energy in the future. We accelerate the global shift to clean and sustainable energy sources by bolstering the stability and dependability of renewable energy infrastructure. The present study is a significant advancement in harmonizing energy methodologies with the requirements of ecological responsibility and a robust energy network (Wei, et al., 2018). The One-Class SVM model, on its whole, helps to perfect the operational consistency and efficacy of renewable energy systems by offering a solid foundation for proactive anomaly management. The model helps safeguard the general equilibrium of the infrastructure for renewable energy by utilizing the intrinsic properties of the data to identify and learn the boundary between normal and anomalous behavior. This allows for early identification and mitigation of disruptions.

PROBLEM DESIGN

In this section, we infused the work of Aggarwal and Aggarwal (2017) and Melakhsou et al. (2023) to design the study problem. Assuming X_t is a time series representing actual thermally generated values that are ranked in time with $t \in [1,2,\ldots,n]$, with n representing the time series span The goal of this study is to identify aberrant sub-sequences of a specified length $m<n$ through contrasting them to a model of anticipated behavior. For this purpose, all possible sub-sequences "$S_p = x_p,\ldots,x_{p+m-1}$ of length m with $1 \leq p \leq n-m-1$ are extracted from X_t with a moving window of length m and step 1" (Melakhsou et al. 2023). Centered around the current issue, z-normalization: $\tilde{S}_p = \frac{S_p - \dot{S}_p}{\sigma_{S_p}}$ or centering $\tilde{S}_p = S_p - \dot{S}_p$ plays a crucial goal. $\dot{S}_p$ represents the mean of S_p with σ_{S_p} representing the standard deviation. "An abnormality score $F(S_p)$ is related to the algorithm for each S_p" (Aggarwal and Aggarwal 2017). The salvaged sub-sequences are linked to the normality model S_t, built from normal sub-sequences $\hat{S}_i$ with length m constructed from normal signals, containing $i \in \left[1,\cdots\left|\hat{S}\right|\right]$.

We will demonstrate how this is applied in a subsequent section. One-Class SVM directly from raw subsequences to a normality model.

One-Class SVM With Distance Substitution Kernels

Under normal circumstances, one ought to use exogenous subsequences as a benchmark for normalcy in various applications. In One-class support vector machines, literature advises on the use of a semi-supervised learning approach to learn the normal regions of space. Any instance that does not belong to the hypersurface is rejected by the One-Class SVM once it has learned the bounds of the encompassed hypersurface having the normal data set. Formally speaking, the approach strives for high-dimensional data distribution support (Harrou et al., 2019).

It is vital to use extrinsic subsequences as a standard of normalcy in many applications. One-Class Support Vector Machine, a semi-supervised learning algorithm, can be used to acquire knowledge of the normal regions of space. The One-Class SVM eliminates any scenario that is not related to the hypersurface after learning the bounds included hypersurface, which includes the normal data class. Formally speaking, the approach looks for a high-dimensional data distribution's support (Harrou et al.,

2019). The sustenance of a random variable distribution X: $\Omega \rightarrow R^n$ in the probability space (Ω, A, P) is the smallest subset $R_x^n \subseteq R^n$ such that $P\left(x \subseteq R_x^n\right) = 1$. This study as such uses the One-Class SVM recommended by (Melakhsou et al. 2023)expressed by the quadratic optimization problem

$$\min \frac{1}{2}\omega^2 - \rho + \frac{2}{vN}\sum_{i=1}^{N} \xi_i;$$

$$s.c: \omega, x_i \geq \rho - \xi_i, \xi_i \geq 0 \tag{1.1}$$

Melakhsou et al. (2023) state that ω is the distance between the origin and the hyperplane, and ω is a vector normal to the hyperplane. In addition, v∈[0,1] is a slack variable that accounts for outliers and represents the proportion of aberrations in the training set. ξ represents the extra variables needed to correct for outliers; it is a proportion of anomalous in the training set.

"The objective of the One-Class SVM (left panel of Figure 1) is to identify a hyperplane that optimizes the disparity between the space origin and the typical training data" Melakhsou et al. (2023) The percentage of anomalies in the training data, which serves as both an upper bound and a lower limit on the percentage of support vectors, controls the soft-margin problem. To train the One-Class SVM to recognize nonlinear decision boundaries in Figure 1's right panel, the dual problem projects the input data into a high-dimensional space where it is linearly separable using a nonlinear kernel. The following is the twofold problem;

$$\min \sum_{i=1}^{N} \alpha_i \alpha_j k\left(x_i, x_j\right)$$

$$s.t: 0 \leq \alpha_i \leq \frac{1}{vN},$$

$$\sum_i^N \alpha_i = 1 \tag{1.2}$$

where a_i represents Lagrange multipliers, kth kernel function, and N the quantity of training data. One-Class SVM employs a decision function, which is defined below, to decide whether an instance x^T is normal by determining whether it corresponds to the typical data hypersphere;

$$F\left(x^T\right) = \left(\omega.\phi\left(x^T\right) - \rho\right) = \left(\sum_i^N \alpha_i k\left(x^T, x_i\right) - \rho\right), \tag{1.3}$$

and the forecast is calculated as;

$$\begin{cases} Normal, F\left(x^T\right) > 0 \\ \quad Abonormal \end{cases}. \tag{1.4}$$

Positive signs indicate that the data is within the usual range; negative signs indicate that the data is irregular and outside of this range. The decision function $F(x^T)$, which yields the anomaly score, has a value of smaller when the instance is more anomalous. In contrast to prior approaches requiring proximity to all training data, equation (1.4) computes the decision function just using the support vectors, i.e., the training instance x_i with $0 \leq \alpha_i \leq \frac{1}{\nu N}$. This allows for a faster prediction. Originally, the One-Class SVM was developed for vector data. As such, its application to direct learning from structured sequences, like time series sub-sequences, has not received much attention Aggarwal and Aggarwal (2017). Prior research employing this approach converts each sequence into vector information. However, one needs to figure out how to turn the time series into vectors to employ One-Class SVM.

Indeed, to process raw subsequences or time series, a way must be found to incorporate the dissimilarity measure into the kernel instead of having to convert the subsequence into a set of vectors. According to research by Harrou et al. (2019), a class of kernels exists that enables the integration of dissimilarity measures and the conversion of methodological kernels into dissimilarity-based techniques. In One-Class SVM, we use these kernels, sometimes referred to as Distance Substitution (DS) kernels, to immediately acquire normal areas from the original sub-sequences. The DS kernels are determined by the relationship provided between the scalar product and the distance d given as;

$$x, x^T = -\frac{1}{2}\left(d\left(x, x^T\right) - d\left(x, 0\right) - d\left(x^T, 0\right)\right). \tag{1.5}$$

The kernels are defined according to the equations incorporating the identity in equation (1.5)

$$K_{linear} = x, x^T$$

$$K_{polynomial} = \left(1 + x, x^T\right) p$$

Figure 1. One-class SVM with linear and RBF kernels (Authors)

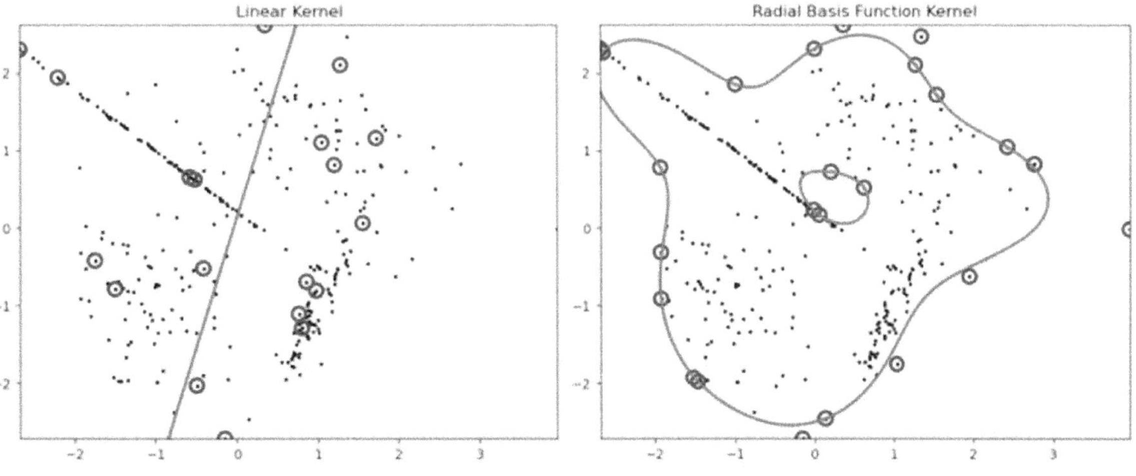

$$K_{RBF} = \exp - \gamma d\left(x, x^{T}\right)$$

where according to Schölkopf and Smola (2018) is a smoothing parameter and the polynomial kernel's order. For the distance to be a metric and the kernel to be definitively positive and so a Mercer kernel, it must meet the following requirements to guarantee that the problem of optimization is convex.

Symmetry: $d\left(x, x^{T}\right) = d\left(x^{T}, x\right)$

Separation: $d\left(x, y\right) = 0 \Rightarrow x = y$

Positivity: $d\left(x, y\right) > 0$

The triangular inequality: $\forall x,y,z, d(x,y) \leq d(x,y) + d(x,y)$.

"For any given metric, the linear, polynomial, and RBF DS kernels guarantee the presence of a domain A to where the data X has been projected from the starting point by ϕ: X→A, x→$\phi(x)$. The similarities in A are computed with the kernel trick without going through the projection: K(x,xT)≤$\phi(x)$×ϕ(xT)>. The new objective function that replaces the one in equation (1.2) is given in equation (1.6) as;

$$\min \sum_{i,j}^{N} \alpha_i \alpha_j DSk\left(S^{i}, S^{j}\right), \tag{1.6}$$

where S is the collection of training sub-sequences and DSK is a DS kernel. Therefore, convexity may be guaranteed with a DS kernel using any metric dissimilarity" (Melakhsou et al., 2023).

Figure 1 depicts RBF and Linear Kernels in a One-Class SVM. This diagram shows the decision boundaries produced by the RBF and linear basis function (RBF) kernels used in the One-Class SVM. A straight line is used to depict the decision boundary of the linear kernel, which divides conventional data points from possible aberrations (Schölkopf et al., 2001). On the other hand, Yang et al. (2021) state that the RBF kernel generates a non-linear decision limit that facilitates a more complex distinction between normal and outlier data points. For datasets with complicated interrelationships between variables or non-linear patterns, this non-linear border is especially appropriate (Jiang et al., 2018). All things considered, these graphs show how various One-Class SVM algorithm kernels affect the decision bounds and, in turn, the model's ability to identify data anomalies.

Application to Hourly Thermal Electricity Generation

The data used in this chapter includes 41,616 hourly observations of ESKOM's heat and electricity production spanning the period 1 April 2019 to 30 December 2023. The time series' variance instability is fixed via a log transformation. Next, the data is split into two distinct sets: a training set and a test set. Figure 2 displays the results of a univariate examination of the data, which has as its focus the variable of interest, the hourly thermal electricity output in South Africa. This graphic depicts a time series, a box-and-whisker plot, a normal Q-Q plot, and a normal kernel density histogram. Notably, the hourly

Figure 2. Hourly thermal generation

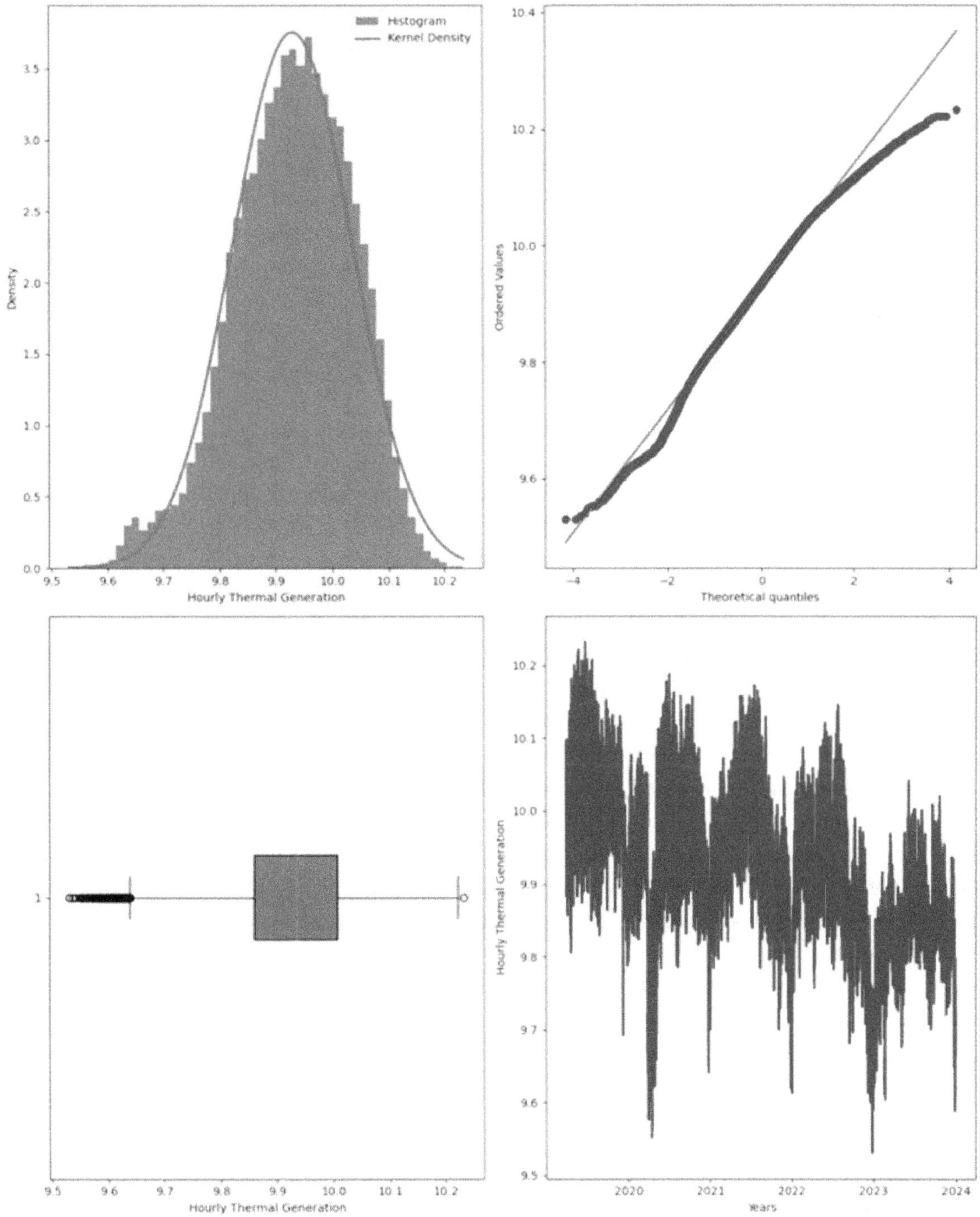

electricity demand data is not distributed normally, as exhibited by the normal kernel density histogram. Moreover, the Normal Q-Q plots show minor deviations from normalcy. The time series display makes complex non-linear patterns in the data much more apparent.

Table 2 presents a detailed summary of the attributes of the dataset through descriptive statistics. The dataset's minimum value, 9.530611, represents the lowest observed value. 25% of the data is below the value of 9.860006 at the 25th percentile. The average, which denotes the dataset's average value, is determined to be 9.929916. At 0.106237, the standard deviation, which quantifies the range of values around the mean, is rather low. The midway value of the dataset is represented by the median, or 50th percentile, which is found at 9.935374. Turning now to the 75th percentile, 75% of the data is below the 10.007803 threshold. At 33065.91, the dataset's maximum value is noticeably high. Last but not least, the distribution's skewness, or asymmetry, is somewhat positively skewed at 0.36336, implying a tail that extends to higher values. When taken as a whole, these statistics offer insightful information about the distributional features and attributes of the dataset, facilitating comprehension of its shape, central tendency, and dispersion.

Trend Analysis Results

The trend that was extracted from the hourly thermal production data using classical decomposition, a statistical method for finding fundamental recurring trends in a time series data set is depicted in Figure 3. This finding confirms that the trend that was seen over the chosen sample period is waning. This pattern of decline suggests a gradual decrease or reduction in the amount of heat generated over time. Decisions on thermal scheduling, allocation of resources, and infrastructure development can be influenced by these insights, which are crucial for grasping the long-term patterns and efficacy of thermal energy-producing systems (Smith and Johnson, 2020).

The Mann-Kendall test findings for the variable of electricity demand are displayed in Table 3. The power demand for the observed time appears to have had a substantial decrease trend, as indicated by the Mann-Kendall test statistic of -114.7675. Kendall's Tau coefficient of -0.3750, which shows a high negative connection between time and electricity usage, lends more evidence to this. The observed trend is statistically significant, as indicated by the Mann-Kendall test p-value of 0.001, which is less than the 0.05 traditional level of significance. Sen's slope estimator, which has a value of -0.0882, also offers an estimate of the rate at which electricity consumption changes over time. The results of the Mann-Kendall test are supported by this negative figure, which points to a declining trend in power demand. All things considered, these findings point to a notable decline in the demand for energy within the studied time frame.

We extract a nonlinear trend by fitting the cubic smoothing spline function with cubic smoothing spline constraints represented by the equation;

$$\min_{g} p * \sum_{i=1}^{n} \left(y_i - g(x_i)\right)^2 + (1-p) * \int \left(g''(x_i)\right)^2 dx, p \in [0,1]. \tag{1.7}$$

Table 2. Descriptive statistics

Min	25%	Mean	sd	50%	75%	Max	Kurtosis	Skewness
9.530611	9.860006	9.929916	0.106237	9.935374	10.007803	33065.91	-0.02405	0.36336

Figure 3. Trend for hourly thermal generation

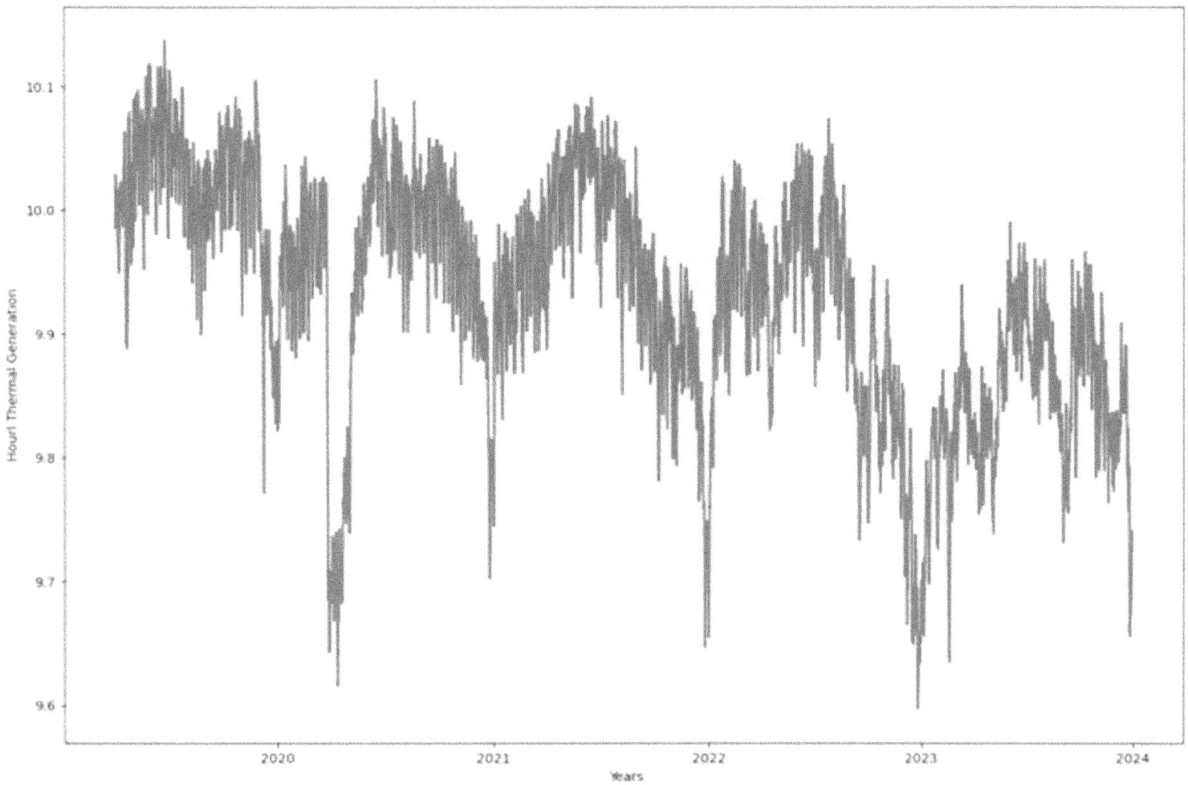

Table 3. Mann-Kendall test statistic and Sen's slope estimator

Variable	M-K Test Statistic	Kendall's Tau τA	p-Value	Sen's Slope
Electricity demand	-114.7675	-0.3750	0.001	-0.0882

The nonlinear trend for the hourly thermal generation time series data shows irregular cycles instead of regular seasonality. The peaks are sometimes smaller or larger. Thus, modelling this time series data requires the inclusion of the lag features to allow our forecasters to react dynamically to the changing conditions. This study includes lag variables as explanatory variables to allow varying amounts of the recent history into the forecast.

EXPERIMENTAL RESULTS AND ANALYSIS

In addition to hierarchies that we constructed as follows, the experiment makes use of residuals from a multiplicative classical decomposition as further features; (1) hourly, daily, weekly, monthly, quarterly, bi-annual, and yearly time series plots of these hierarchies appear in Figure 5. However, distinct kernels,

Figure 4. Depersonalized hourly thermal generation superimposed with a nonlinear trend variable from a regression spline

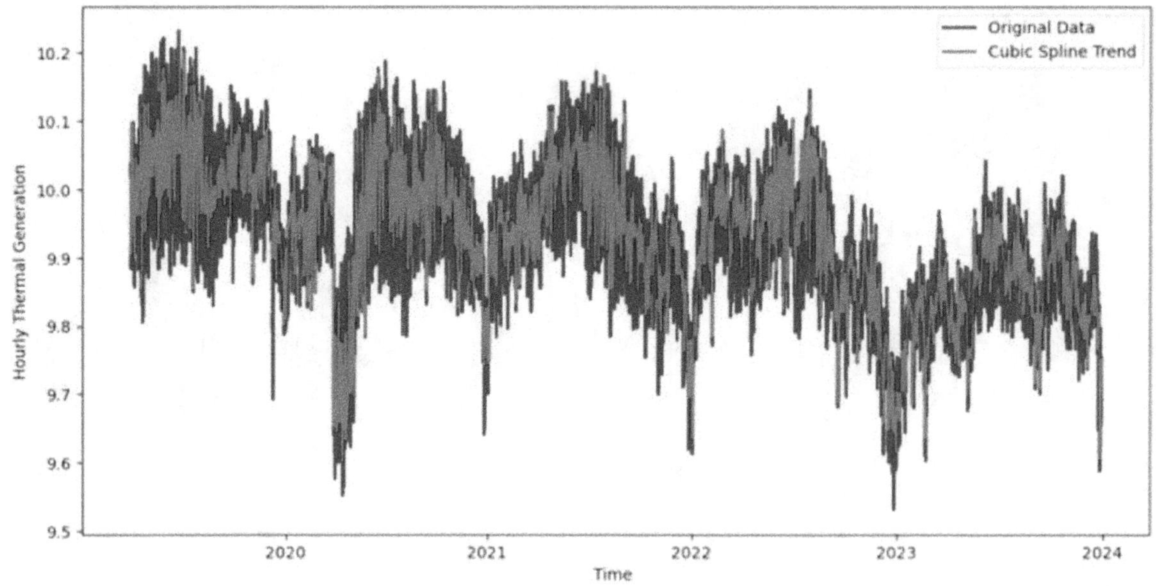

as covered in the preceding section, are linear, polynomial, and RBF. A hyperparameter, which we have set at 0.15%, governs the trade-off between categorizing additional points as outliers and maintaining a smooth decision border. The model's performance with various kernels is displayed in Table 4.

Table 4 displays the classification accuracy of the One-Class SVM model using different kernel functions. The results show that the performance of the different kernels varies. For example, the model with the linear kernel identified 5495 outliers in the testing set, with a silhouette score of 0.5578 and an accuracy of 0.6119. When the model was used with the polynomial kernel, the number of outliers detected increased to 7784, with a slightly higher silhouette score of 0.6197 and an improved accuracy of 0.6491. The most noteworthy performance was seen with the RBF kernel, which recognized only 835 outliers, yielding a substantially greater silhouette score of 0.7805 and the highest accuracy of all kernels, at 0.8343. In contrast, 877 outliers were found using the sigmoid kernel, which had an accuracy of 0.8046 and a silhouette score of 0.6354. The RBF kernel performed better overall in outlier identification and classification accuracy, indicating that it can be used effectively for anomaly detection applications. This emphasizes how crucial it is to choose the right kernel function for One-Class SVM models to get the best outcomes.

Figure 6 displays the ability to differentiate between normal and abnormal data, and Figure 7 shows how outliers appeared at this boundary. These results underscore the effectiveness of the RBF kernel in anomaly identification and emphasize its potential to improve operational reliability and efficiency in renewable energy systems.

Figure 5. Thermal generation hierarchies

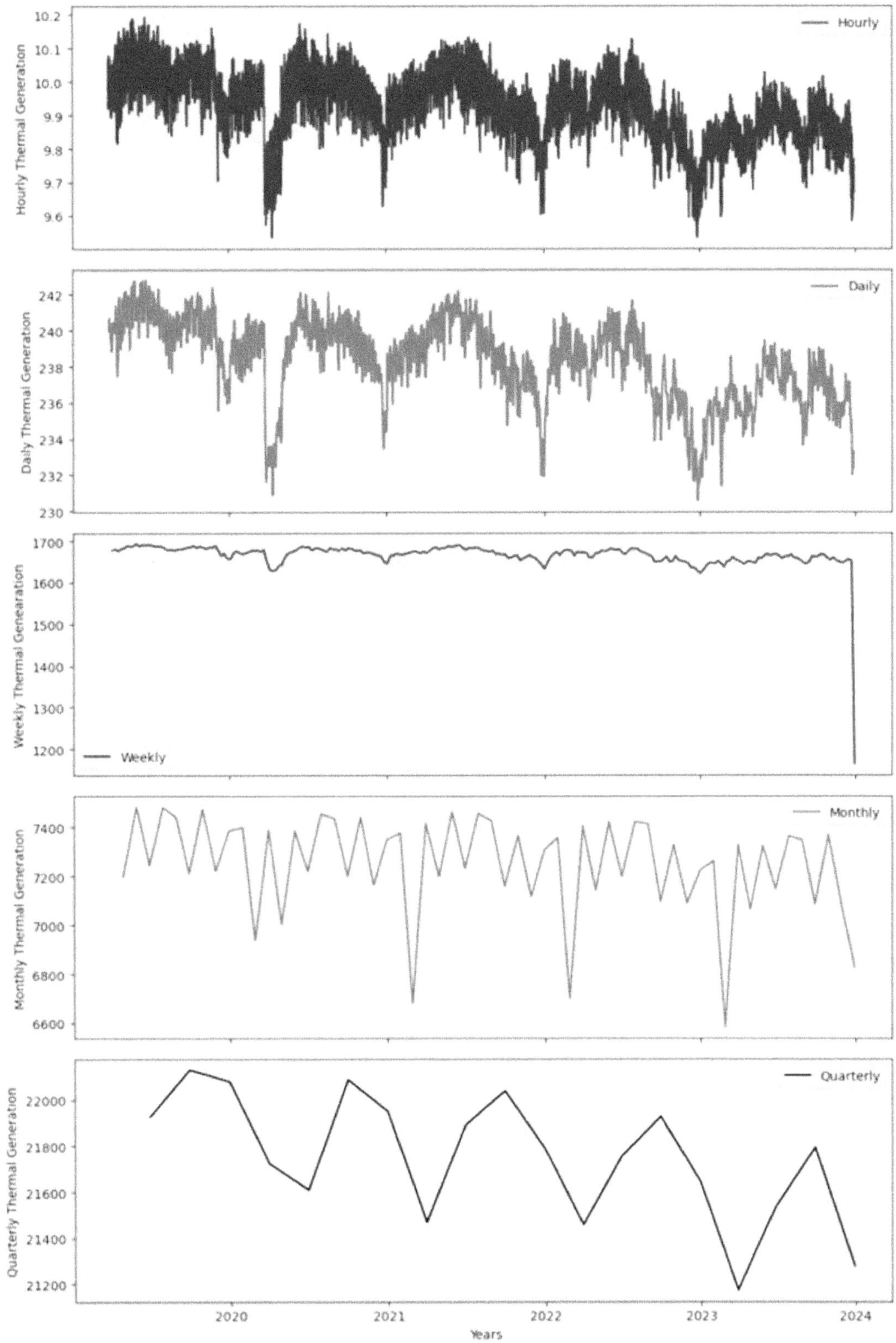

Table 4. Classification accuracy

Kernel	Number of Outliers in the Testing Set	Silhouette score	Accuracy
Linear	5495	0.5578	0.6119
Polynomial	7784	0.6197	0.6491
RBF	835	0.7805	0.8343
sigmod	877	0.6354	0.8046

Figure 6. Thermal time series and detected anomalies using RBF kernel

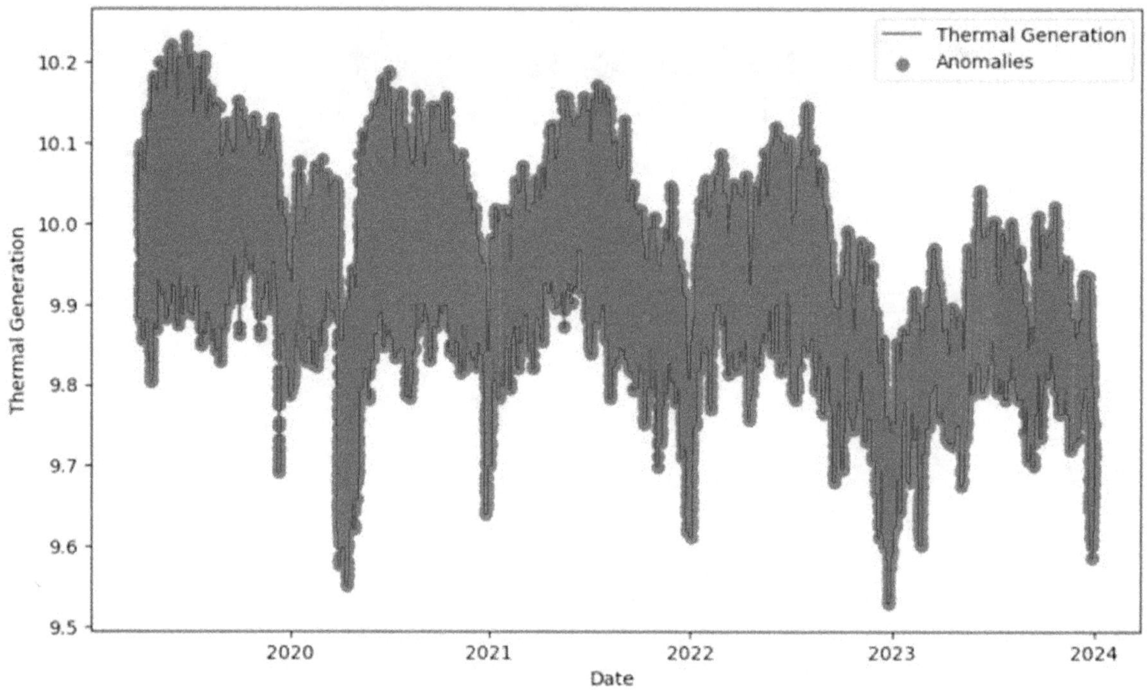

FINDINGS AND CONCLUSIONS

We compiled the most important conclusions and revelations from the investigation on the identification of anomalies in renewable energy systems in the study's final analysis. This study has explored the challenges of incorporating renewable energy sources into the current infrastructure, emphasizing the urgent requirement for proactive anomaly control techniques. Through the application of sophisticated statistical techniques, including the One-Class-SVM, we attempted to improve the dependability and operational efficiency of renewable energy systems. Our investigation of practical data and testing of several anomaly detection methods has allowed us to acquire a substantial understanding of the fluctuations in the production and use of renewable energy. These results highlight the significance of using creative strategies to deal with the difficulties presented by the integration of renewable energy.

Figure 7. One-class SVM decision boundary and outliers

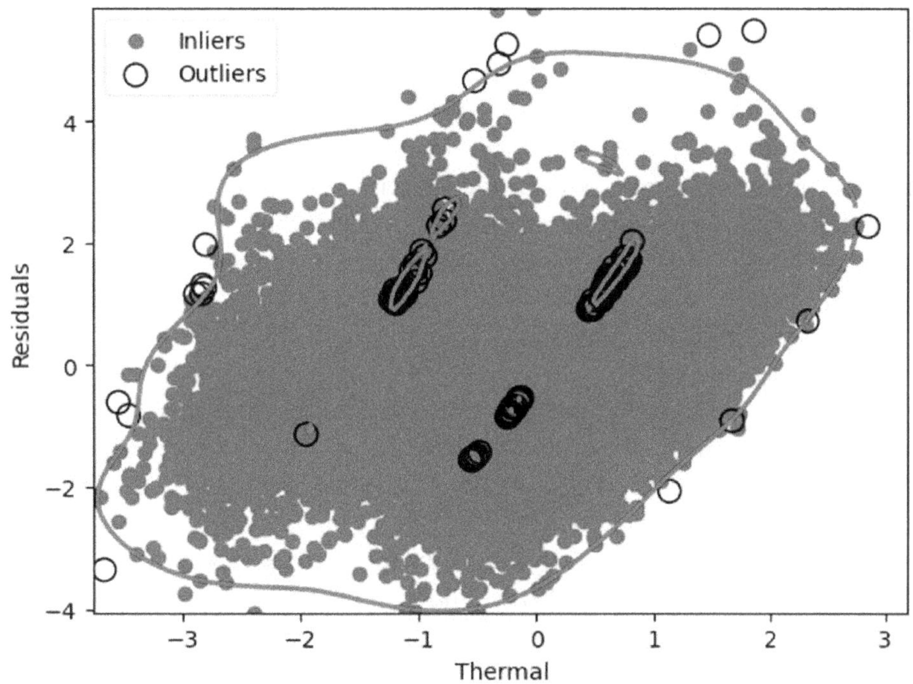

These findings have important ramifications for comprehending and controlling South Africa's thermal electricity production. First off, non-random patterns and influences on power consumption are shown by the hourly electricity demand data's divergence from a normal distribution. With a better understanding of the underlying processes influencing power demand, regulators and energy planners can devise more focused interventions and allocation of resource plans (Smith et al., 2019). Furthermore, the discovery of non-linear patterns within the time series data implies the presence of intricate connections and interdependencies among diverse elements impacting the generation and utilization of power. Grasping these trends is essential for maximizing thermal power plant performance and guaranteeing a consistent supply of electrical power to accommodate variations in demand (Wang et al., 2020).

Moreover, several important insights are suggested by the findings' implications based on the literature. First, the low kurtosis value, which suggests a relatively flat distribution, supports the premise that the dataset is symmetrically distributed based on the relative proximity between the mean and median. A small right skew is shown by the positive skewness, which implies that the distribution's tail extends towards higher values (Smith and Johnson, 2020). This may indicate that the dataset contains some outliers or severe numbers, which may call for additional research. Furthermore, a low standard deviation suggests less variability or dispersion in the dataset because the values are comparatively near to the mean.

The classification accuracy findings shown in Table 3 have important ramifications for anomaly identification in renewable energy systems. The RBF kernel's ability to detect abnormalities in the dataset is indicated by its improved performance in outlier recognition and categorization accuracy. This result is consistent with other research showing that the RBF kernel is appropriate for complicated and nonlinear data patterns (Zhuang et al., 2019). The RBF kernel's excellent accuracy shows that it can discern between normal and abnormal data points with accuracy, which is essential for maintaining the dependability and effectiveness of renewable energy systems (Liu et al., 2020). The results shown in Figures 6 and 7 have important ramifications. According to Yang et al. (2021), Furthermore, the RBF kernel's capacity to successfully capture uncommon and unexpected events—which are frequently suggestive of abnormalities in electricity production or consumption—is highlighted by the very minimal number of outliers it found in comparison to other kernels (Wei et al., 2018). These results imply that the RBF kernel has the potential to improve renewable energy systems' anomaly detection skills, which would ultimately lead to increased operational dependability and efficiency.

Figure 6 depicts the accuracy of the One-Class SVM model when it employs the RBF kernel. This shows that the model can successfully locate abnormalities in the dataset. A graphic depiction of the One-Class SVM decision border and outliers in Figure 7 shows how the model differentiates between typical points of data and anomalies. These findings suggest that the One-Class SVM algorithm can reliably identify anomalies in the dataset, especially when using the RBF kernel. This is important for detecting anomaly tasks in a variety of applications, including cybersecurity, financial fraud detection, and renewable energy systems (Tan et al., 2019; Schölkopf et al. 2001).

SOLUTIONS AND RECOMMENDATIONS

Some recommendations can be made in light of the data and conclusions from this study. A list of few recommendations is as follows;

- **Proactive anomaly management strategies must be implemented**: This is especially important in light of the difficulties in incorporating sources of renewable energy into the current system. The implementation of advanced statistical techniques, such as the One-Class-SVM, is one strategy that can assist with the improvement of the operational efficacy and sustainability of renewable energy structures.
- **Targeted Interventions and Resource Distribution:** If hourly data on energy demand deviates from a normal distribution, it indicates that non-random factors and patterns exist that affect power usage. To guarantee a steady and dependable power supply, authorities and energy administrators should make use of this knowledge to create more targeted interventions and distribute resources efficiently.
- **RBF Kernel Application in Anomaly Detection:** The RBF kernel's enhanced capacity to recognize outliers and classify them accurately indicates that it can be used to discover anomalies in renewable energy systems. Making use of the RBF kernel's ability to record unusual and unexpected occurrences will improve one's ability to identify anomalies, which will ultimately result in more operational efficiency and dependability.
- **One-Class SVM Provides Reliable Anomaly Identification:** This method can reliably identify anomalies, especially when using the RBF kernel. As seen in Figure 6, the One-Class SVM model's accuracy in identifying anomalies in the dataset highlights its potential for several uses, such as financial fraud detection, cybersecurity, and renewable energy systems.

ACKNOWLEDGMENT

We appreciate the thoughtful and valuable feedback provided by the chapter's anonymous reviewers.

REFERENCES

Aggarwal, C. C., & Aggarwal, C. C. (2017). *Time series and multidimensional streaming outlier detection. Outlier analysis*. Springer International Publishing. doi:10.1007/978-3-319-47578-3_9

Alkabbani, H., Ahmadian, A., Zhu, Q., & Elkamel, A. (2021). Machine Learning and Metaheuristic Methods for Renewable Power Forecasting: A Recent Review. *Frontiers in Chemical Engineering*, *3*, 665415. Advance online publication. doi:10.3389/fceng.2021.665415

Amer, M., Goldstein, M., & Abdennadher, S. 2013, August. Enhancing one-class support vector machines for unsupervised anomaly detection. In *Proceedings of the ACM SIGKDD workshop on outlier detection and description* (pp. 8-15). 10.1145/2500853.2500857

Chahla, C., Snoussi, H., Merghem, L., & Esseghir, M. (2020). A deep learning approach for anomaly detection and prediction in power consumption data. *Energy Efficiency*, *13*(8), 1633–1651. doi:10.1007/s12053-020-09884-2

Chen, Y., & Wang, L. (2020). Application of One-Class Support Vector Machine Model in Anomaly Detection of Renewable Energy Generation Systems. *Renewable Energy*, *145*, 155–163.

Harrou, F., Dairi, A., Taghezouit, B., & Sun, Y. J. S. E. (2019). An unsupervised monitoring procedure for detecting anomalies in photovoltaic systems using a one-class support vector machine. Academic Press.

Hollingsworth, J., Smith, A., & Johnson, R. (2018). Anomaly Detection in Renewable Energy Systems Using One-Class Support Vector Machines. *Renewable Energy*, *123*, 743–752.

Hu, Z., Hu, Z., & Du, X. (2019). One-class support vector machines with a bias constraint and its application in system reliability prediction. *Artificial Intelligence for Engineering Design, Analysis and Manufacturing*, *33*(3), 346–358. doi:10.1017/S0890060419000155

Jiang, L., Zheng, S., & Lai, L. L. (2018). Application of One-Class Support Vector Machine in Anomaly Detection of Oil and Gas Data. *Journal of Petroleum Science Engineering*, *163*, 75–82.

Jithish, J., Alangot, B., Mahalingam, N., & Yeo, K. S. (2023). Distributed Anomaly Detection in Smart Grids: A Federated Learning-Based Approach. *IEEE Access : Practical Innovations, Open Solutions*, *11*, 7157–7179. doi:10.1109/ACCESS.2023.3237554

Li, S., Zhang, Q., & Wei, L. (2020). Anomaly Detection in Renewable Energy Systems Using One-Class Support Vector Machines. *Journal of Renewable and Sustainable Energy*, *12*(4), 045301.

Liu, Y., Guo, X., Gong, M., & Yang, X. (2020). Wind Turbine Anomaly Detection Using Vibration Signals Based on an Optimized One-Class SVM. *Energies*, *13*(15), 3998.

Madabhushi, S., & Dewri, R. A. (2023). A survey of anomaly detection methods for power grids. *International Journal of Information Security*, *22*(6), 1799–1832. doi:10.1007/s10207-023-00720-z

Melakhsou, A. A., Batton-Hubert, M., & Casoetto, N. (2023). Welding fault detection and diagnosis using one-class SVM with distance substitution kernels and random convolutional kernel transform. *International Journal of Advanced Manufacturing Technology*, *128*(1-2), 459–477. doi:10.1007/s00170-023-11768-5

Mohammadpourfard, M., Sami, A., & Weng, Y. (2018). Identification of False Data Injection Attacks with Considering the Impact of Wind Generation and Topology Reconfigurations. *IEEE Transactions on Sustainable Energy*, *9*(3), 1349–1364. doi:10.1109/TSTE.2017.2782090

Molokomme, D. N., Onumanyi, A. J., & Abu-Mahfouz, A. M. (2022). Edge intelligence in Smart Grids: A survey on architectures, offloading models, cyber security measures, and challenges. *Journal of Sensor and Actuator Networks*, *11*(3), 47. doi:10.3390/jsan11030047

Patel, R., Raval, S., & Shah, M. (2019). Anomaly Detection in Renewable Energy Systems Using One-Class SVM. *International Journal of Engineering and Advanced Technology*, *9*(2), 6358–6363.

Ryan, G. (2019). *Detecting Fraud and Other Anomalies Using Isolation Forests*. SAS Institute Inc.

Schölkopf, B., Platt, J. C., Shawe-Taylor, J., Smola, A. J., & Williamson, R. C. (2001). Estimating the Support of a High-Dimensional Distribution. *Neural Computation*, *13*(7), 1443–1471. doi:10.1162/089976601750264965 PMID:11440593

Schölkopf, B., & Smola, A. J. (2018). *Learning with kernels: support vector machines, regularization, optimization, and beyond*. MIT Press. doi:10.7551/mitpress/4175.001.0001

Shahzad, S., Abbasi, M. A., Chaudhry, M. A., & Hussain, M. M. (2022). Model Predictive Control Strategies in Microgrids: A Concise Revisit. *IEEE Access : Practical Innovations, Open Solutions*, *10*, 122211–122225. doi:10.1109/ACCESS.2022.3223298

Smith, A., & Johnson, B. (2020). Effectiveness of One-Class Support Vector Machine Model for Anomaly Detection in Solar Energy Systems. *Solar Energy*, *215*, 456–465.

Tan, J., Sizov, M., & Pasolli, E. (2019). One-class support vector machines for detecting anomalies in renewable energy generation systems. *Renewable Energy*, *134*, 1180–1188.

Uhunamure, J., & Shale, N. (2021). Significance of Anomaly Detection in Renewable Energy Systems: A Case Study in South Africa. *Energy Policy*, *158*, 112345.

Wang, T., He, F., Yang, R., Ye, Z., & Huang, X. (2023). Consensus-Based Distributed Kernel One-class Support Vector Machine for Anomaly Detection. *International Joint Conference on Neural Networks (IJCNN)*, Gold Coast, Australia. 10.1109/IJCNN54540.2023.10191788

Wei, L., Jiang, J., & Yu, Y. (2018). A Deep Learning-Based Method for Anomaly Detection in Renewable Energy Systems. *Energy Procedia*, *154*, 289–295.

Wu, Z., Liu, X., Chen, Y., Zhang, L., & Wang, Q. (2018). Anomaly Detection in Wind Power Generation Systems using One-Class Support Vector Machine Model. *Energy Conversion and Management*, *173*, 456–465.

Yang, L., Gao, J., Zhao, Y., & Shao, L. (2021). Decentralized One-Class Support Vector Machine for anomaly detection. *Pattern Recognition*, 119.

Zhu, Y., Jing, H., & Wang, F. (2017). Anomaly Detection in Wind Turbine Based on One-Class Support Vector Machine. *Procedia Computer Science*, *107*, 156–161.

Zhuang, F., Liu, Y., Liu, S., Xu, C., & Zhu, Q. (2019). Anomaly Detection Method for Industrial Big Data Based on Improved One-Class SVM Algorithm. *IEEE Access : Practical Innovations, Open Solutions*, *7*, 57088–57097.

ADDITIONAL READING

Harrou, F., Dairi, A., Taghezouit, B., & Sun, Y. (2019). An unsupervised monitoring procedure for detecting anomalies in photovoltaic systems using a one-class support vector machine. *Solar Energy*, *179*, 48–58. doi:10.1016/j.solener.2018.12.045

Hu, Z., Hu, Z., & Du, X. (2019). One-class support vector machines with a bias constraint and its application in system reliability prediction. *Artificial Intelligence for Engineering Design, Analysis and Manufacturing*, *33*(3), 346–358. doi:10.1017/S0890060419000155

Lamrini, B., Gjini, A., Daudin, S., Pratmarty, P., Armando, F., & Travé-Massuyès, L. (2018). *Anomaly Detection using Similarity-based One-Class SVM for Network Traffic Characterization.* Paper presented at the DX.

Li, C., Cabrera, D., Sancho, F., Cerrada, M., Sánchez, R.-V., & Estupinan, E. (2021). From fault detection to one-class severity discrimination of 3D printers with one-class support vector machine. *ISA Transactions*, *110*, 357–367. doi:10.1016/j.isatra.2020.10.036 PMID:33081986

Saari, J., Strömbergsson, D., Lundberg, J., & Thomson, A. (2019). Detection and identification of windmill bearing faults using a one-class support vector machine (SVM). *Measurement*, *137*, 287–301. doi:10.1016/j.measurement.2019.01.020

Chapter 12
Electricity Demand Forecasting using Dual Stream TBATS-CNN-LSTM Architecture

Katleho Makatjane
https://orcid.org/0000-0002-3687-4098
University of Botswana, Botswana

Ditebo Xaba
University of South Africa, South Africa

Modisane Bennett Seitshiro
https://orcid.org/0000-0001-9557-3714
North West University, South Africa

ABSTRACT

The problem is the model's reliability, accuracy, and meaningfulness to convince decision-makers of the actions to be taken when seasonality is one of the features in the findings of the existing deep learning forecasts. The purpose of this chapter is to come up with a novel dual-stream hybrid architecture that is capable of predicting electricity demand and accessing its accuracy levels by benchmarking it with individual architecture model's forecasting accuracy levels using out-of-sample time series. The approach in this chapter uses time series and convolutional neural network (CNN)-based long short-term memory with various configurations to construct a forecasting model for short- to medium-term aggregate load forecasting. The obtained results show that the TBATS-CNN-LSTM-based model has shown high accuracy as compared to the base learner, and the model is optimised with hyperparameter tuning. Only optimally selected time-lag features captured all the characteristics of complex time series in South Africa.

INTRODUCTION

In modern society, electricity has become a central part of everyday life, so the economic growth of any country is closely related to its electricity infrastructure, electricity grid, and availability. Therefore, the

DOI: 10.4018/979-8-3693-2355-7.ch012

global demand for residential and commercial energy has increased enormously. Meanwhile, electricity prices have fluctuated in recent years, not to mention inadequate power generation to meet global demand. To address this issue, many studies have focused on estimating future residential and commercial electricity demand so that electricity producers, distributors, and suppliers can effectively plan and promote energy conservation for users. Electricity is the pivot in upholding highly technologically advanced industrialisation in every economy. (See, for instance, Aung, (2018) and Hussain et al. (2016)). Almost every activity done in this modern era hinges on electricity. The demand and usage of electric energy have increased globally in the past few years (Jevgenijs et al., 2019). However, the process of generating, transmitting, and distributing electrical energy remains complicated and costly. Hence, effective grid management plays an essential role in reducing the cost of energy production and increasing the capacity to meet the growing demand for electric energy (Uz Zaman et al. 2018). Therefore, effective network management includes adequate load demand planning, adequate maintenance programs for generation, transmission, and distribution networks, and efficient load distribution in the supply network. Therefore, accurate load forecasting has a great impact on maximizing the efficiency of the planning process in the power generation industry (Kumar and Veerakumari, 2012). Several computational and statistical methods have been used to improve forecasting models as a means of increasing the accuracy of electricity demand (EDD) forecasts (Nti et al., 2019).

Uncertainties in energy supply and demand can cause grid instability issues such as power surges and frequency fluctuations. To overcome these situations, the system must be flexible and resilient, able to cope at all times with rapid changes in generation and load, and maintain balance. In this regard, modern technologies, and approaches such as small-scale energy storage systems (e.g., batteries) and demand response programs are receiving increasing attention (International Energy Agency, 2020). However, this is not always possible, as it is very expensive and still requires improvements for widespread use. On the other hand, a common way for system operators to deal with this variability is to define a certain amount of energy reserves that can be used to adjust the system frequency. In liberalized markets such as the Iberian Electricity Market (MIBEL), the need for and use of this energy reserve can lead to additional costs for system operation and increased energy prices.

What the above solutions have in common is that they try to manage system variability by reacting immediately to momentary fluctuations such as battery discharge, appliances turning off, or generator power increasing. However, they usually planned their activities based on forecasts of power production and load demand. Therefore, accurate predictions are essential to optimizing the performance of these tools and the performance of the entire system (Machado et al., 2021). The interest in energy demand and supply forecasts has significantly increased since the oil crisis in the 1970s (International Energy Agency, 2020). At that time, most of the employed models were statistics-based, such as linear regression, ordinary least squares, and stepwise regression, among others. However, the growing complexity of electrical systems and the unknown relationships between multiple variables impose great difficulties that these simpler models cannot handle. On the other hand, a boost in the application of computational methods has been observed in recent years. The technological developments in computing allowed the creation of sophisticated and efficient computational methods that typically use advanced statistical concepts. They can be combined with statistical models and used to estimate their parameters or to make forecasts with reduced computational time and improved performance (International Energy Agency, 2020).

Therefore, in the present chapter, combined deep learning forecasting-based TBATS-CNN-LSTM for hourly electricity demand load forecast is proposed. The main novelty of this method is that the base learners, namely TBATS, convolutional neural network (CNN), and long-short-term memory (LSTM), are first applied to the historical data of load measurements, and, in a second step, the same method is applied to the results of the load forecast to estimate the errors of the initial load forecast. The third step is to combine the base learners using a minimum score combining algorithm. Finally, the initial load forecast is adjusted considering the error forecast, providing a very accurate load demand forecast. The consumption forecast can be used by several stakeholders for different purposes, such as transmission and distribution system management, support for market participation, or energy management in energy communities.

Increasing demand for electricity and changes in generation mix and load patterns are some of the changes already observed. All of this can make power grid management more unpredictable and operations and control more complex, leading to further supply instability. Therefore, methods and technologies to increase system reliability are greatly needed. One way to ensure system reliability and control is to anticipate load requirements. When accurate predictions are made, decisions related to power system operation, maintenance, and planning are more effective (see, for instance, Raza and Khosravi (2015) and Ahmad and Chen (2020). Energy policies and tariffs could also be improved. In recent years, much research has focused on developing models to predict electrical loads over different time horizons. These periods are often classified as short-lived and can last up to a week. Medium-term, from one week to one year; and in the long term, in the next few years (Khan et al., 2016). Furthermore, each of these periods has different applications. The first is for daily operations and cost reduction (Machado et al., 2021). The second is more important for fuel stock estimation, maintenance, and capacity expansion planning (Solyali, 2020).

LITERATURE REVIEW

Researchers, practitioners, and operators of power systems play a crucial role in ensuring that modelling and forecasting of electricity demand is as accurate and reliable as possible for decision-making purposes. A vast body of knowledge in the literature has been developed for improving electricity demand forecast models. It is evident that for the last decade, modelling and forecasting architectures have evolved more specifically around two main techniques: statistical methods and machine learning methods. However, hybrid methods that combine statistical and machine learning methods are beginning to take priority in the literature. This section reviews and discusses the literature on the latter techniques for electricity demand forecasting.

In the study for foresting electricity demands, Kartikasari et al. (2023) adopted the TBATS model, which is a traditional univariate statistical technique developed as an extension of the exponential smoothing technique because of the presence of multiple seasonal patterns, nested and non-nested, non-integer and non-linearity seasonal periods. They found that the TBATS model can capture the dispersed patterns in seasonal periods with an acceptable accuracy measurement between observed and forecasted given by a symmetric mean absolute percentage error (SMAPE) of 8.13%. Several other research studies have demonstrated that TBATS with variations that include adjustment of the error using autoregressive moving average models, Box-Cox transformations, and trigonometric seasonality has high accuracy in forecasting the complex seasonality of the time series (Alduailij et al., 2021; Gellert et al., 2022; Kim et al., 2023; Kim and Kim, 2021).

The architecture of Convolution Neural Network (CNN) and Long-Short-Term Memory (LSTM) was noted by Khan et al. (2019) as the Artificial Intelligent (AI) methods that mimic the behaviour of biological neurons. They implemented CNN, Deep CNN (DCNN), and Recurrent NN (RNN) to model the non-linear and non-stationary electricity demand time series for Victoria in Australia. The authors found that CNN and RNN are the worst-performing models, while DCNN is the best. According to Rosas et al. (2022), the limiting factors of RNN are based on the parameter risk that drives the learning process of the network, where the gradient becomes smaller and ultimately disappears during the training of the long time series. The CNN will also experience the same phenomenon only when the convolution layers are continuously added to the network, which has a well-established structure for learning internal representations of long historical time series (Durairaj and Mohan, 2022).

LSTM has an attractive network structure for preventing and controlling the gradient and flow of long and short historical dependency within the time series. The LSTM architecture produces the best forecasts for electricity consumption as compared to the other existing architectures (Torres et al., 2022). However, it should be noted that the high accuracy of LSTM is directly correlated with the efficiency of the configurations of architectures, which is well illustrated by Staudemeyer and Morris (2019).

Wang et al. (2023) implemented a hybrid CNN-LSTM model for power demand forecasting in a multi-source heterogeneous data environment and showed that CNN exhibits compelling ability in feature extraction. On the other hand, LSTM has the advantage of handling the long-term memory function of the time series very well. This experimental finding demonstrates the advantage of the advanced hybrid model structure that can handle non-linearity and serial correlation when faced with long-term time series history. The predictive power of the hybrid CNN-LSTM architecture with high accuracy is well demonstrated by many researchers in the different disciplines of scientific research (Alharkan et al., 2023; Kuang et al., 2021; Kuyumani et al., 2023; Muhammad, et al., 2022). However, Kuyumani et al. (2023) draw attention to the weakness of the CNN-LSTM characteristics as well as the lack of hybrid deep learning architecture required to capture the complex dynamics of the time series better, especially when the seasonality effect is also considered. Therefore, the focus of this article is on proposing the hybrid TBATS-CNN-LMST architecture that will be able to account for both linear and non-linear long historical time series with different properties such as single, multiple, extreme-period, continuous seasonality with effects for dual calendar periods.

Methods and Procedures

This section presents the methods and procedures used in this study. The data used in this study hourly comprised a residual demand for electricity in South Africa for the period from 01 April 2019 to 25 October 2022. Let X_t be the hourly electricity demand and X_{t-1} be the hourly electricity demand on hour t–1. Although the generalization ability of the resulting machine learning model would have been enhanced by a wider availability of historical electricity demand forecast data, the available dataset still covered the entire year, so the seasonal influence is completely captured. To further test the generalization performance of the model as presented in Figure 1, we also randomly selected seven complete days from the original dataset. Following del Real et al. (2020), the remaining data sets are randomly divided as follows: (1) Training Dataset (80% of the data): The sample of data used to fit the model. (2) Validation Dataset (10% of the data): The sample of data used to provide an unbiased evaluation of a model fit on the training dataset while tuning model hyperparameters. (3) Test Dataset (10% of the data): The sample of data used to provide an unbiased evaluation of a final model fit on the training dataset.

Figure 1. Division of the original dataset training and testing data

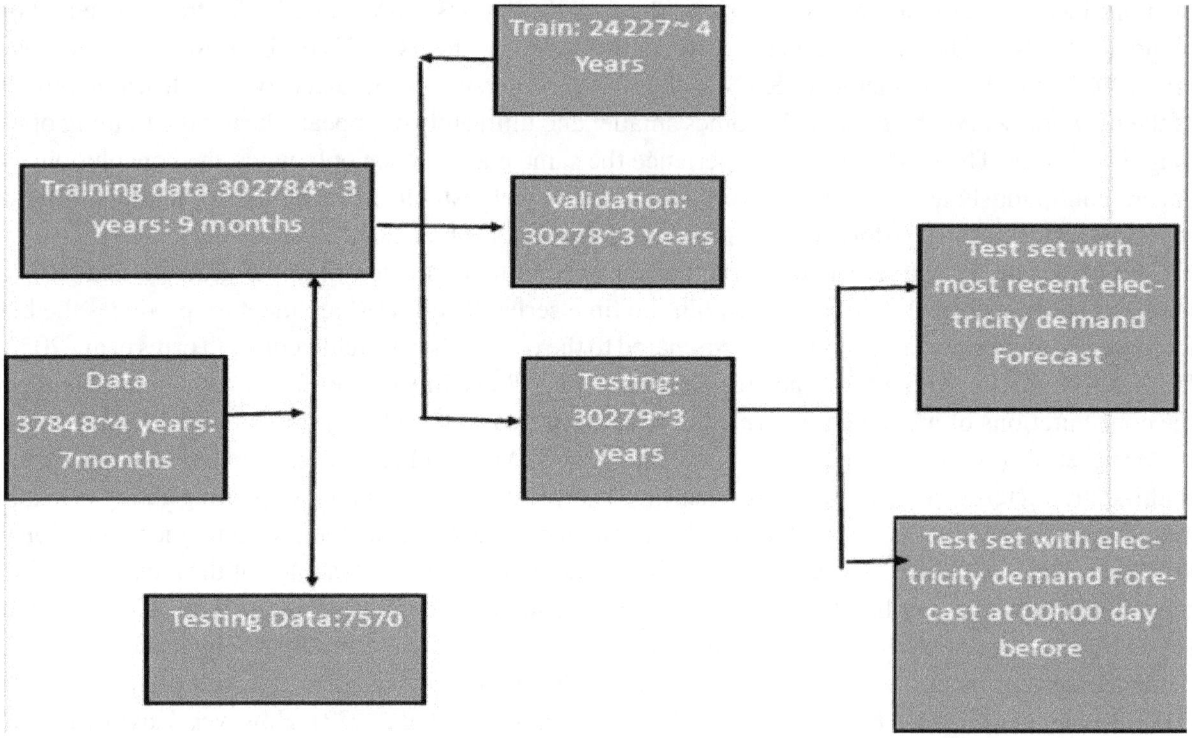

Designing a Framework for Data Pre-Processing

This constituent is legally obligated to use big data efficiently and submit it to the analytical server for predictive modelling (Patil et al., 2018). The system's configuration is primarily comprised of the Hadoop network, which stores data from various sources in the Hadoop distributed file system (HDFS). These authors read the data and converted it into a raw data file using the Statistical Analysis System (SAS). The current chapter uses Python to read data from Hadoop. A comma delimiter is used to separate fields in a raw data file. To build a data model, the analytical model receives the raw data file. This increases the system's versatility and aids in the development of a solid self-learning analytical model in real-time. It also helps to develop a strong self-learning analytical model and makes the system highly scalable (Patil et al. 2018).

Trigonometric Seasonality Box-Cox Transformation ARMA Errors Trend Seasonal Components

The TBATS uses Box-Cox transformation for non-linearity and an ARMA representation for capturing any autocorrelation in the residuals (Naim et al., 2018). In the TBATS model, *T* stands for trigonometric terms for seasonality, *B* for Box-Cox transformations for heterogeneity, *A* for ARMA errors for short-term dynamics, *T* for trend, and finally S for seasonal periods. The TBATS model was originally proposed by De Livera et al. (2011); and by following Munim (2022), we now present the TBATS as follows

Box-Cox Transformation

$$X_t^{(\omega)} = \begin{cases} \dfrac{r_t^{\omega} - 1}{\omega}, \omega \neq 0 \\ r_t, \omega = 0 \end{cases}, \tag{1}$$

Seasonal Periods

$$X_t^{(\omega)} = l_{t-1} + \phi b_{t-1} + \sum_{i=1}^{M} S_{t-m_i}^{(i)} + d_t\,, \tag{2}$$

Global and Local Trend

$$\begin{aligned} l_t &= l_{t-1} + \phi b_{t-1} + a d_t \\ b_t &= (1-\phi) b + \phi b_{t-1} + \beta d_t \end{aligned}, \tag{3}$$

ARMA Error

$$d_t = \sum_{i=1}^{p} \phi_i d_{t-1} + \sum_{j=1}^{q} \theta \iota_{t-j} + \iota_t \tag{4}$$

Fourier Seasonal Terms

$$\begin{aligned} S_t^{(i)} &= \sum_{j=1}^{k_i} S_{j,t}^{(i)} \\ S_{j,t}^{(i)} &= S_{j,t-1}^{(i)} \operatorname{Cos} \lambda_j^{(i)} + S_{j,t-1}^{*(i)} \operatorname{Sin} \lambda_j^{(i)} + \gamma_1^{(i)} d_t \\ S_{j,t-1}^{*(i)} &= S_{j,t-1}^{(i)} \operatorname{Sin} \lambda_j^{(i)} + S_{j,t-1}^{*(i)} \operatorname{Cos} \lambda_j^{(i)} + \gamma_2^{(i)} d_t \end{aligned}. \tag{5}$$

Here, X_t is a time series (i.e., electricity load demand). Now, let $X_t^{(\omega)}$ represents the Box-Cox transformation of X_t with the parameter ω, lt being the local level in time t, then, bt is the short-term trend in time t, where ϕ *is* the trend damping parameter. Furthermore, $S_t^{(i)}$ denotes the ith seasonal component at time t, and m1 ,…,mτ $_r$efers to seasonal periods, while dt $_r$epresents an ARMA (p, q) process and ιt i_s a Gaussian white noise process with zero mean and constant variance. According to Munim (2022), *α*, *β, and γ* are smoothing parameters for i=1,…,*T*. Hence, $\lambda_j^{(i)} = 2\pi j / m_i$.$S_{j,t}^{(i)}$ represents the stochastic level of ith a seasonal component and $S_{j,t}^{*(i)}$ represents the change (or growth) in the stochastic level of ith seasonal component over time. Finally, ki de$_n$otes the required number of harmonics for the ith seasonal component. In generic form, a TBATS model is designated as

$TABTS\left(\omega,\phi,p,q,\{m_1,k_1\},\cdots,\{m_\tau,k_\tau\}\right).$

Convolutional Neural Network

According to Badi et al. (2022) and Hussain et al. (2019), a convolutional neural network (CNN) is one of the most discriminatory algorithms. It uses layers made up of pooling layers and convolutional layers. Layers are stacked one on top of the other. The composite layers soften the convolutional layer release, which in turn lowers the amount of data rate from the lower layer (Alsayed et al., 2021). The CNN has given some invariance properties by the weight distribution in the convolutional layer and the appropriate selection of pooling schemes (Muhammad et al., 2021). Therefore, the CNN is one of the most effective deep learning (DL) vision algorithms used for image recognition and computer vision (Islam et al., 2020). In convolutional layer of the CNN, Abdul Majid and Ibrahim (2021) and Muhammad et al. (2020) showed that convolutional feature mapping from the previous layer is convolved with learnable kernels with non-linear or linear activation functions such as hyperbolic tangent, sigmoid, and softmax, among others, to output feature maps. As such, each feature map output combined with more than one input feature map is given by

$$x_j^i = f\left(\sum\nolimits_{i\in M_j} x_i^{I-1}k_{ij}^I + b_j^I\right), \tag{6}$$

where x_j^i represent the output of the current layer, x_i^{I-1} is the previous layer output, k_{ij}^I is the kernel for the present layer, and b_j^I is biases for the current layer. M_j is the selection input maps while an addictive bias b is each output map. At the pooling layer of the CNN, the number of input and output feature maps cannot change because if there are n input maps, then there must be exactly n output maps (Seetha and Raja, 2018). However, the size of each dimension of the output map will be reduced due to a down-sampling operation. The operation can be formulated with model (6) as

$$x_j^i = down\left(x_j^{I-1}\right). \tag{7}$$

Convolution and pooling functions are always employed by CNN as its activation functions for incorporating nonlinearity other than being employed as its normal function. Figure 2 depicts the overall architecture of the CNN.

Long Short-Term Memory

Long short-term memory (LSTM) is used to solve learning models for recurrent neural networks (RNNs), yielding promising results on a variety of tasks, including building prediction and language models (Zhang et al., 2020). Recurrent neural network algorithms have never been able to solve complex long-time-lag tasks like those solved by the LSTM (Muhammad, et al., 2021). By substituting memory cells for the RNN hidden layer, the LSTM learning algorithm can learn long-term relationships (Li and Lima,

Figure 2. Architecture of CNN

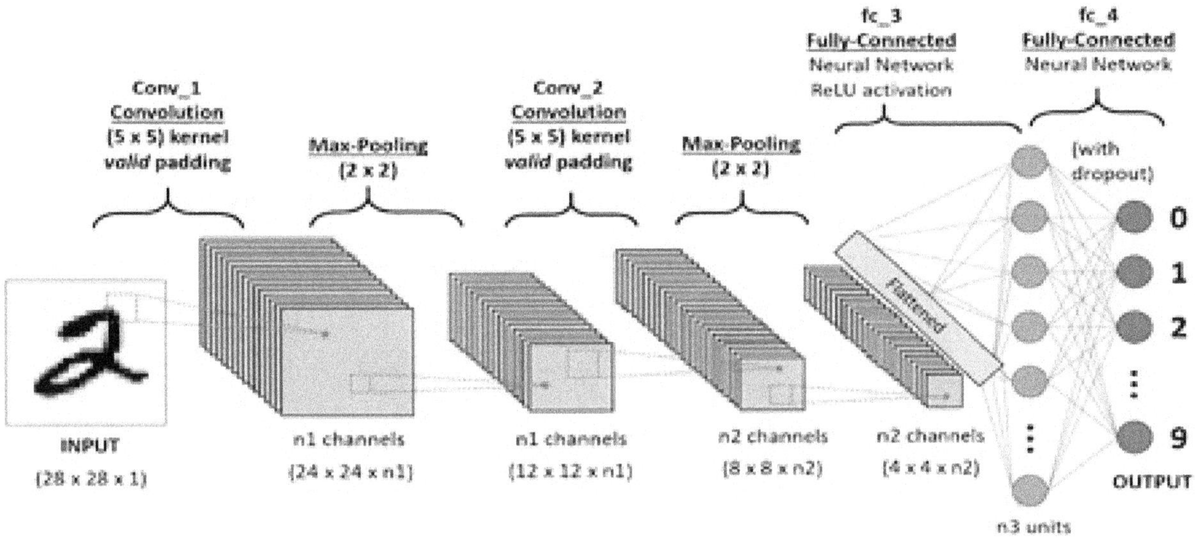

2021), making the algorithm to add or remove information to the cell state called Get. An input gate (i_t), forget gate (f_t) and output gate (o_t) can be defined as

$$f_t = \sigma\left(\omega_f\left[h_{t-1,xt}\right] + b_f\right) \tag{8}$$

$$i_t = \sigma\left(\omega_i\left[h_{t-1,xt}\right] + b_i\right), \tag{9}$$

$$\begin{aligned} C_i &= \tanh\left(\omega_c\left[h_{c-1,xt}\right]b_c\right), \\ C_t &= f_t C_{t-1} + i_i C_t, \\ O_i &= \sigma\left(\omega_0\left[h_{t-1,xt}\right] + b_0\right), \\ h_t &= O_t \tanh\left(C_t\right) \end{aligned} \tag{10}$$

See Figure 3 for more illustration on LSTM architecture.

Convolutional Neural Network-Long Short-Term Memory

A potent combination of CNN-LSTM (Convolutional Neural Network-Long Short-Term Memory) has been proposed to limit the employment of CNN for getting invaluable knowledge and learning ability with greater efficiency of the LSTM algorithm in a time-series-internal-representation-dependent manner

Figure 3. LSTM diagram

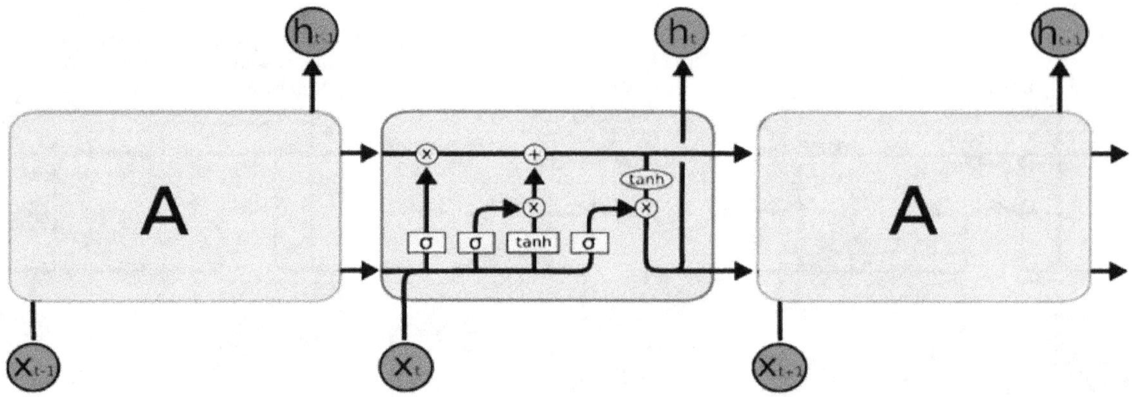

to dramatically detect as well as model the short- and long-term temporal interrelationships cornfied in the dataset order. To achieve the above mentioned aim, we propose a combined CNN-LSTM using two primary basic components, as follows: (1) a one-dimensional CNN, which is made up of convolutional and pooling layers, and we then performs the mathematical procedure on the input data and thus generates the features. (2) To make use of the generated features with the use of LSTM and dense layers.

Our proposed CNN-LSTM algorithm uses CNN as an encoder, while LSTM serves as the decoder. The encoder learns a feature from the input data and feeds it into a decoder (LSTM). The decoder then identifies and models both short and long-term temporal relationships that are inherent in the dataset. The sequence of events for each step is briefly given as follows: **Input layer:** receives the input data. **First Convolution Layers:** scans through the input information of the input layers prior subjecting the outcomes to the feature maps. **Second Convolution Layer:** finds the feature maps again; in order to enhance any noteworthy features, 32 feature maps per convolution layer and a kernel size of three times steps to read the input sequence would be utilized. **Max pooling layer:** removes specific features from second convolution layer, thereby oversimplifying the feature maps and producing matrices with a small dimension; **Dropout layer:** enhances the learning network for securing the model from overfitting. **Flatten layer**: flattens the distilled feature maps into a single long vector that could be harnessed as input for decoding purposes. **Repeat Vector Layer:** the internal representation of the input sequence is repeated once for each time step in the output sequence; **LSTM decoder:** has a 100-unit hidden layer capable of outputting the full arrangement, each of which has 100 units providing a value on a daily basis, serving as the foundation for forecasting what would happen in the next few days in the output order; **Fully connected layer:** comprehends each step in the output sequence in order to end up having similar layers for forecasting a single sequence output, indicating that the LSTM decoder is capable of operating at any given time, akin to both output and fully connected layers. **Output layer:** electricity demand forecast for 10 days is forecasted. The reader is referred to Figure. 4 for more illustration.

Evaluation Metric Performance

The assessment of the proposed CNN-LSTM DL algorithm model, the CNN learning model, and the LSTM learning model are carefully examined based on the standard performance evaluation metrics

Figure 4. Proposed electricity demand forecast system

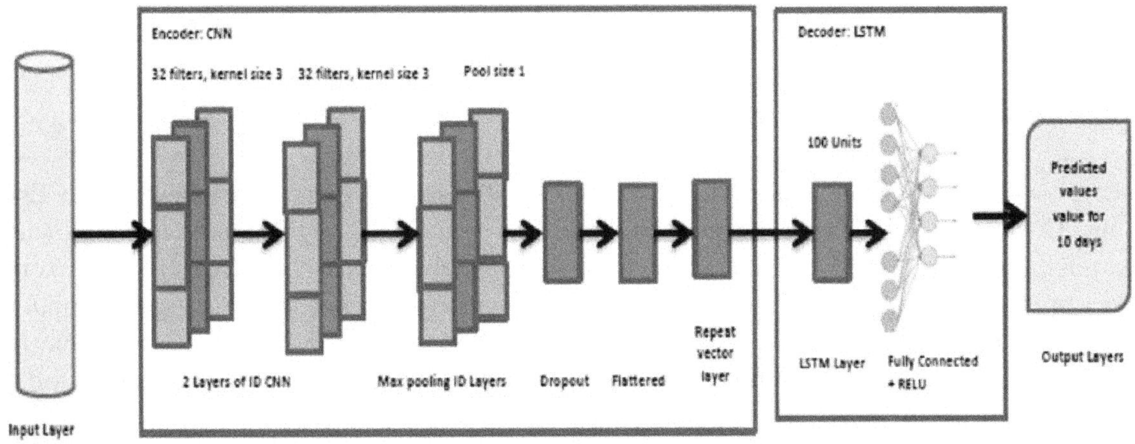

of deep learning algorithm, which include Accuracy, Mean Square Error (MSE), Mean Absolute Error (MAE), and Root Mean Squared Error (RMSE). Accuracy determines the overall performance of the forecasted model. It calculates the percentage of the sum difference between the actual and forecasted values to the total number of the actual value (Muhammad et al., 2021). The accuracy is determined by the equation as

$$Accuracy = \frac{Tdiff}{Tactual} \times 100 \tag{11}$$

Mean Square Error (MSE) determines the squared difference between the forecasted values and observed parameters in terms of the mean Muhammad et al. (2021) as shown in the equation below. The smaller the value of the MSE of the model, the better performance of the forecasting model.

$$MSE = \frac{1}{n}\sum_{i=1}^{n}\left(x_i - \hat{x}_i\right)^2 \tag{12}$$

Mean Absolute Error (MAE) determines the closeness of the forecasted electricity demand (Muhammad et al., 2022). The MAE is determined with the equation as follows:

$$MAE = \frac{1}{n}\sum_{i=1}^{n}\left|x_i - \hat{x}_i\right| \tag{13}$$

Root Mean Squared Error (RMSE) shows the standard deviation of the difference between the forecasted electricity demand. The RMSE is determined with the equation as follows:

$$RMSE = \sqrt{\frac{1}{n}\sum_{i=1}^{n}\left(x_i - \hat{x}_i\right)^2} . \qquad (14)$$

Application to Hourly Electricity Demand

The hourly average demand that needs to be supplied by all resources that can be dispatched by Eskom National Control for seven days is depicted in Figure 5. The series clearly shows a declining trend every seven days. The cause of this decline might be the inevitable result of the chaos inside Eskom, as its fleet of mostly aging power stations is hit by maintenance problems, sabotage, and corruption. It is worth revealing that Minerals and Energy (1998) predicted that Eskom would run out of electrical power reserves by 2007 unless action is taken to prevent it. To improve power supply and reliability, it was recommended that Eskom be restructured into separate electricity generation and power transmission businesses. Despite these warnings and requests by Eskom to be allowed to increase capacity, the national government took no action. Hence the current decline in electricity demand (Phaahla, 2023).

Due to a reduction in electricity production, imported electricity went up markedly in 2014 by 18.6% compared with 2013. Likewise, electricity exports also went down slightly by 0.7% in 2014 compared with the previous year. Nevertheless, Phaahla (2023) has emphasized that even though the production of electricity by Eskom showed a declining trend, so did distribution. Table 1 below shows indications of the amount of electricity distributed between 2010 and 2014 as Eskom struggled to keep up with demand.

Exploratory Data Analysis

Table 2 confirms the results reported in Figure 5. The outcome of the Mann-Kendall test results revealed that hourly average electricity demand has significant monotonic decreasing long-term trends (p < 0.05

Figure 5. Day of the week trend analysis

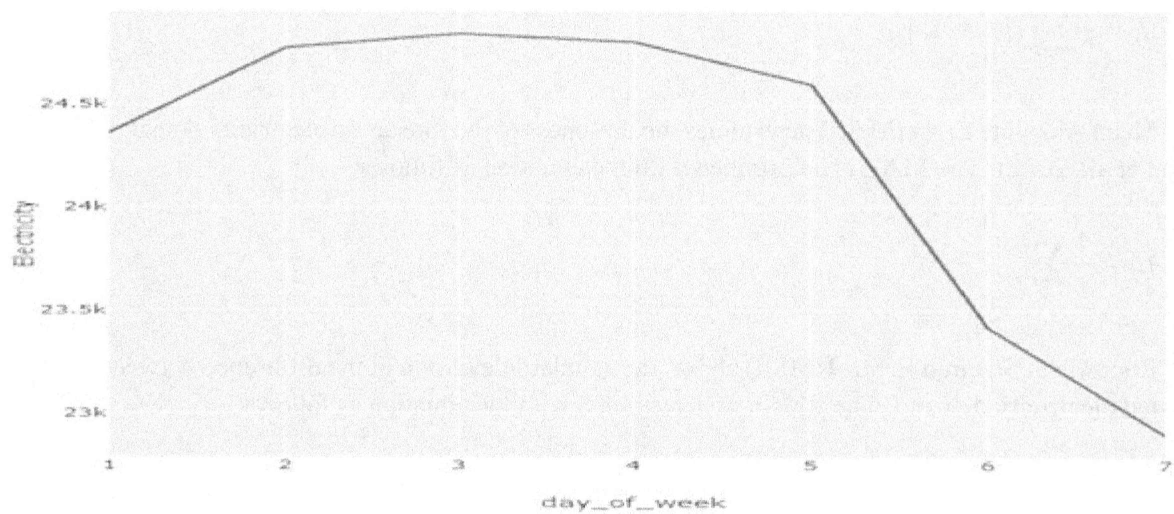

Table 1. Electricity distributed – Gigawatts-hours (GWH), 2010 – 2014

Year	Generated (GWH)
2010	238 272
2011	240 528
2012	234 174
2013	233 105
2014	231 449

Source: Stats SA, 2015

and τ is negative). Sen's slope value shows significant decreasing magnitudes of trends, which correspond with the Mann-Kendall test results.

Combined Forecast Model Using Minimum Score Combining

Eskom hourly data is split into the training, validation, and test data sets as described in the previous section in Figure 1. We therefore proceed and fit a TBATS, and the selection of the best TBATS model is done automatically by using the TBATS function in Python for hourly electricity demand load. The parameters of the fitted model are $\hat{\lambda}$ =0.027603, $\hat{\alpha}$ =1.890744, $\hat{\beta}$ = -0.374124 and the damping parameter is $\hat{\phi}$ =0.800000.

Using the minimum score combination proposed by Trucíos and Taylor (2020), CNN and LSTM are combined with the TBATS. Based on the estimates of combined models, an out-of-sample analysis evaluates the ability of deep learning methods coupled with Box-Cox transformations, Fourier representations that have time-varying coefficients, and autoregressive moving average (ARMA) error correction. Figure 6 presents the loss predicted by individual models. It can be seen that the TABTS-CNN-LSTM has less loss function compared to counterpart models. We therefore conclude that the best model is the TBATS-CNN-LSTM. This model's key performance indicators are significantly lower than the others. In contrast to Gasparin et al. (2022), who used three error metrics in their deep learning for time series forecasting: in the electric load case study, we employ four error metrics in this study. Each model's performance is assessed using these metrics, and the results are summarized in Table 2. This table leads to a few speculative inferences. The findings demonstrate that combining point forecasts from various models enhances the model's predictive capability. The best model is in Table 2.

Figure 7 presents the fitted conditional mean of the estimated TBATS-CNN_LSTM. By visual inspection, the model has a reasonable fit for the electricity demand load by Eskom. This model combined

Table 2. Summary statistics, Mann-Kendall test statistic, and Sen's slope estimator

Variable	Mean	Standard Dev	Skewness	Kurtosis
Electricity Demand	25113.378	2909.06	-0.137	-0.377
	Mann-Kendall test statistic and Sen's slope estimator			
	M-K Test Statistic	**Kendall's Tau**	**p-Value**	**Sen's Slope**
	-19.127	-0.066	0.001	-0.0274

Figure 6. Performance of individual models

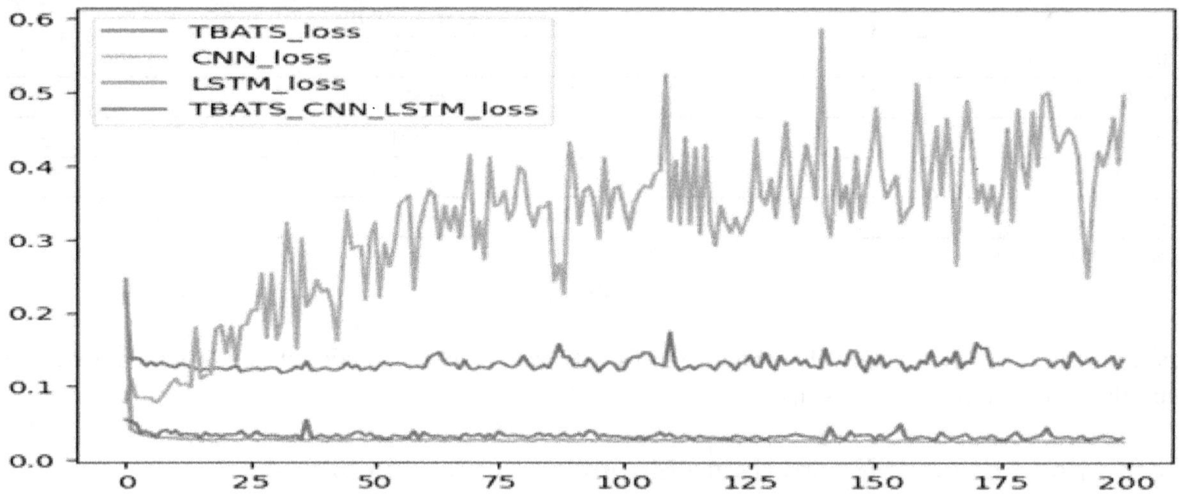

Table 2. Key performance indicators for the fitted models

Model	MSE	MAE	RMSE	Accuracy
TBATS	2.774	12.457	1.665	5.211
CNN	1.180	16.607	1.086	3.171
LSTM	1.324	10.725	1.151	1.396
TABTS-CNN-LSTM	1.132	9.935	1.064	1.263

forecasting model is found to be the best forecasting model for the short and long-run term for electricity hourly demand load by Eskom.

Short- and Medium-Term Forecasting Results

To check the model performance on short (i.e., accuracy) and medium-term (i.e., stability) horizons, a random sample of 700 data points were selected uniformly from various time ranges from one week to several months, as shown in Table 3. The mean of the performance metrics values is computed for each range. Results show that error values are consistent with a low standard deviation. The accuracy measured by CV (RMSE) is 0.60% for the short term and an average of 0.56% for the medium term. With very comparable performances for both the short-term and medium-term horizons, it is obvious that the estimated TBATS-CNN_LSTM is accurate and stable.

Discussion

The construct validity in this chapter is concerned with two threats. The first one is related to the sufficiency of time series data to well predict the hourly electricity demand load. This question was considered from the beginning of the methodology process and other sources of data; in particular,

Figure 7. A comparison of the predicted values versus the observed values from the test set

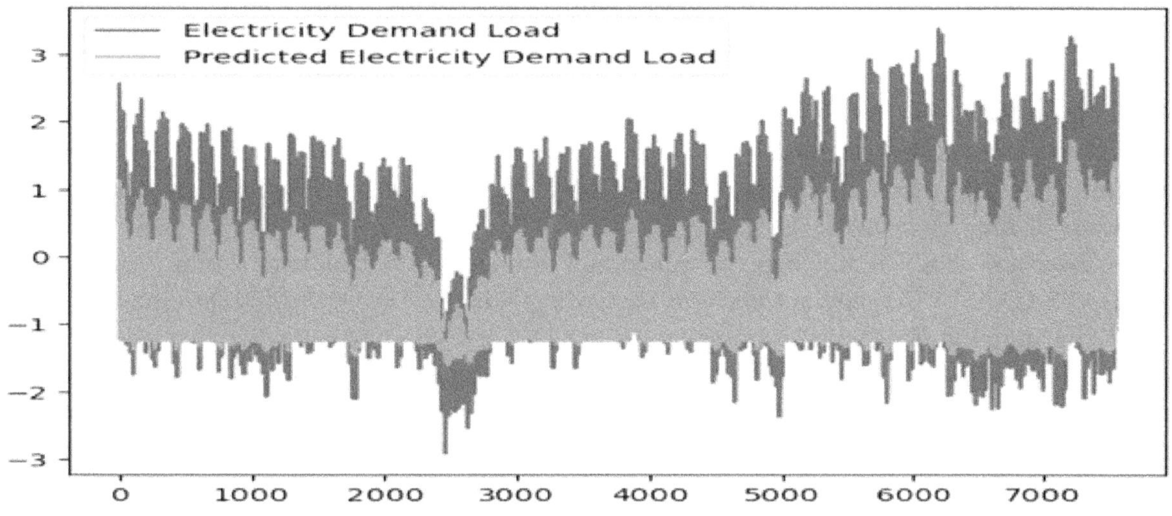

Table 3. Performance metrics of LSTM-RNN on various time horizons

Forecasting Horizon	MSE	MAE	RMSE	CV (RMSE) %
2-Weeks	11.49	2.51	3.39	0.60
Between 2–4 weeks	6.66	2.14	2.58	0.56
Between 2–3 Months	8.64	2.25	2.94	0.63
Between 3–4 Months	7.56	2.08	2.75	0.5
Mean-Medium term	7.60	2.156	2.756	0.56
Std. Dev.	3.58	8.6	1.8	0.06

intra-day, weekly, and monthly information were added as input rather than excluded by different feature selection techniques. The second threat is concerned with the algorithms used in the different steps and components of the methodology, which could use some unrealistic assumptions inappropriate for the current chapter's case context. To avoid such a threat, all the algorithms are developed in this chapter.

The internal validity in the case of this chapter is concerned with three main threats. The first one is related to how much the data set is suitable to train a TBATS-CNN-LSTM model that needs a large amount of data to mine historical patterns. The ESKOM hourly electricity demand load data set ranges from April 2019 until October 2023. This data set is large enough to allow deep learning, which is particularly targeted by TABTS-CNN-LSTM. The second threat is that the TBATS-CNN-LSTMS error starts increasing while the training error still decreases. Too much training with a complex forecasting model leads to bad generalizations. To prevent overfitting for the TBATS-CNN-LSTM model, the training examples at each epoch were randomly shuffled. This helps to improve generalization. Furthermore, early stopping is also used, whereby training is stopped at the lowest error achieved on the test set. This approach is highly effective in reducing overfitting.

The diagnostic plot of the estimated TBATS-CNN-LSTM model in Figure 8 shows that the train and validation losses decrease as the number of epochs increases. The loss stabilizes around the same point, showing a good fit. The third threat is related to the choice of alternative machine learning for comparison. The spectrum of techniques used for the energy prediction problem may be misrepresented. For this concern, the most successful techniques, according to the literature exploration, are selected. A stronger strategy to avoid this threat is to carry out a specific study to identify the top techniques for each class of prediction circumstances. This latter is one of the objectives of future works.

The external validity of this chapter is concerned with the generalization of results. In other words, the TBATS-CNN-LSTM may not perform well when it is used in very different circumstances of energy consumption. This threat is mitigated by two choices. The first is the fact that a deep learning approach is used in this chapter that is known to recognize good patterns from a large amount of data and from far-reaching history in the case of time series. Deep learning is a variation of ANN that produces a robust model. The second choice is the use of a large amount of training data (30,274) out of 40,523 observations with a rigorous validation method that evaluates the robustness of the models against being trained on different training sets. Besides, one of future works will be applying this methodology, in particular TBATS-CNN-LSTM modelling, to many data sets from distant energy contexts and training more parameters of the CNN-LSTM model via stacking ensembles in a distributed manner. For instance, this approach could be applied to forecasting several decision variables, such as the energy flow and the load, in the context of urban vehicle networks (Scellato et al., 2010). Many alternatives to validation methods for this approach can be inspired by the extended discussion on validation in Bouktif (2005).

Figure 8. TBATS-CNN-LSTM learning curve

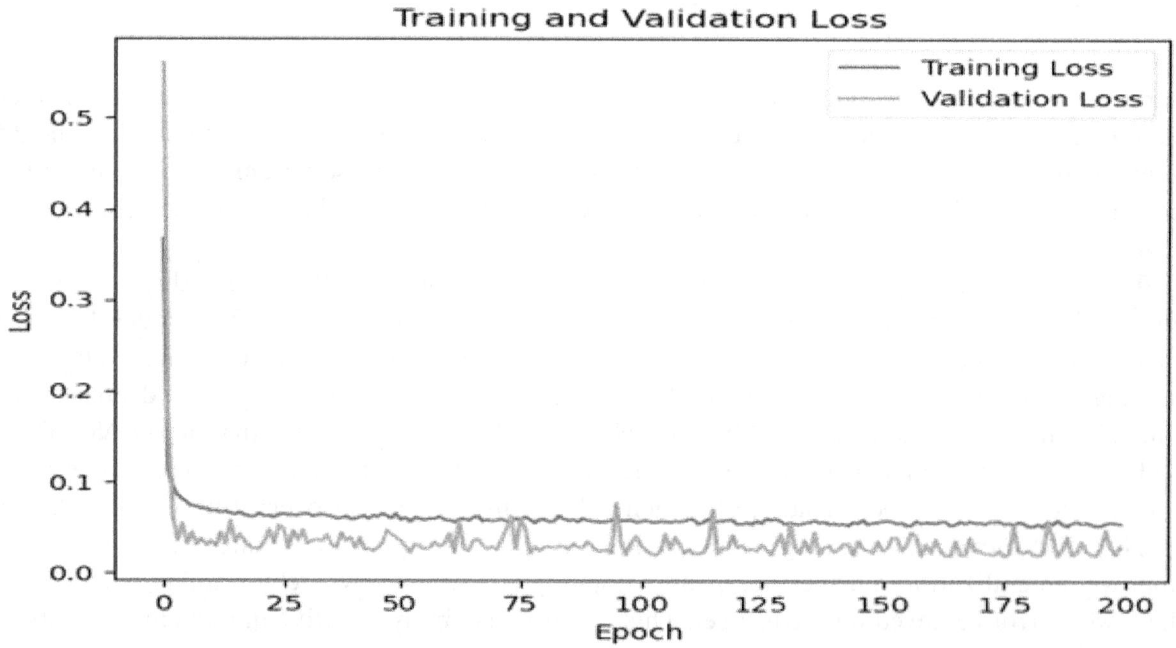

CONCLUSION AND RECOMMENDATIONS

With the growth of smart grids, load forecasting is becoming more and more important. The prediction of electrical loads with high accuracy is a difficult problem. The non-linearity and volatility of real-time energy consumption make it difficult to estimate energy demand and consumption. Accurate load forecasting is essential for power system planning and rapid decision-making. Accurate energy consumption prediction is an important part of energy management for sustainable systems.

In this chapter, TBATS-CNN-LSTM is built and optimized for load demand forecasting, providing accurate short-term (days to two weeks) and medium-term (weeks to months) forecasts. For comparison, seven prediction techniques were implemented, representing the spectrum of machine learning approaches commonly used in the field of energy prediction. The best-performing model from the studied consumption data set is used as the reference model. Encapsulated and nested feature selection techniques, namely recursive feature elimination, and incremental tree regression, are used to test the significance of model inputs. Hyperparameter tuning is used to further improve the performance of ensemble-based machine-learning models. These methods remove the unpredictability of quarterly characteristics and focus more on historical lags for hourly electricity demand load forecasts. This is because temporary load parameters encapsulate the information originally contained in the workday specification. For time-lagged features, a common problem is choosing the appropriate lag length, which is solved using a genetic algorithm (GA).

The validation of results using simple train test splits, multiple train test splits, and various time horizons showed that the TBATS-CNN-LSTM-based forecasting method has lower forecast errors in the challenging short- to medium-term electric load forecasting problem compared to the base learners. Electricity load forecasting has been modelled as a univariate time series problem, and thus the approach can be generalized to other time series data. Future works will include applying this methodology and, in particular, TBATS-CNN-LSTM modelling many data sets from distant energy contexts. Another objective is to carry out an empirical study to identify the top techniques for each class of prediction circumstances. Examining the results using single-train tests, multiple-train tests, and different time horizons, the TBATS-CNN-LSTM-based forecasting method is found to have low forecast errors for short- and medium-term electric load forecasting problems. Since power load forecasting is modelled as a univariate time series problem, the approach can be generalized to other time series data. Future work may include applications of this methodology, particularly TBATS-CNN-LSTM modelling on multiple datasets in distant energy contexts. Another goal is to conduct empirical research to determine which method is most appropriate for each class of forecasting situations.

It is also anticipated that the results from this research can open roads toward future implementation of advanced calculation methods regarding energy demand and consumption. It is expected that such models will help energy planners to accurately plan for the future and utilize sustainable and renewable energy resources to a greater extent. The models will help policymakers and administrators make decisions for a greener tomorrow. In addition, the results of this study are expected to lay the foundation for the introduction of advanced computational techniques related to energy demand and consumption in the future. These models are expected to help energy planners accurately plan for the future and use more sustainable and renewable energy sources. The model helps decision-makers and managers make decisions for a greener future.

ACKNOWLEDGMENT

The authors appreciate the positive and constructive feedback and ideas provided by the chapter's anonymous reviewers.

Data Description: The data used in this chapter was provided by the South Africa Electricity Distribution, ESKOM and it is available from the authors upon request.

REFERENCES

Abdul Majid, M. H., & Ibrahim, K. (2021). On Bayesian approach to composite Pareto models. *PLoS One*, *16*(9), e0257762.

Ahmad, T., & Chen, H. (2020). A review on machine learning forecasting growth trends and their real-time applications in different energy systems. *Sustainable Cities and Society*, *54*, 102010.

Alduailij, M. A., Petri, I., Rana, O., Alduailij, M. A., & Aldawood, A. S. (2021). Forecasting peak energy demand for smart buildings. *The Journal of Supercomputing*, *77*, 6356–6380.

Alharkan, H., Habib, S., & Islam, M. (2023). Solar power prediction using dual stream cnn-lstm architecture. *Sensors (Basel)*, *23*(2), 945.

Alsayed, A. O., Rahim, M. S. M., AlBidewi, I., Hussain, M., Jabeen, S. H., Alromema, N., & Jibril, M. L. (2021). Selection of the Right Undergraduate Major by Students Using Supervised Learning Techniques. *Applied Sciences (Basel, Switzerland)*, *11*(22), 10639.

Aung, S. (2018). Electric power is the main driving force for industrialization. Academic Press.

Badi, I., Muhammad, L., Abubakar, M., & Bakır, M. (2022). Measuring sustainability performance indicators using FUCOM-MARCOS methods. Operational Research in Engineering Sciences: Theory and Applications, 5(2), 99-116. doi:10.31181/oresta040722060b

Bouktif, S. (2005). Improving software Quality prediction by combining and adapting predictive models. Montreal University Montreal. doi:10.1198/jasa.2011.tm09771

del Real, A. J., Dorado, F., & Durán, J. (2020). Energy demand forecasting using deep learning: Applications for the French grid. *Energies*, *13*(9), 2242.

Durairaj, D. M., & Mohan, B. K. (2022). A convolutional neural network-based approach to financial time series prediction. *Neural Computing & Applications*, *34*(16), 13319–13337.

Gasparin, A., Lukovic, S., & Alippi, C. (2022). Deep learning for time series forecasting: The electric load case. Transactions on Intelligence Technology, 7(1), 1-25. doi:10.1049/cit2.12060

Gellert, A., Fiore, U., Florea, A., Chis, R., & Palmieri, F. (2022). Forecasting electricity consumption and production in smart homes through statistical methods. *Sustainable Cities and Society*, *76*, 103426.

Hussain, A., Rahman, M., & Memon, J. A. (2016). Forecasting electricity consumption in Pakistan: The way forward. *Energy Policy*, *90*, 73–80.

Hussain, S., Muhammad, L., Ishaq, F., Yakubu, A., & Mohammed, I. (2019). Performance evaluation of various data mining algorithms on road traffic accident datasets. Paper presented at the Information and Communication Technology for Intelligent Systems: Proceedings of ICTIS 2018, 1.

International Energy Agency. (2020). Power Systems in Transition. Retrieved from https://www.iea.org/reports/power-systems-in-transition

Islam, M. Z., Islam, M. M., & Asraf, A. (2020). A combined deep CNN-LSTM network for the detection of novel coronavirus (COVID-19) using X-ray images. *Informatics in Medicine Unlocked*, *20*, 100412.

Jevgenijs, S., deW, J., Kochnakyan, A., and Vivien, F. (2019). Forecasting electricity demand: an aid for practitioners. Academic Press.

Kartikasari, P., Warsito, B., Yasin, H., Utami, I. T., & Suhermi, N. (2023). Forecasting double seasonal electricity consumption with TBATS model. Paper presented at the AIP Conference Proceedings.

Khan, A. R., Mahmood, A., Safdar, A., Khan, Z. A., & Khan, N. A. (2016). Load forecasting, dynamic pricing and DSM in smart grid: A review. *Renewable & Sustainable Energy Reviews*, *54*, 1311–1322.

Khan, S., Javaid, N., Chand, A., Khan, A. B. M., Rashid, F., & Afridi, I. U. (2019). Electricity load forecasting for each day of week using deep CNN. Paper presented at the Web, Artificial Intelligence and Network Applications: Proceedings of the Workshops of the 33rd International Conference on Advanced Information Networking and Applications (WAINA-2019).

Kim, H., Park, S., & Kim, S. (2023). Time-series clustering and forecasting household electricity demand using smart meter data. *Energy Reports*, *9*, 4111–4121.

Kim, Y., & Kim, S. (2021). Forecasting charging demand of electric vehicles using time-series models. *Energies*, *14*(5), 1487.

Kuang, H., Guo, Q., Li, S., & Zhong, H. (2021). Short-term wind power forecasting model based on multi-feature extraction and CNN-LSTM. Paper presented at the IOP Conference Series: Earth and Environmental Science.

Kumar, C. J., and Veerakumari, M. (2012). Load forecasting of Andhra Pradesh grid using PSO, DE algorithms. International Journal of Advanced Research in Computer Engineering and Technology, 1(9), 179-184.

Kuyumani, E., Hasan, A. N., & Shongwe, T. (2023). A Hybrid Model Based on CNN-LSTM to Detect and Forecast Harmonics: A Case Study of an Eskom Substation in South Africa. *Electric Power Components and Systems*, *51*(8), 746–760.

Machado, E., Pinto, T., Guedes, V., & Morais, H. (2021). Electrical Load Demand Forecasting Using Feed-Forward Neural Networks. *Energies*, *14*(22), 7644.

Minerals, D. o., and Energy. (1998). White paper on the energy policy of the Republic of South Africa. In. Pretoria: Department of Minerals and Energy. doi:10.1007/s42979-021-00731-4

Muhammad, L., Amshi, J. M., Usman, S. S., Badi, I. A., Mohammed, I., Dada, O., & Haruna, A. A. (2021). Deep learning models for classification and localization of COVID-19 abnormalities on chest radiographs. Paper presented at the 2021 4th International Conference on Computing and Information Sciences (ICCIS).

Muhammad, L., Badi, I., Haruna, A. A., Mohammed, I., & Dada, O. (2022). Deep learning models for classification of brain tumor with magnetic resonance imaging images dataset. In Computational Intelligence in Oncology: Applications in Diagnosis, Prognosis and Therapeutics of Cancers (pp. 159-176). Springer.

Muhammad, L., Haruna, A. A., Sharif, U. S., & Mohammed, M. B. (2022). Cnn-lstm deep learning-based forecasting model for COVID-19 infection cases in Nigeria, South Africa, and Botswana. *Health and Technology*, *12*(6), 1259–1276.

Muhammad, L. J., Algehyne, E. A., & Usman, S. S. (2020). Predictive Supervised Machine Learning Models for Diabetes Mellitus. *SN Computer Science*, *1*(5), 240.

Muhammad, L., & Algehyne, E. A. (2021). Fuzzy-based expert system for diagnosis of coronary artery disease in Nigeria. Health and Technology, 11(2), 319-329. doi:10.1007/s12553-021-00531-z

Munim, Z. H. (2022). State-space TBATS model for container freight rate forecasting with improved accuracy. Maritime Transport Research, 3, 100057. doi:10.1016/j.martra.2022.100057

Naim, I., Mahara, T., & Idrisi, A. R. (2018). Effective short-term forecasting for daily time series with complex seasonal patterns. *Procedia Computer Science*, *132*, 1832–1841.

Nti, I. K., Samuel, A.-A., & Michael, A. (2019). Predicting Monthly Electricity Demand Using Soft-Computing. *International Research Journal of Engineering and Technology*, *6*(6), 1967–1973.

Patil, S., Nemade, V., & Soni, P. K. J. P. s. (2018). Predictive modelling for credit card fraud detection using data analytics. *Procedia Computer Science*, *132*, 385–395.

Phaahla, E. (2023). The State of Electricity in South Africa-Part I: The problems in Eskom. Retrieved from https://hsf.org.za/publications/hsf-briefs/the-state-of-electricity-in-south-africa-part-i-part-i-the-problems-in-eskom

Raza, M. Q., & Khosravi, A. (2015). A review on artificial intelligence-based load demand forecasting techniques for smart grid and buildings. [doi:https://doi.org/10.1016/j.rser.2015.04.065]. *Renewable & Sustainable Energy Reviews*, *50*, 1352–1372.

Rosas, M. A. T., Pérez, M. R., & Pérez, E. R. M. (2022). Itineraries for charging and discharging a BESS using energy predictions based on a CNN-LSTM neural network model in BCS, Mexico. *Renewable Energy*, *188*, 1141–1165.

Scellato, S., Fortuna, L., Frasca, M., Gómez-Gardeñes, J., & Latora, V. (2010). Traffic optimization in transport networks based on local routing. *The European Physical Journal B*, *73*, 303–308.

Seetha, J., & Raja, S. S. (2018). Brain tumor classification using convolutional neural networks. *Biomedical & Pharmacology Journal*, *11*(3), 1457–1461.

Solyali, D. (2020). A Comparative Analysis of Machine Learning Approaches for Short-/Long-Term Electricity Load Forecasting in Cyprus. *Sustainability*, *12*(9), 3612.

Staudemeyer, R. C., & Morris, E. R. (2019). Understanding LSTM—a tutorial into long short-term memory recurrent neural networks. arXiv preprint arXiv:1909.09586.

Torres, J. F., Martínez-Álvarez, F., & Troncoso, A. (2022). A deep LSTM network for the Spanish electricity consumption forecasting. *Neural Computing & Applications*, *34*(13), 10533–10545. doi:10.1007/s00521-021-06773-2

Uz Zaman, M., Islam, A., & Sultana, N. (2018). Short-term load forecasting based on Internet of things (IoT). BRAC University. Retrieved from http://hdl.handle.net/10361/10170

Wang, D., Gan, J., Mao, J., Chen, F., & Yu, L. (2023). Forecasting power demand in China with a CNN-LSTM model including multimodal information. *Energy*, *263*, 126012.

Zhang, Y., Satapathy, S. C., Zhu, L.-Y., Górriz, J. M., & Wang, S. (2020). A seven-layer convolutional neural network for chest CT-based COVID-19 diagnosis using stochastic pooling. *IEEE Sensors Journal*, *22*(18), 17573–17582.

ADDITIONAL READING

Baker, M. R., Jihad, K. H., Al-Bayaty, H., Ghareeb, A., Ali, H., Choi, J.-K., & Sun, Q. (2023). Uncertainty management in electricity demand forecasting with machine learning and ensemble learning: Case studies of COVID-19 in the US metropolitans. *Engineering Applications of Artificial Intelligence*, *123*, 106350. doi:10.1016/j.engappai.2023.106350

Chapagain, K., Gurung, S., Kulthanavit, P., & Kittipiyakul, S. (2023). Short-term electricity demand forecasting using deep neural networks: An analysis for Thai data. *Applied System Innovation*, *6*(6), 100. doi:10.3390/asi6060100

Choi, E., Cho, S., & Kim, D. K. (2020). Power demand forecasting using long short-term memory (LSTM) deep-learning model for monitoring energy sustainability. *Sustainability (Basel)*, *12*(3), 1109. doi:10.3390/su12031109

Cui, Z., Wu, J., Lian, W., & Wang, Y.-G. (2023). A novel deep learning framework with a COVID-19 adjustment for electricity demand forecasting. *Energy Reports*, *9*, 1887–1895. doi:10.1016/j.egyr.2023.01.019

Deina, C., dos Santos, J. L. F., Biuk, L. H., Lizot, M., Converti, A., Siqueira, H. V., & Trojan, F. (2023). Forecasting electricity demand by neural networks and definition of inputs by multi-criteria analysis. *Energies*, *16*(4), 1712. doi:10.3390/en16041712

Kim, H., Park, S., & Kim, S. (2023). Time-series clustering and forecasting household electricity demand using smart meter data. *Energy Reports*, *9*, 4111–4121. doi:10.1016/j.egyr.2023.03.042

Ofori-Ntow, E., & Ziggah, Y. Y. (2023). Electricity demand forecasting based on feature extraction and optimized backpropagation neural network. e-Prime-Advances in Electrical Engineering. *Electronics and Energy*, *6*, 100293.

Pierre, A. A., Akim, S. A., Semenyo, A. K., & Babiga, B. (2023). Peak electrical energy consumption prediction by ARIMA, LSTM, GRU, ARIMA-LSTM and ARIMA-GRU approaches. *Energies*, *16*(12), 4739. doi:10.3390/en16124739

Saglam, M., Spataru, C., & Karaman, O. A. (2023). Forecasting electricity demand in Turkey using optimization and machine learning algorithms. *Energies*, *16*(11), 4499. doi:10.3390/en16114499

Sekhar, C., & Dahiya, R. (2023). Robust framework based on hybrid deep learning approach for short term load forecasting of building electricity demand. *Energy*, *268*, 126660. doi:10.1016/j.energy.2023.126660

Wang, Z., Chen, Z., Yang, Y., Liu, C., Li, X., & Wu, J. (2023). A hybrid autoformer framework for electricity demand forecasting. *Energy Reports*, *9*, 3800–3812. doi:10.1016/j.egyr.2023.02.083

Yuniarti, E., Nurmaini, N., Suprapto, B. Y., & Rachmatullah, M. N. (2019). Short term electrical energy consumption forecasting using rnn-lstm. Paper presented at the 2019 International Conference on Electrical Engineering and Computer Science (ICECOS). 10.1109/ICECOS47637.2019.8984496

Chapter 13
Electricity Production Prediction by Microsoft Azure Machine Learning Service and Python User Blocks

Vladyslav Pliuhin
O.M. Beketov National University of Urban Economy in Kharkiv, Ukraine

Yevgen Tsegelnyk
https://orcid.org/0000-0003-1261-9890
O.M. Beketov National University of Urban Economy in Kharkiv, Ukraine

Maria Sukhonos
O.M. Beketov National University of Urban Economy in Kharkiv, Ukraine

Ihor Biletskyi
O.M. Beketov National University of Urban Economy in Kharkiv, Ukraine

Sergiy Plankovskyy
O.M. Beketov National University of Urban Economy in Kharkiv, Ukraine

Illia Khudiakov
O.M. Beketov National University of Urban Economy in Kharkiv, Ukraine

ABSTRACT

In this chapter, the forecasting of electricity consumption and production is conducted by analyzing indicators from previous years. The problem is addressed using machine learning within Microsoft Azure Machine Learning Studio. The outcome is an independent service integrated into Excel, enabling consumption forecasting for specified dates. The Excel user interface is developed using Visual Basic for Applications. Python was used to create user blocks for modifying input data pools and forming graphical dependencies, seamlessly integrated into the original modules of Microsoft Azure Machine Learning Studio. An additional aspect of the forecast results involves evaluating the quality of the predicted electricity consumption indicators. The materials used for this chapter were sourced with the support of Ukraine's National Power Company UKRENERGO.

DOI: 10.4018/979-8-3693-2355-7.ch013

INTRODUCTION

Electricity consumption by the civilian and industrial enterprises depends on many factors: temperature, time of day (light/dark), weather conditions (in cloudy weather, people are more likely to stay at home), etc. Thus, communications are affected by different loads. If the network is overloaded, it may fail, causing severe consequences. The constant need to supply electricity to both civilian and industrial enterprises create the need for strong protection against breakdowns. Stopping the supply of energy to vital objects can cause a catastrophe and lead to such huge financial losses that in most cases it is more appropriate to spend money on preventing crisis situations (Wang et al., 2022a).

Solving this task can be done in two ways. The first one consists in studying and constantly monitoring the state of the equipment (Aksonov et al., 2021; Kombarov et al., 2021; Tsegelnyk et al., 2022), as well as in forecasting the state several periods ahead. This approach requires the creation of an information system and long-term work of experts and analysts. The result is achieved after a rather long period of time and requires significant financial costs. However, it is stable and allows to fully protect from risks in the long term.

An alternative way is to forecast electricity consumption. As a rule, experts are aware of the potential power grids and consumption information will be sufficient for them to expertly predict possible breakdowns. This way requires comparatively low financial expenditure or considerable time to collect representative samples. The information needed for forecasting is usually collected by sensors at power plants or utilities. This approach gives less stable results, but they appear for a much shorter period.

Formally, the researcher is faced with several tasks that are not determined in advance. In general, the goal of the research can be formulated as: "to obtain as much information as possible from the available data and to build the most accurate forecast possible." Specific actions are chosen by the analyst based on his experience and available data. The lack of formalization forces preliminary research, which is called "descriptive analysis". The organization of information, therefore, plays an important role in research planning and task formulation.

Forecasting the long-term needs of society in electricity helps to determine the electricity generating capacities for the future (Wang et al., 2022b; Wen et al., 2023). Such forecasts are also used to analyze the scope and composition of power supply expansion projects that include nuclear power. In the conditions of scientific and technical progress and improvement of the energy system of the state, forecasting becomes one of the decisive scientific factors in the formation of the strategy and tactics of energy development. To solve the problem of uncertainties, it is necessary to carry out comprehensive forecasting, which includes a number of forecasts of energy demand in the future.

The analysis should be conducted using up-to-date and consistent macroeconomic and microeconomic data so that electricity demand forecasts are more reliable and linked to demographic, economic and industrial development forecasts. Similarly to the analysis of demand for electricity, when determining the feasibility of implementing an energy project, it is necessary to conduct an analysis of the supply in the electricity sector. Such forecast shall be based on historical data, including consideration of past and present power supply trends, study of power generation systems and power transmission and distribution systems, as well as available resources for power generation and existing and planned connections to adjacent systems.

Currently, there are a number of works devoted to the practical application of machine learning (ML) technology in real life. A framework for regression-based ML that provides researchers with a common language and abstraction to aid in their study presented by Bojer (2022). To demonstrate the utility of the framework, is shown how it can be used to map and compare ML methods used in the M5 Uncertainty competition. In Tarallo et al. (2019) exploratory research presents the benefits of ML in sales forecasting for short shelf-life and highly-perishable products, as it surpasses the accuracy level of traditional statistical techniques and, as a result, improves inventory balancing throughout the chain, reducing stockout rates at points of sale, improving availability to consumers and increasing profitability.

In the series of publications (Hofmann & Rutschmann, 2018; Ke et al., 2017; Martins & Galegale, 2023; Zhou et al., 2017) research aims to evaluate the main methods and identify the one with the greatest accuracy in sales prediction. Based on an integrative literature review, three main methods were evaluated: time series, artificial neural networks and ML algorithms. The results show that ML is more suitable in terms of accuracy, particularly when models contain exogenous and endogenous variables, in addition to allowing the identification of hidden patterns in demand that can be used to identify market trends.

In Rohaan et al. (2022); Theoharidis et al. (2023) considered deep learning research direction and proposed a hybrid deep learning model that merges Variational Autoencoders and Convolutional LSTM Networks (VAE-ConvLSTM) to forecast inflation. Such results suggest that macroeconomic forecasting could take advantage of deep learning models when tackling nonlinearities and nonstationarity, potentially delivering superior performance in comparison to traditional econometric approaches based on linear, stationary models.

Many studies use ML to predict atmospheric pollutant levels, prioritizing accuracy over interpretability (Ampomah et al,, 2020; Balaji et al., 2021; Houdou et al., 2024; Htun et al., 2023; Kureljusic and Karger. 2023; Rahimi et al., 2023). These methods identify important atmospheric features, enhancing researchers' understanding and making ML outcomes more accessible to non-experts. This can enhance prediction and prevention of adverse weather events and air pollution, benefiting public health. Some of papers devoted to predicting article quality scores with ML that is outside of technical issue (Akella et al., 2021; Alchokr et al., 2023; Chen et al., 2022; Thelwall et al., 2023). In the area of electrical energy production and electrical networks power grid machine learning is used for considering challenging problems in smart grids (Akhtar et al., 2023; Bomfim, 2020; Duchesne et al., 2020; Eskandarpour and Khodaei, 2016; Kumbhar et al., 2021; Mira et al., 2023; Srinivasan et al., 2023). These challenges lead to make fundamental theoretical contributions in the areas of artificial intelligence and smart grids, building new bridges between them. These models can be classified as AI- or mathematically inspired models and include dynamical systems, Bayesian inference, stochastic differential equations, ML methods, deep learning, reinforcement learning, and reservoir computing.

Different papers examines the current state of applying artificial intelligence and machine learning algorithms in energy conversion and management evaluation and optimization tasks (Entezari et al., 2023; Franki et al., 2023; Huang et al., 2022; Xu et al., 2022). It highlights the latest developments and the most promising algorithms and assesses their merits and drawbacks, encompassing specific applications and relevant scenarios. Furthermore, the authors propose recommendations to emphasize the prioritization of acquiring real-world experimental and simulated data and adopting standardized, explicit reporting in research publications.

The concept of artificial intelligence (AI) and machine learning (ML) in the field of power energy was studied in Entezari et al. (2023). Mentioned work investigate the usage of AI and ML to determine the energy system the optimum point for supply and demand. But this works as the row of same by Dayupay et al. (2024); Forootan et al. (2022); Huotari et al. (2024) rather describes obtained result than the method details, so such works cannot be reproduced.

In Moradzadeh et al. (2021) authors shows selected applications of machine learning and deep learning based on their learning, structure, mode of operation, and application in the load forecasting of power systems. To analyze the condition of power systems they used different prediction models based on linear regression, classification, and clustering. To analyze the behavior of power systems, advanced metering, and condition monitoring devices and systems are utilized, which generate a huge amount of data. This chapter is useful to understand practical application of ML in power energy field and males links with energy consumption and development of industry and technology systems.

In Kirubakaran et al. (2023) the hybrid electricity production was investigated to show how produce and use electrical energy from many sources as part of an integrated setup. This study optimizes geographic locations with various wind and sun renewable energy input potentials. Convolutional neural networks with long short-term memory are designed to maximize electricity production while consumer load, demand, and weather conditions are considering. This paper is close to current authors work and beside of all describes in details application of neural networks algorithms. Almost the same work but with more details in datasets creation and optimization metrics determinations is studied in Yang (2022). The main outcome of this research can support decision optimization for an ML based decision support system. The proposed system is very promising for potential applications in wind (power) energy management. This study optimizes three distinct geographic locations with various wind and sun renewable energy input potentials. Convolutional neural networks with long short-term memory are designed to maximize electricity production while taking consumer load, demand, and weather conditions into account. Finally, in Balal et al. (2023) was shown a study that utilizes various machine learning models to predict solar photovoltaic (PV) power generation in Lubbock, Texas. The developed machine learning models describes how to aid solar PV investors in streamlining their processes and improving their planning for the production of solar energy.

Despite the wide range of works devoted to the topic of machine learning, the analysis of trends in the consumption of electricity networks within the country has not been performed or has not been publicly announced. The authors of this work have already carried out research on the implementation of machine learning in the field of electrical engineering (Pliuhin et al., 2018; Pliuhin et al., 2019a; Pliuhin et al., 2019b; Sukhonos et al., 2023). The practical use of machine learning technology Azure Machine Learning (AML) is also embodied in the research works of master's scientists and in the topics of master's works of students of O.M. Beketov National University of Urban Economy in Kharkiv. The goal of the work performed by the students together with teachers was to develop an innovative approach to the optimization of electric machines and power industry facilities. In addition, a team of university researchers (consisting of two professors, two master's students, and two first-year graduate students) took part in Microsoft's worldwide competition "The Next AI Guardians" (Microsoft, 2020) with a project to introduce standard calculation methods into the database blocks of Microsoft AML Studio own modules, which are capable of fully automatic design of electrical machines with their further optimization by means of machine learning.

The purpose of this paper is to determine the ways of development of society and the economic resources that ensure their achievement, to identify the most likely and economically effective options for long-term, medium-term and current plans, to substantiate the main directions of energy, economic and technical policy, to predict the future decisions made and measures implemented at this moment.

BACKGROUND

Predictive experimentation in the field of ML plays a key role in creating accurate and efficient models. The success of this process depends on the variety and volume of input data that determine the target parameter. The more and more diverse the data, the more competence and accuracy can be achieved in the development of predictive models. In this paper, for the prediction of electricity consumption and production, test data sets are considered, which include:

1) Time and date of data entry. Fundamental data collected during a certain period are key to considering the dynamics of electricity consumption and production.
2) Energy production by different types of power plants:
    - Nuclear power plant. Information about energy production using nuclear reactors.
    - Thermal power plant. Data on energy supply based on thermal energy.
    - Block thermal power plant. Guidelines for energy production from block thermal power plants.
    - Hydroelectric power station. Data on electricity production using hydropower technologies.

The results of the predictive experiment are the basis for identifying the causes of atypical electricity consumption, and serve as the basis for building a decision-making model for managing energy infrastructure reconstruction programs and projects. In this work, the predictive experiment was developed using the Microsoft Azure Machine Learning (AML) studio. The use of this service is determined by the following factors:

- ease of use – the components of the ML system are represented by blocks with inputs and outputs, respectively, the output data of one block is the input for another; thus, there is no need to write code to create a model;
- wide functionality – the AML studio has a large number of prediction and data classification methods for model training;
- the possibility of adding custom scripts to the model – additional data processing and analysis tools are implemented using scripts in the *Python* and *R* programming languages;
- convenience of analyzing the quality of model training – the studio provides the ability to analyze the quality of the model using a few metrics: root mean square error, relative absolute error, coefficient of determination, etc.

Value of final energy consumption by users Q_{end} is suggested to be used as a reference value when determining model constraints. Thus, we accept:

$$Q_{end} = kQ, \tag{1}$$

where Q is the total heat capacity of energy sources, MW; k is the coefficient that considers energy losses during transportation to end users.

Application of the parameter Q_{end} in such a definition will allow to simulate an engineering system, in which generating capacities will be selected in accordance with the volumes of final consumption. This will prevent the occurrence of excessive energy production and minimize the consumption of energy resources for the further functioning of the system. Defining model constraints is a user function, and it is possible to define constraints for both quantitative and qualitative indicators. Thus, for the existing system to be reconstructed within the framework of the implemented program, the existing disturbances of the external environment, it is possible to adapt individual parameters of the model. Possible limitations for generating subsystems of utility power facilities are represented by equations (2)–(5).

$$Q = 4.5, \tag{2}$$

$$Q_{los} \leq 0.01, \tag{3}$$

where Q_{los} is the total loss of thermal energy in the subsystem, MW.

$$P_{out} \geq 1.4, \tag{4}$$

where P_{out} is the coolant pressure at the exit from the subsystem, MPa.

$$FCon = 1, \tag{5}$$

where *FCon* is the presence of a frequency converter.

The list of limitations given above is not exhaustive and may be supplemented as necessary. The operation of the model involves the setting of all parameter values acting as constraints. This is due to the limitations of the machine learning technology used and the Microsoft AML Studio. The input data for training the model is provided as a dataset created in *csv* format. The data of the dataset is fed to the input of the block that implements the selection of columns used for further training. In this case, the block selects all columns except the first one, which contains the serial numbers of the objects. Columns are selected using the "*Column selector*" tool, the interface of which is shown in Figure 1. With the help of a script written in the Python programming language, columns of the dataset containing parameters of objects of a type that are not considered in a particular experiment are removed. For example, in the current task there was a deletion from further studies, the columns that related to the production of energy by wind and solar power plants, the analysis of which was transferred to the second stage of studies.

The "*Clean Missing Data*" block deletes the terms of the dataset in which there are empty values. The cleaned dataset enters the input of the "*Split Data*" block, which divides the dataset into data for training the model and data for testing. In this model, 75% of the time is allocated for model training.

Regression using a neural network is used to predict the output values. The neural network approach has advantages over traditional methods in three cases:

- when the task under consideration due to specific features does not lend itself to adequate formalization, as it contains elements of uncertainty that cannot be formalized by traditional mathematical methods;
- when the task under consideration is formalized, but there is currently no apparatus for solving it;
- when there is a suitable mathematical apparatus for the considered, well-formalized problem, but the implementation of calculations with its help on the basis of computer systems does not satisfy the requirements for obtaining solutions in terms of time, size, weight, energy consumption, etc.

Figure 1. Interface of the "column selector" tool of Microsoft Azure Machine Learning Studio

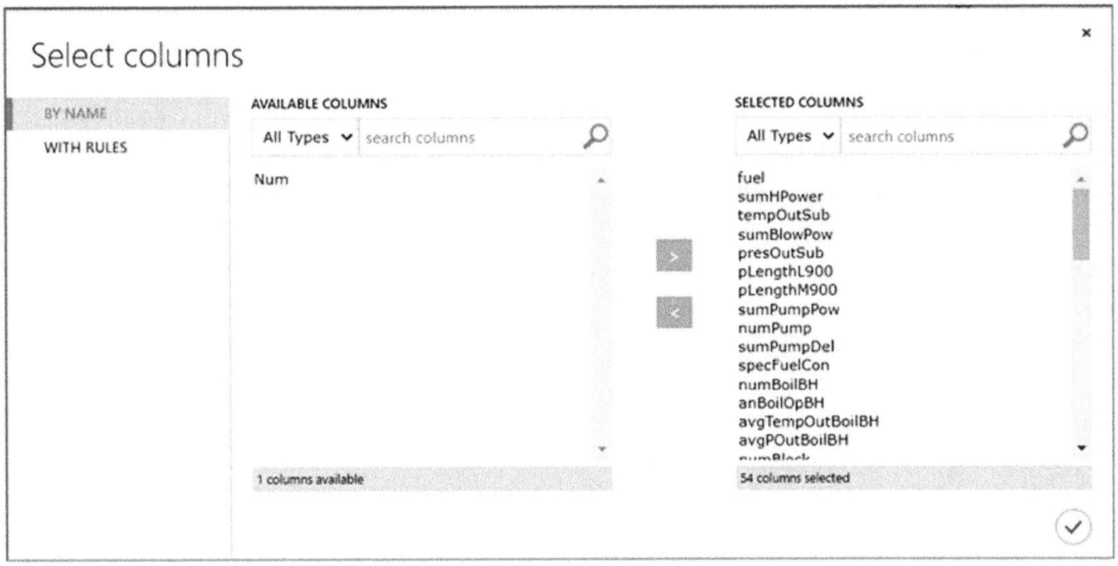

In such a situation, it is necessary to either simplify calculation methods, which reduces the quality of decisions, or apply a suitable neural network approach, if it will ensure the necessary quality of the task.

Due to the lack of the possibility of a clear complex formalization of engineering systems, including heat supply systems, it is appropriate to choose regression using a neural network. The parameters of the neural network used in these models are listed in Table 1. The neural network parameters specified in the Table 1 are used in the AML Studio for advanced configuration of the module responsible for the prediction algorithm using a neural network. The feedforward neural network that is used in Microsoft AML Studio, is the simplest type of artificial neural network. In a feedforward network, information moves in only one direction from input layer to output layer. Feedforward neural networks transform an input by putting it through a series of hidden layers. Every layer is made up of a set of neurons, and each layer is fully connected to all neurons in the layer before. The last fully connected layer (the output layer) represents the generated predictions.

Table 1. Neural network parameters

Parameter	Value
Number of hidden layers	60
Learning speed	0.005
Number of training cycles	60
Diameter of primary weights	0.1
Gradient	0

After training the model using the "*Train Model*" block and testing it using the "*Score Model*", it is evaluated in the "*Evaluate Model*" block. Also, from the "*Score Model*" block, data is fed to the six inputs of "*Execute Python Script*" block, which are responsible for generating graphs of the dependence of the values of the objective function on the values of some basic constraints. The function of generating charts is an element of post-processing of model data and will be described in detail below in the method of practical use of the ML service. After completion of training, testing and evaluation, the web service is set up based on the model using the option "*Set Up Web Service*" – "*Predictive Web Service*". The architecture of the created web service is shown in Figure 2.

To use the prediction model outside the Azure ML web-site service and to get this model ready for deployment, it is needed to convert the training experiment to a predictive one. To get this model ready for deployment. This involves three steps:

1. Save the model you've trained and then replace our training modules
2. Trim the experiment to remove modules that were only needed for training
3. Define where the web service will accept input and where it generates the output

One could do this manually, but fortunately all three steps can be accomplished by clicking Set Up Web Service at the bottom of the experiment canvas (and selecting the Predictive Web Service option).

Appropriate simplification of the architecture by removing some blocks and adding a transformed dataset. The simplified architecture of the web service is shown in Figure 3. The service includes: a trained model, a transformed dataset, a model testing block, as well as the input and output of the web service. The transformed dataset is the input dataset to which all data processing operations have been applied. The saved trained model is added back into the experiment. Web service input and Web service output modules are added (these identify where the user's data will enter the model, and what data is returned, when the web service is accessed).

When the web service is accessed, the user's data enters through the Web service input module where it's passed to the Score Model module and scored. The way you've set up the predictive experiment, the model expects data in the same format as the original credit risk dataset. The results are returned to the user from the web service through the Web service output module.

After the work on the web service is completed, it is deployed. In the future, it is possible to use it both in the browser and with the help of a Microsoft Excel file with the AML Studio add-on. During model creation and training, AML Studio creates a polynomial that is used to further predict system parameter values based on input data.

Machine Learning Studio makes it easy to call web services directly from Excel without the need to write any code. When entering input data, Microsoft Excel makes a call to this polynomial deployed as a web service, after performing calculations, the output data is returned to the user in numerical and textual form. The interface of the AML Studio add-on is shown in Figure 4. Once macros are enabled, a table is generated. Columns for initial data are required as input into the RRS web service, or PARAMETERS. Note the output of the RRS service, PREDICTED VALUES in separate column. When all columns for a given row are filled, the workbook automatically calls the scoring API, and displays the scored results.

Point 1 in Figure 4 presents information about the structure of the model – a list of input and output parameters, as well as a list of global parameters. In point 2, the data for forecasting is determined: the addresses of the cells in which the values of the input parameters are presented and the cells in which the output data will be output are specified. The "*Auto-predict*" checkbox allows to enable the function

Figure 2. Architecture of a model-based web service

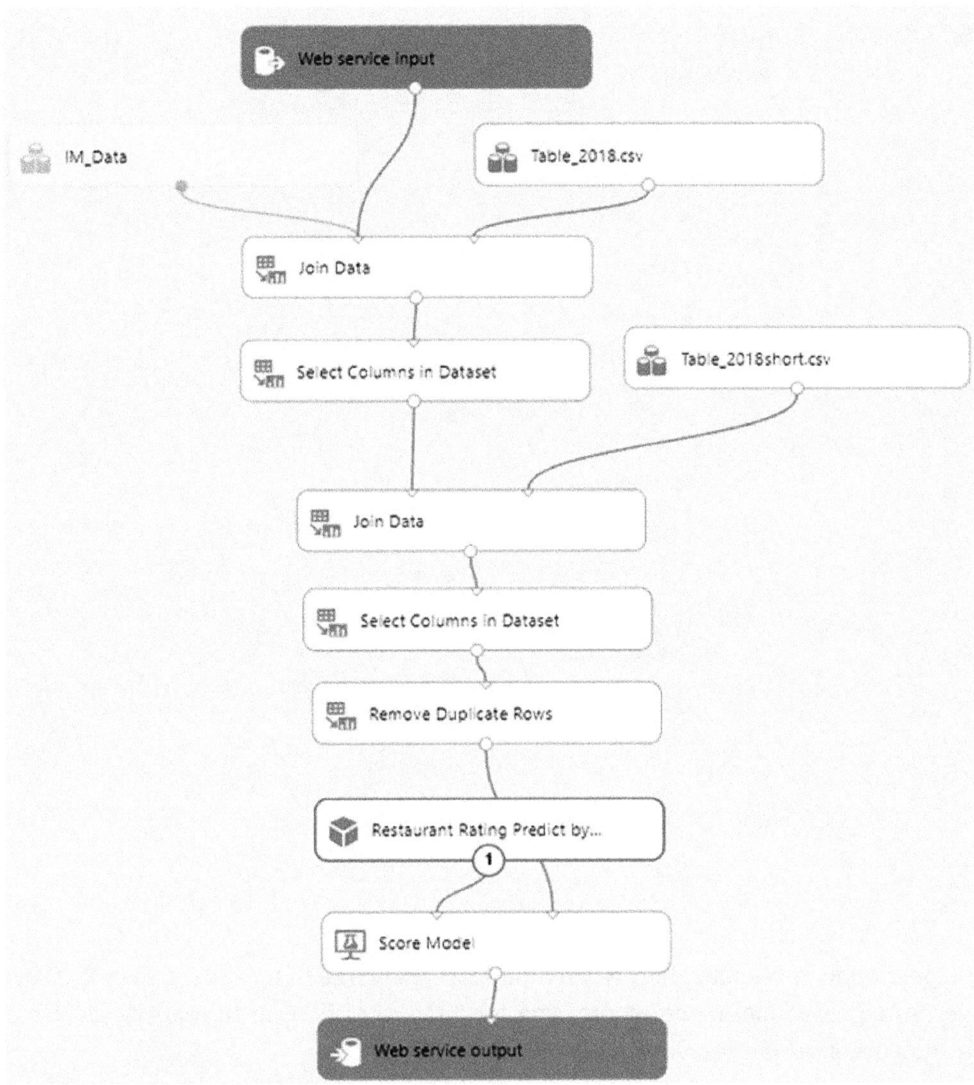

Figure 3. Simplified web service architecture

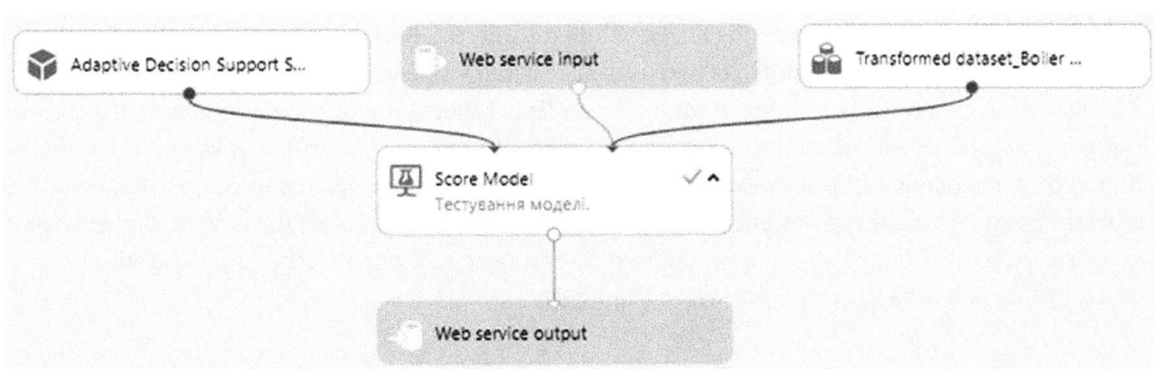

Figure 4. Azure Machine Learning add-on interface for Microsoft Excel

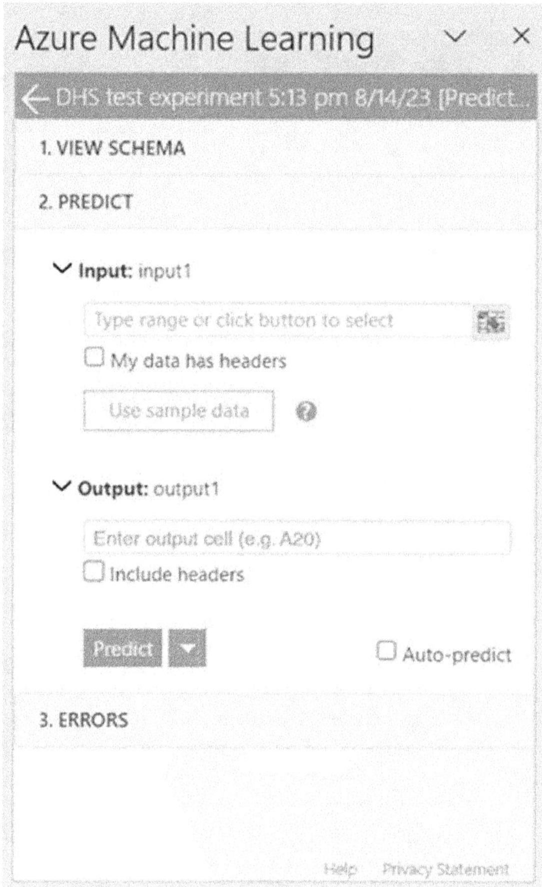

of automatic prediction of output values when input data is entered in the specified cells. The prediction procedure is started in manual mode by pressing the *"Predict"* button. In point 3, error messages are displayed if they occur during forecasting.

The user interface can be presented in a separate Microsoft Excel file. The choice of this software as the environment for using the decision support model is reasonable because of the ease of use for the end user-decision maker. Also, an important advantage of this approach is the absence of the need to create a separate software product with a graphical interface for integrating the trained model into it. Data entry and model management functions are developed using the built-in *Visual Basic for Applications (VBA)*. A Microsoft Excel workbook has two user-accessible worksheets used for data input and output.

The data entry interface is a series of tables on an Excel sheet containing parameter names, divided by category, and fields for data entry. Each of the tables, therefore, each of the parameter categories is highlighted in a separate color for ease of use. Within each of the tables are common parameters and parameters unique to each type of energy generation facility. After successfully passing the predicted value of the objective function, the data entered by the user is passed to the Data Output sheet. The interface of the data output sheet is shown in Figure 5.

Figure 5. Data output interface

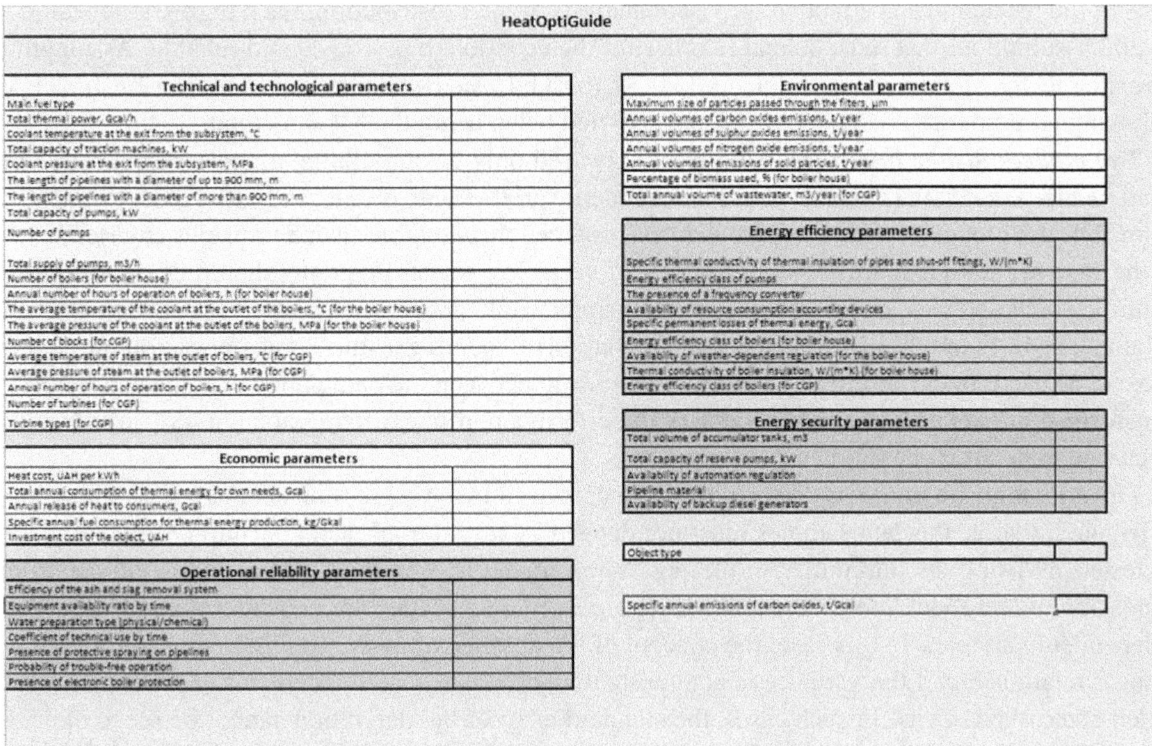

After obtaining the forecasting results, the user has the opportunity to change the values of the indicators in the cells on the "*Data Entry*" sheet and, by pressing the "*Calculation*" button, obtain the corresponding value of the objective function. Accordingly, the user can change the parameters of the simulated system, focusing on the obtained value.

FEATURES OF REGRESSION ANALYSIS IN MACHINE LEARNING

When forming the structure of the experiment in the Microsoft AML studio, the predicative analysis algorithm block is added to the "*Train Model*" block. Available algorithms include neural networks, search trees, clusters, and various types of regression algorithms. In this work, among a number of proposed ones, we focused on the *multiple regression* algorithm. First, as it can be seen from the name of multiple linear regression, it is assumed that the relationship between the variables is linear. In fact, this conjecture can, in fact, be proven at any time. Multiple regression analysis procedures are slightly affected by small deviations from this assumption. However, it always makes sense to look at two-dimensional scatterplots of the variables of interest. If the non-linearity of the relationship is obvious, it is possible to consider whether the transformation of the variables or explicitly allow the inclusion of non-linear terms. In multiple regression, it is assumed that the residuals (predicted values minus the observed values) are normally distributed (that is, follow the law of normal distribution). Although most tests (especially the F-test) are quite robust to deviations from this assumption, it is always worth considering the distributions of the variables of interest before drawing final conclusions. Histograms or normal probability plots of the residuals can be plotted for visual analysis of their distribution. Each regression method has several assumptions that must be met in order for the equation to be considered reliable. Assumptions, for example, for Method of Least Squares (MLS), must be tested when building a regression model. The following assumptions must be verified and satisfied when using the MLS method.

The model must be linear. MLS regression is used only when building a linear model. The linear relationship between the dependent and independent variables can be checked using a dot (scatter) diagram. The dot plot matrix can test all variables, provided that no more than 5 variables are used in total. In the case of multiple regression of each pair of variables, scatter diagrams should be constructed. In Figure 6 shown how a data set of linear relationships would appear on a scatterplot. If such a graph is obtained, as in Figure 7, which has an exponential form, then logarithmic transformation into a linear one can be used. In the absence of a linear relationship between the data points or if the data cannot be transformed into a linear one, it is necessary to perform a non-linear regression analysis, which will be discussed in detail in the following lectures.

The data must be randomly distributed. The data used in the regression analysis must be randomly distributed, that is, the data samples must not depend on an external factor. Arbitrary distribution can be tested by using discontinuities in the regression model. Covariances calculated as output from a regression model should not be correlated when plotted on a scatterplot or scatterplot matrix with the independent variables. In this case, the concept of the absence of *heteroskedasticity* is considered. For a linear relationship, if the variance is not constant and increases as the predictor (X) increases, this is called heteroskedasticity. In such cases, the standard error of the derivation cannot be relied upon, but the coefficients will still be unbiased.

Figure 6. An example of linear dependence of regression analysis

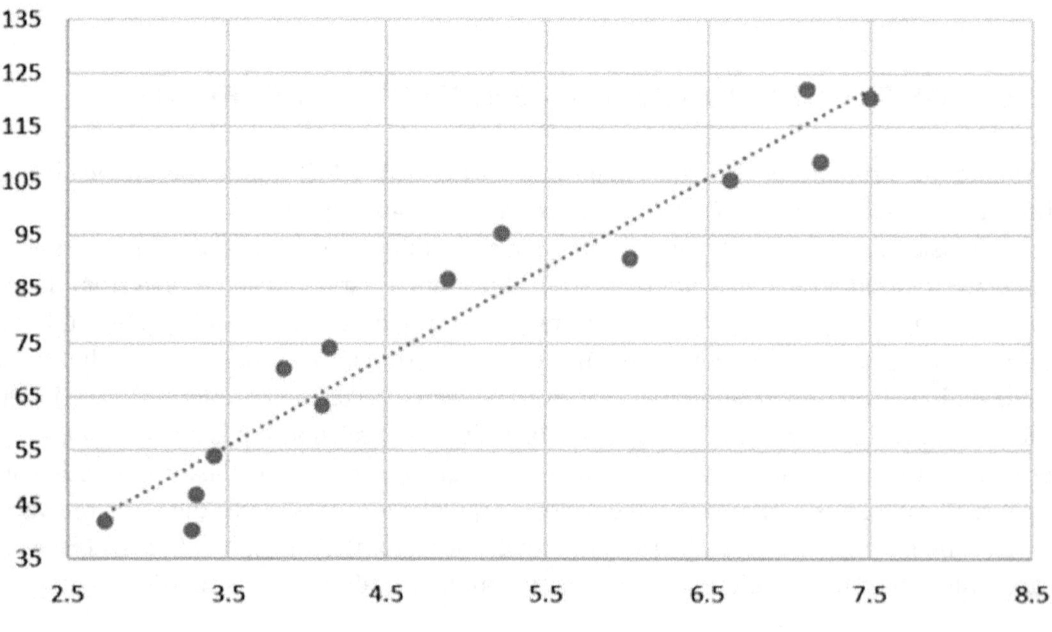

Figure 7. An example of the exponential nature of the regression dependence

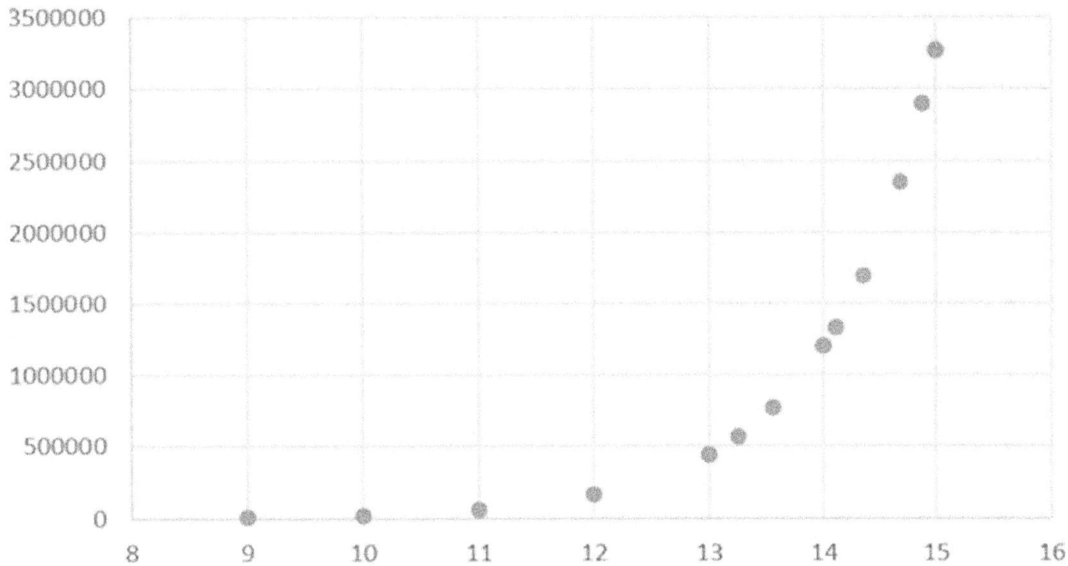

The best way to detect heteroskedasticity is to use a scatterplot of the residuals against the value of *X*. Heteroscedasticity can be avoided by performing a log transformation or by examining the bias of the omitted variables, or by identifying outliers and trying to remove them.

Independent variables should not be collinear. Collinearity is a linear relationship between independent variables that creates redundancy in the model. In some cases, the model is created with collinearity. However, if one of the collinear variables depends on the other, it may be worth removing it from the model. Collinearity can be estimated using a dot plot or dot plot matrix of independent variables. Predictors should not be correlated with the error term. This can be represented by the equation below:

$\sigma X\varepsilon=0\colon \forall x,\varepsilon.$

If the predictors are correlated, this is called omitted variable elimination. This occurs when a predictor is excluded, and this is reflected in the error term, leading to biased and counterintuitive estimates. It can be found by examining the correlation between predictors. Thus, when estimating the amount of wages based only on the number of years of education, the variable error is not taken into account. Because the salary depends on many other factors, such as the type of education, additional socio-economic status. But still, it can be used for predictions.

A concomitant feature of the correctness of regression analysis is the absence of multicollinearity. If the multiple linear regression is represented as

$$Y=\beta_0+\beta_1\left(X_1\right)+\beta_2\left(X_2\right)+\beta_3\left(X_3\right)+\beta_4\left(X_4\right)+\varepsilon,$$

then ideally no predictor should be explained by another predictor.

Multicollinearity occurs when the predictors themselves are correlated. To determine whether multicollinearity exists between predictors, the correlation between all pairs of predictors must be examined.

If the correlation coefficient is very high, it clearly indicates that the predictor is explained by another predictor. Multicollinearity can also be detected using the variance inflation factor (VIF). To find the VIF, it is necessary to perform an auxiliary regression on all predictors. For the above equation, the auxiliary regression equation for X_1

$$X_1 = \beta_0^* + \beta_2^*(X_2) + \beta_3^*(X_3) + \beta_4^*(X_4) + \varepsilon^*.$$

This will help to understand how much X_1 is explained by other predictors. If the R^2 of this model is equal to RX_1, then the variance inflation factor,

$$VIF = \frac{1}{1 - R - X_1^2}.$$

This means that the higher $R - X_1^2$, the higher it will be VIF. A larger VIF value indicates that X_1 is explained by other predictors. Likewise, the VIF of all other predictors should be calculated to test whether this predictor is explained by another predictor. To overcome this, it will be necessary to find out whether two predictors give the same information, in which case one of them is removed. When deleting, care must be taken to eliminate missing variables. Another method is to convert correlated predictors into a single predictor.

Independent variables should have a small measurement error. The accuracy of the regression model corresponds to the accuracy of the input data. If the independent variables have many errors, the model cannot be considered accurate. When performing regression analysis, it is very important to use data sets only from known and trusted sources to be sure that errors are small. It is assumed that the error ε has a normal distribution, that is, the mean value is zero (the sum of all errors will be zero (0) or almost 0), and the variance $\sigma 2$ error terms is equal to e~ N(0,2).

Normality is violated when the error variance is not consistent across observations. For example, when a linear regression model is used to express the insurance payout as a function of the customer's age. Insurance will not be claimed by all individuals who have chosen insurance. So there will be a lot of zero claims alongside a few very high claims. If the sample size is very large, the central limit theorem will apply and the error variance will be consistent across all observations. But for a small sample size, this will affect the standard errors of the inference. This can be easily detected by constructing a histogram of residuals.

The expected sum of non-connections should be equal to zero. The variance is the difference between the expected and observed values in a regression analysis. Values observed above the regression curve have a positive lag value, and values below the regression curve have a negative value. The regression curve should pass through the center of the data points. Accordingly, the sum of non-connections should tend to zero. The sum of the values can be determined in the summary table.

Disjoints must have uniform variability. The value of the variability must be the same for all non-connections. This assumption is tested using a scatterplot of correlations (y-axis) and estimated values (x-axis). The final dot plot is displayed as a horizontal bar with randomly scattered points across the area.

The distribution of non-connections should be normal. The normal distribution – a curve in the form of a parabola – is a natural distribution, where a high frequency of the phenomenon is observed close to the mean value, where the frequency decreases as the distance from the mean increases. In statistical analysis, the normal distribution is often used as a *null hypothesis*. If the distribution of linkage disequilibrium is normal, the line of best fit passes through the center of the observed data points, rather than diverging, approaching some and deviating from others. This assumption can be checked by constructing a histogram of the connections. The normal distribution curve may not fit on the card and the skewness and kurtosis are carried over to the back of the histogram card.

Adjacent lags should not exhibit autocorrelation. This assumption is based on the chronology of the data. If the data follows a timeline, each data point must be independent of the previous or subsequent data point. Therefore, when performing regression analysis, it is important to make sure that the chronological order of the data corresponds to the normal course of time. This assumption is calculated using the Durbin-Watson test.

The Durbin-Watson test measures the autocorrelation of the lags in a regression model. The Durbin-Watson test uses a scale from 0 to 4 where values from 0 to 2 indicate positive autocorrelation, 2 indicates no autocorrelation, and 2 to 4 indicates negative autocorrelation. That is, to meet the assumption of no autocorrelation of the links, a value close to 2 should be obtained. In general, values between 1.5 and 2.5 are considered acceptable, and values less than 1.5 or greater than 2.5 indicate that the model does not meet the assumption of no autocorrelation. The values of the error terms must not have any identifiable relationship: $\sigma \varepsilon i \varepsilon_j = 0_; \forall i \neq j$.

If there is any relationship between the values of the error term, then autocorrelation appears. Autocorrelation is also known as serial correlation, which affects the standard errors without affecting the coefficients, which will be unbiased. This is not observed in cross-sectional data, but is reflected in time-series data such as stock prices. For stock price analysis, the phenomenon of the day of the week effect suggests high returns on Fridays and low returns on Mondays. Monday's errors will be shifted down, and Friday's errors will be shifted up.

The main cause of autocorrelation is related to missing variables or incorrect functional forms. A common way to detect autocorrelation is to plot a scatterplot of all residuals and look for patterns. If there are no patterns, it means there is no autocorrelation. The Durbin-Watson test method can be used to find autocorrelation. If there is a pattern, autocorrelation exists and it is better to avoid linear regression models and use an autoregressive model, or a moving average model, or an autoregressive moving average model, or an autoregressive integrated moving average model.

Suitability of the model. The accuracy of the regression equation is the basis of regression analysis. All models will have some error, but understanding these statistics will help you determine whether the model can be used for your analysis, or whether additional transformations are necessary. There are two methods of testing the correctness of a regression model: **exploratory analysis** and **confirmatory analysis**.

Exploratory analysis is the technology of data analysis using various statistical and visual methods. As part of the exploratory analysis, the MLS regression assumption is tested and the performance of various independent variables is compared. Exploratory analysis allows to compare the performance and accuracy of different models, but it cannot determine whether you should use or reject a particular model. Exploratory analysis must be conducted before the analysis of confirmation of each regression model, possibly several times, comparing different models.

The following charts and statistics may be used as part of research analysis:

- dot diagram (scattering) and dot diagram matrix;
- histogram and analysis of normal distribution;

- regression equation and prediction of new observations;
- coefficient of determination, R^2 and adjusted R^2;
- standard error of the lag;
- dot chart.

Exploratory analysis begins when the independent variables are chosen and before the regression model is constructed. Since MLS is a linear regression method, the main assumption is that the model should be linear. A scatter plot (scatter) and a scatter plot matrix can be used to analyze the linear relationship between the dependent variable and the independent variables. A scatterplot matrix can display up to 4 independent variables with a dependent variable, allowing comparisons between all variables at once. A simple scatter plot can display only two variables: one dependent and one independent. Viewing a scatterplot with one dependent variable and one independent variable allows for a more accurate estimate of the relationship between the variables. Linearity can be tested before building a regression model to determine which independent variables should be used to create a suitable model. Several output statistics are also available after creating the regression model, including:

- egression equation;
- value R^2;
- Durbin-Watson criterion.

After creating the regression model, it is needed to use the output, as well as charts and tables, to test other assumptions of the MLS regression. If the created model satisfies the assumptions, the research analysis can be continued. The regression equation makes it possible to estimate the effect of each independent variable on the predicted values, including the regression coefficient of each independent variable. It is possible to compare the values of the slope of determining the impact of each independent variable on the dependent variable: the farther from zero the value of the slope (no matter whether in the positive or negative direction) – the greater the impact. A regression equation can also be used to predict the values of the dependent variable through the values of each independent variable.

The coefficient of determination, denoted as R^2, measures how well the regression equation models the actual data points. The R^2 value is a number in the range from 0 to 1, and the closer the value is to 1, the more accurate the model. If R^2 is 1, this indicates a perfect model, which is highly unlikely in real-world situations given the complexity of interactions between various factors and unknown variables. Therefore, one should strive to create a regression model with the maximum possible value of R^2, understanding that the value cannot be equal to 1.

When performing regression analysis, there is a risk of creating a regression model that has an acceptable R^2 value by adding independent variables that randomly show a good fit. The adjusted R^2 value, which should also be between 0 and 1, considers additional independent variables, reducing the role of randomness in the calculation. Adjusted R^2 should be used in a model with many independent variables or when comparing models with different numbers of independent variables.

The standard error of the variance measures the accuracy with which a regression model can predict values with new data. The smallest values indicate a more accurate model, respectively, when comparing several models, the one where this value is the smallest of all is the model in which the standard error of the discrepancy is minimized. A scatterplot can be used to analyze the independent variables to identify clustering or outliers that may affect model accuracy.

Confirmatory analysis is the process of evaluating a model against the null hypothesis. In regression analysis, the null hypothesis states that there is no relationship between the dependent and independent variables. For a model with no relationships, the value of the slope is 0. If the elements of the confirmation analysis are statistically significant, the null hypothesis can be rejected (in other words, the existence of a relationship between the dependent and independent variables is statistically confirmed). The following statistical indicators are used to determine significance as a component of the analysis:

- F-statistics and the associated p-value;
- t-statistics and the associated p-value;
- confidence intervals.

The *F-statistic* is a global statistic returned by the F-test that shows the predictive power of the model by calculating regression coefficients in the model that are significantly different from 0. The F-test analyzes the combined effect of independent variables rather than evaluating each one individually. Associated with the F-statistic is the corresponding p-value, which is a measure of the probability that the deterministic relationship between the variables is due to chance. A small p-value, usually 0.05 or less, indicates that there are real relationships between variables in the model (that is, the revealed regularity is not random), which gives the right to reject the null hypothesis. In this case, the probability that the relationship in the model is random is 0.05, or 1 in 20. Or the probability that the relationship is real is 0.95, or 19 in 20.

The *t-statistic* is a local statistic returned by the t-test that shows the predictive power of each independent variable separately. Like the F-test, the t-test analyzes regression coefficients in a model that are significantly different from 0. Because the t-test is applied to each independent variable, the model will return a t-statistic value for each independent variable, rather than a single value for of the entire model. Each t-statistic value has an associated p-value that indicates the significance of the independent variable.

As with the F-test, the p-value for each t-test must be 0.05 or less to reject the null hypothesis. If the p-value for an independent variable is greater than 0.05, that variable should not be included in the model and a new model should be built, even if the global probability value for the original model indicates statistical significance.

Confidence intervals visualize the regression coefficients for each independent variable and can be 90, 95, and 99 percent. Therefore, confidence intervals can be used alongside t-test p-values to assess the significance of the *null hypothesis* for each independent variable. The regression coefficients must be equal to 0, only in this case you can reject the null hypothesis and continue using the model. Therefore, for each independent variable, the regression coefficient and the associated confidence interval cannot overlap with 0. If the 99% or 95% confidence intervals for this independent variable overlap with 0, this independent variable does not allow us to reject the null hypothesis. Including this model variable may negatively affect the overall significance of your model. If only the 90% confidence interval overlaps 0, the variable can be included in a model with satisfactory overall statistical significance. Ideally, the confidence intervals for all independent variables should be as far from 0 as possible.

Other inputs, such as predicted values and lags, are also important for MLS regression assumptions. *Expected values* are calculated based on the regression equation and the values of each independent variable. Ideally, the expected values should match the observed (actual values of the dependent variable). The expected values, along with the observed values, are used to calculate the correlations. Bias in regression analysis is the difference between the values in the observed data set and the expected values calculated using the regression equation (Figure 8).

The inconsistencies *A* and *B* for the relations above are calculated as follows:

varianceA = observationA – expectedA · variance

varianceB = observationB – expectedB · varianceB

The residuals are used to calculate the error of the regression equation, as well as to test some assumptions. The main conceptual limitation of all considered methods of regression analysis is that they allow to reveal only numerical dependencies, and not the causal relationships underlying them. For example, a strong positive relationship (correlation) can be found between the destruction caused by fire and the number of firefighters involved in fighting the fire. Of course, the most likely explanation for this correlation is that the size of the fire (an extraneous variable neglected to include in the study) affects both the extent of destruction and the number of firefighters required (i.e., the larger the fire, the more firefighters are called from it quenching). Although this example is quite transparent, for correlations, alternative causal explanations are often even considered.

Figure 8. An example of determining the gap

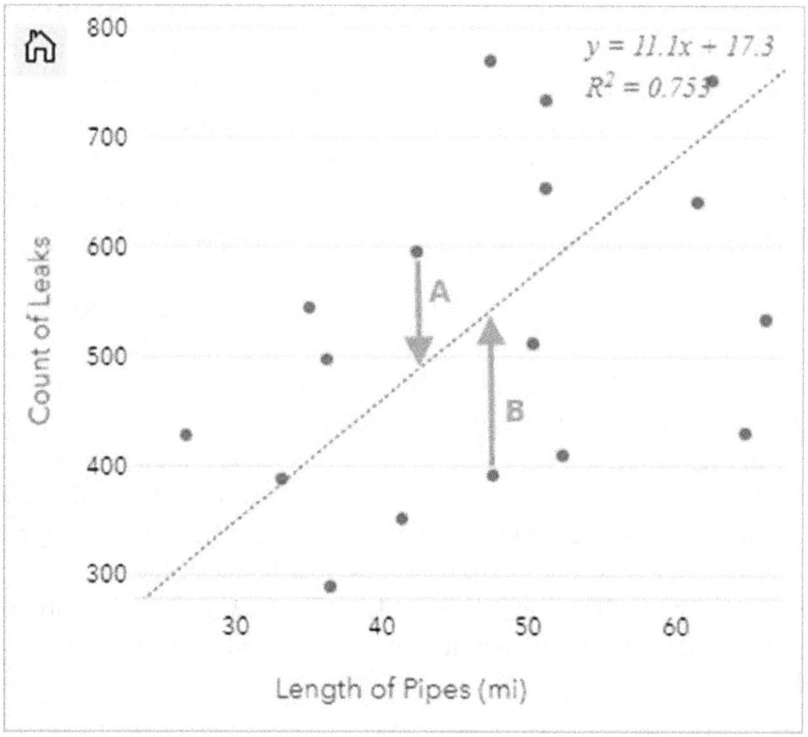

EXPERIMENT DEVELOPMENT IN THE MICROSOFT AZURE MACHINE LEARNING STUDIO

ML is the use of mathematical models of data to help a computer learn without instructions. It is considered one of the forms of artificial intelligence (AI). In ML with the help of algorithms, data patterns are revealed. Based on these regularities, a data model for forecasting is created. The more data such a model processes and the longer it is used, the more accurate the results become. This is very similar to how a person hones skills in practice. The adaptive nature of ML makes it great for scenarios where data is constantly changing, query or problem properties are unstable, or writing code to solve is virtually impossible.

Usually, when solving complex problems of predictive analytics with a large amount of both input data and data categories, a problem arises in the need for large computing capacities. One of the solutions to the problem arising from complex calculations is the use of the computing power of the Microsoft cloud cluster and machine learning technology based on the Microsoft Azure service.

The result of this work is an independent service built into Excel, which allows to perform a consumption forecast by the specified date. Separate results of the forecast are the assessment of the quality of the obtained indicators of electricity consumption.

Any predictive experiment in AML studio is based on a set of input data. The more of them and the more diverse they are in determining the target parameter, the more accurate the forecasting results will be. The Excel file for this work was provided by the forecasting department of NEC "Ukrenergo", Ukraine (Pliuhin et al., 2021) and contains data[1] on the time and date of readings, energy production by various types of power plants (Figure 9): AES is the nuclear power plant (spells as Atomic Electric

Figure 9. Data sampling from the task table

	A	B	C	D	E	F
1	Num	Date	AES	TES	Block TES	HES
2	1	1-01.01.2018	9599	4348	779	1017
3	2	2-01.01.2018	9588	3839	776	772
4	3	3-01.01.2018	9576	3532	776	457
5	4	4-01.01.2018	9603	3513	777	496
6	5	5-01.01.2018	9598	3485	776	816
7	6	6-01.01.2018	9594	3447	777	699
8	7	7-01.01.2018	9607	3405	778	324
9	8	8-01.01.2018	9596	3352	778	144
10	9	9-01.01.2018	9592	3315	777	433
11	10	10-01.01.2018	9607	3400	778	553
12	11	11-01.01.2018	9593	3480	780	905
13	12	12-01.01.2018	9587	3515	783	1012
14	13	13-01.01.2018	9611	3494	782	1008
15	14	14-01.01.2018	9617	3542	782	961
16	15	15-01.01.2018	9626	3512	783	1110

Station); TES – is the thermal power plant (spells as Thermo Electric Station); Block TES is the block thermal power plant (spells as Block Thermo Electric Station); HES is the hydroelectric power station (spells as Hydro Electric Station).

The *Date* column contains data in the following format: time – "dash" – date. Thus, there are 24 measurements for each date. For 2018, the table contains 8,760 rows of measurement data. To use Excel data in ML Studio, the *xlsx* file must be saved as a *csv* (comma separated data) file. But, due to the European localization, such file does not contain commas as data separators, but semicolons (Figure 10), which is a common problem in programming when converting data from a graphic user interface into a program code.

This file can be opened in Windows Notepad and automatically replace the "semicolon" symbol with "comma". The result of such replacement is shown in Figure 11.

The generated data is added as a dataset to the ML studio (Figure 12). The next step is to create a new experiment in Microsoft AML Studio. The first block of such experiment is the dataset input data in the *My Datasets* category, which was prepared earlier. It is enough to drag the dataset from the side menu to the model field (Figure 13).

If to preview the table data (Figure 14), it contains 653 missing values (empty cells or rows). To automatically extract such data, the Clean Missing Data block is used, with the settings shown in Figure 15.

In the data table, as indicated above, one date has 24 measurements, that is, every hour. The use of such format is atypical and cannot be properly analyzed. There are no library units in the AML studio to solve such kind of problem. For such cases, a common issue is to include existing *Python* scripts in AML Studio experiments. The *Execute Python Script* module accepts a *zip* file containing *Python*

Figure 10. csv data file – semicolon separated data

Table_2018.csv

```
Num;Date;AES;TES;Block TES;GES
1;1-01.01.2018;9599;4348;779;1017
2;2-01.01.2018;9588;3839;776;772
3;3-01.01.2018;9576;3532;776;457
4;4-01.01.2018;9603;3513;777;496
5;5-01.01.2018;9598;3485;776;816
6;6-01.01.2018;9594;3447;777;699
7;7-01.01.2018;9607;3405;778;324
8;8-01.01.2018;9596;3352;778;144
9;9-01.01.2018;9592;3315;777;433
10;10-01.01.2018;9607;3400;778;553
11;11-01.01.2018;9593;3480;780;905
12;12-01.01.2018;9587;3515;783;1012
13;13-01.01.2018;9611;3494;782;1008
14;14-01.01.2018;9617;3542;782;961
15;15-01.01.2018;9626;3512;783;1110
16;16-01.01.2018;9615;3645;782;1571
17;17-01.01.2018;9594;3807;783;2230
18;18-01.01.2018;9605;3992;782;1909
19;19-01.01.2018;9608;3992;783;2060
20;20-01.01.2018;9607;3846;785;1923
21;21-01.01.2018;9607;3848;781;1845
22;22-01.01.2018;9607;3908;783;1704
23;23-01.01.2018;9588;3890;784;1079
24;24-01.01.2018;9606;3873;783;591
```

Figure 11. csv data file – comma separated data

	A	B	C	D
1	Num,Date,AES,TES,Block TES,GES			
2	1,1-01.01.2018,9599,4348,779,1017			
3	2,2-01.01.2018,9588,3839,776,772			
4	3,3-01.01.2018,9576,3532,776,457			
5	4,4-01.01.2018,9603,3513,777,496			
6	5,5-01.01.2018,9598,3485,776,816			
7	6,6-01.01.2018,9594,3447,777,699			
8	7,7-01.01.2018,9607,3405,778,324			
9	8,8-01.01.2018,9596,3352,778,144			
10	9,9-01.01.2018,9592,3315,777,433			
11	10,10-01.01.2018,9607,3400,778,553			

Figure 12. Project data uploaded to Machine Learning Studio

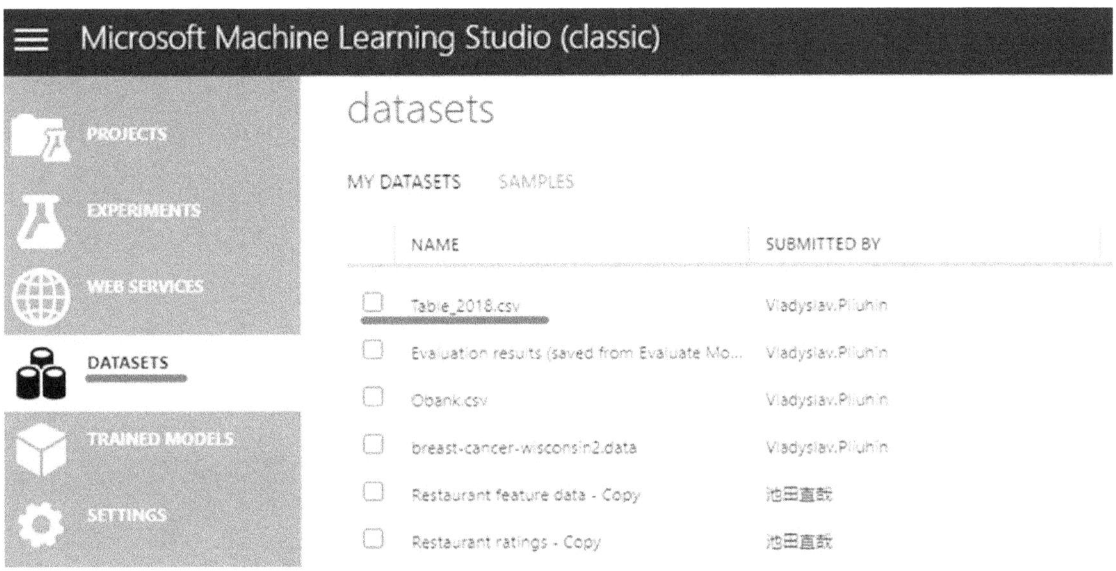

modules on the third input port. The *Anaconda* environment installed in AML Studio includes common packages such as *NumPy*, *SciPy*, and *Scikits-Learn*. These packages can be effectively used to process data in a machine learning pipeline.

Extracting time value from *Date* data series is done using the *Execute Python Script* block, for example: value **01-01.01.2018** change on **01.01.2018**.

Python code and model blocks are shown in Figure 16. In Figure 17 shown the view of the data table after executing the script.

Figure 13. Output table block in MS Azure model field

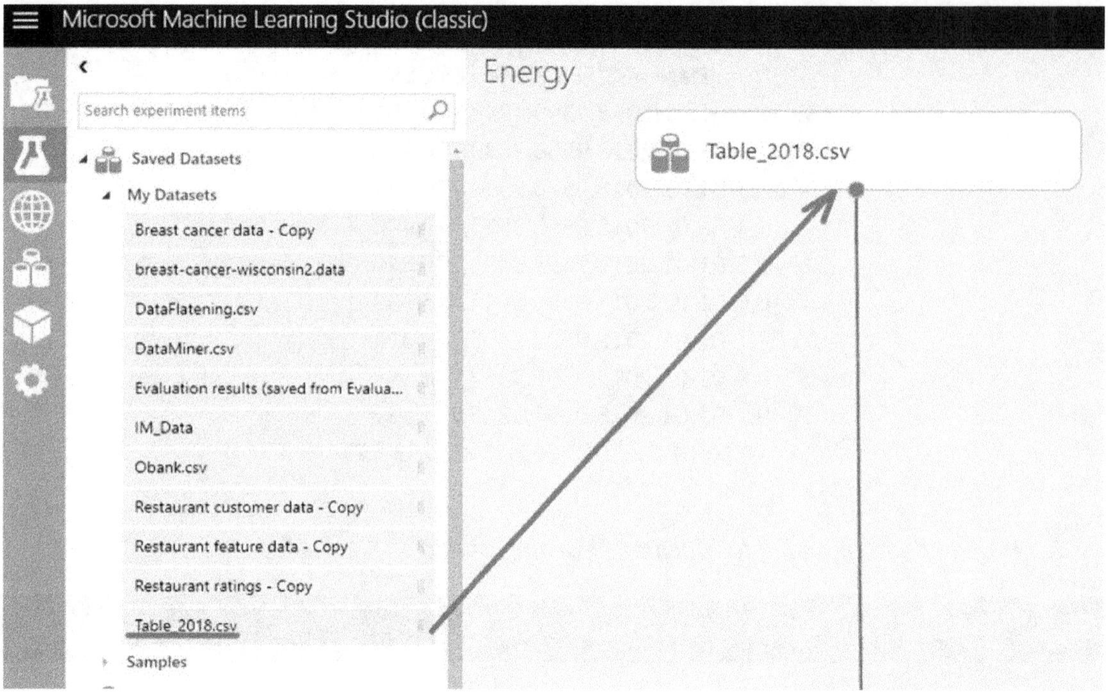

Figure 14. Analysis of input data

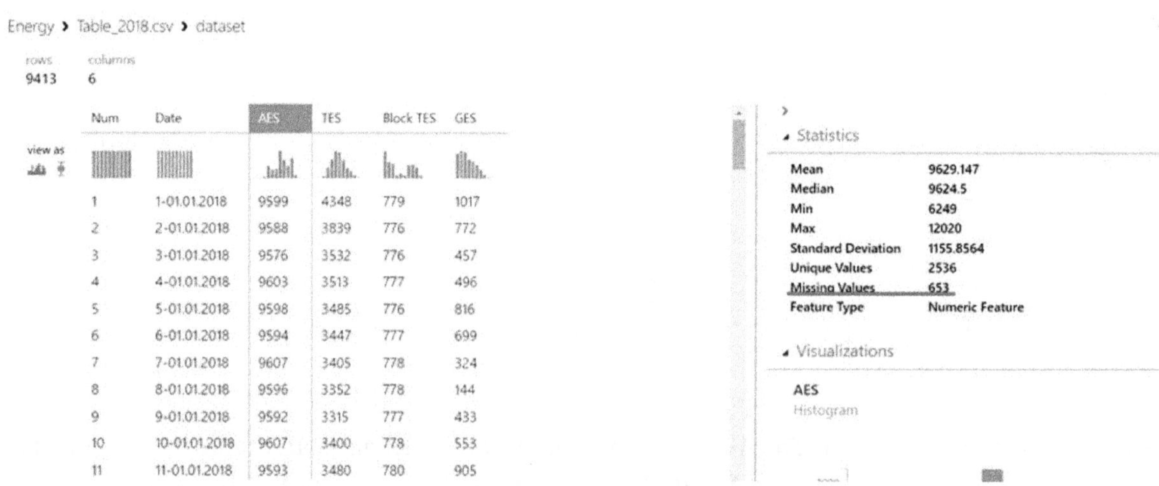

DEVELOPMENT AND INTEGRATION OF USER PYTHON BLOCKS

For further analysis, not all the data of the table will be needed, but only three columns: row number, date, AES, since nuclear power plants were chosen as an example. The *Select Columns in Dataset* block is used to select the required columns and ignore others (Figure 18).

The following consists of two tasks:

Figure 15. Block for removing rows with lost data

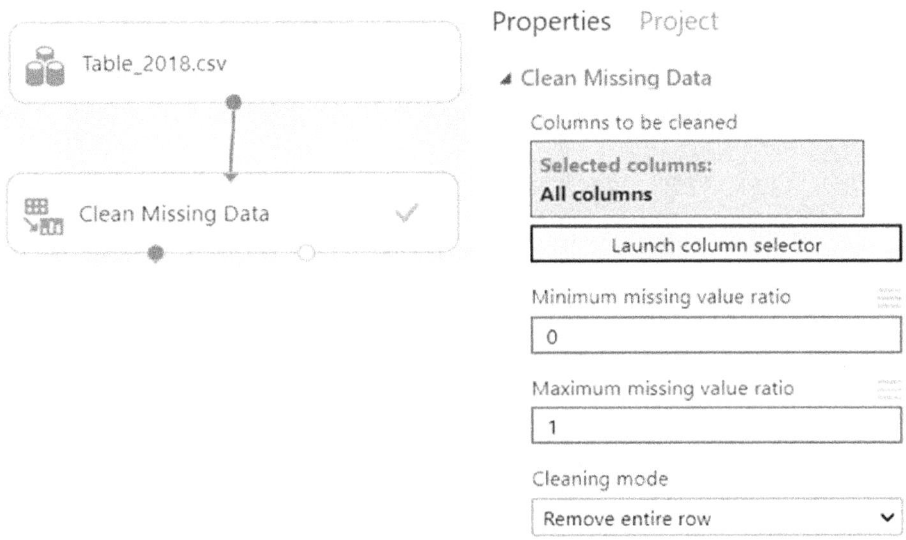

Figure 16. Model and Python script for measuring date correction

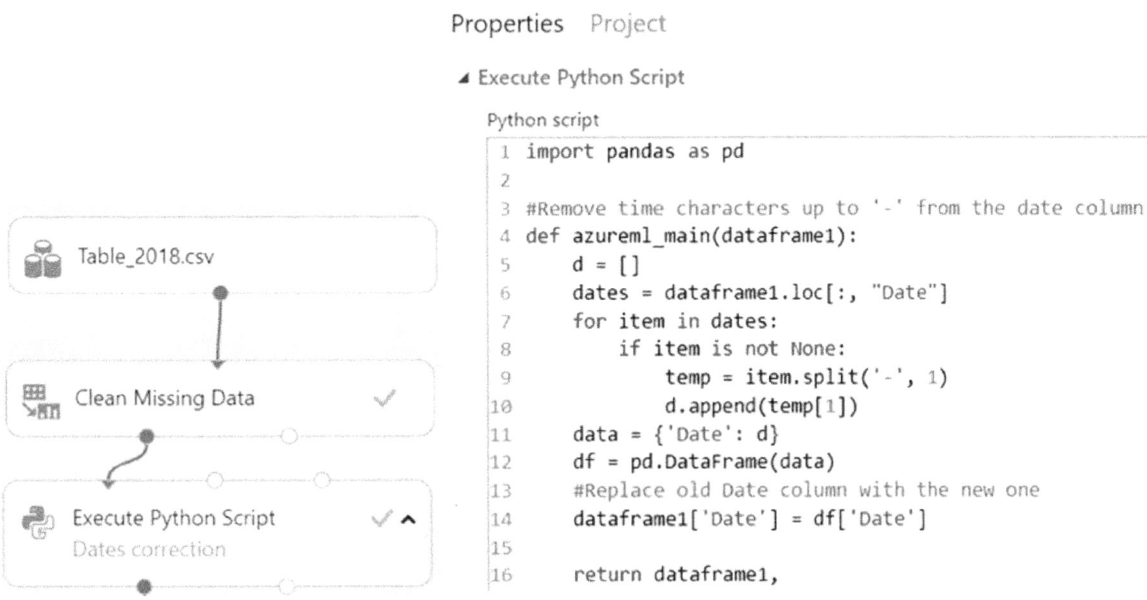

1) create a new column in which to automatically determine the name of the month by date - this will improve the identification of data by an additional feature (or marker);
2) combine readings within one date, as shown in Figure 19. Thus, the dates will not be repeated, and the consumption value for each date is defined as the arithmetic mean of measurements for 24 hours.

Figure 17. Data table after executing the date correction script

Energy › Execute Python Script › Results dataset

rows 8759 columns 6

Num	Date	AES	TES	Block TES	GES
1	01.01.2018	9599	4348	779	1017
2	01.01.2018	9588	3839	776	772
3	01.01.2018	9576	3532	776	457
4	01.01.2018	9603	3513	777	496
5	01.01.2018	9598	3485	776	816
6	01.01.2018	9594	3447	777	699
7	01.01.2018	9607	3405	778	324
8	01.01.2018	9596	3352	778	144
9	01.01.2018	9592	3315	777	433
10	01.01.2018	9607	3400	778	553
11	01.01.2018	9593	3480	780	905

Figure 18. Selection of individual columns from the source table

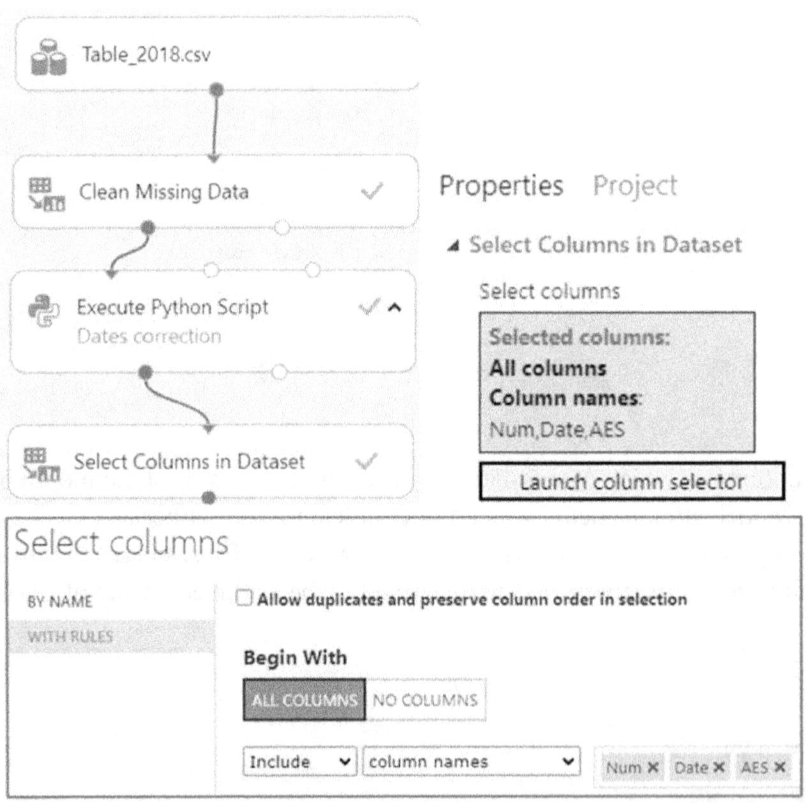

Figure 19. Combining data within one date

Num	Date	AES
1	01.01.2018	9599
2	01.01.2018	9588
3	01.01.2018	9576
4	01.01.2018	9603
5	01.01.2018	9598
6	01.01.2018	9594
7	01.01.2018	9607
8	01.01.2018	9596
9	01.01.2018	9592
10	01.01.2018	9607
11	01.01.2018	9593
12	01.01.2018	9587
13	01.01.2018	9611
14	01.01.2018	9617
15	01.01.2018	9626
16	01.01.2018	9615
17	01.01.2018	9594
18	01.01.2018	9605
19	01.01.2018	9608
20	01.01.2018	9607
21	01.01.2018	9607
22	01.01.2018	9607
23	01.01.2018	9588
24	01.01.2018	9606

AES	Date	Month	Num
9601	01.01.2018	January	1

These operations are shown in Figure 20 and Figure 21 in the *Execute Python Script* block. The result of both scripts is shown in Figure 22 (excerpts from different time periods are shown). In addition, the script shown in Figure 21, allows to plot the distribution of energy production by the measurement date (Figure 23).

The next operation involves dividing the data into training and testing the model in the ratio of 75% and 25%, respectively. For this, the *Split Data* block is used (Figure 24).

The output of the left port of the *Split Data* block and the output of the *Linear Regression* analysis block are fed to the input of the *Train Model* creation block. The level of AES energy production is chosen as the target criterion for the *Train Model* block (Figure 25).

The Score Model block with default settings is used to obtain a prediction of the level of energy production. The forecast result (Scored Labels column) is shown in Figure 26. As can be seen from the figure, the accuracy of the forecast is quite high.

The last block *Evaluate Model* is used to assess the quality of the executed forecast. The complete model together with this block is shown in Figure 27. The value of statistical indicators of the quality of forecasting is shown in Figure 28. In this way, the algorithm of work of a real scientist is implemented – checking the adequacy of calculated data in relation to those obtained experimentally.

As a result of executing the project in the AML Studio workspace, we get an estimate of the deviation of the parameters from the reference parameters.

REPRODUCTION OF THE PREDICTIVE EXPERIMENT

In the educational experiment, all processes of learning and evaluating the model were performed using ready-made data. But after deployment, the model will accept any new data from users and return forecast results to them. Therefore, when turning a training experiment into a predictive experiment and during further deployment, one should remember how this model will be used.

The first step in the transformation of the model and it performed automatically due to the *Setup Web Service* AML studio function (Figure 29). As a result of performing the selected operation, the model will turn from an educational experiment (see Figure 27) to a predictive one (Figure 30).

But in this form, the model is too complex and contains extra blocks that are not needed at the testing stage. First, it is needed to save as a new dataset the data from the port of the *Execute Python Script* block, which is attached to the *Score Model* block. It serves only one purpose - to give names for data input fields when testing a model.

The automatically generated dataset will contain the correct format of this data. Second, it is needed to remove all blocks except *Web Service, Trained Model*, and *Score Model*. The final view of the forecast model is shown in Figure 31.

Figure 20. Model and Python script to add the additional month column

operties Project

Execute Python Script

Python script

```
import pandas as pd
import datetime as dt


def azureml_main(dataframe1):
    month = ['January', 'February', 'March', 'April', 'May', 'June', 'July',
             'August', 'September', 'October', 'November', 'December']
    date_format = '%d.%m.%Y'
    dates = dataframe1.loc[:, "Date"]
    m = []
    for text in dates:
        d = dt.datetime.strptime(text, date_format)
        index = d.month - 1
        m.append(month[index])
    data = {'Month': m}
    dataframe1['Month'] = data['Month']

    return dataframe1,
```

Figure 21. Python model and script to merge data within a date range

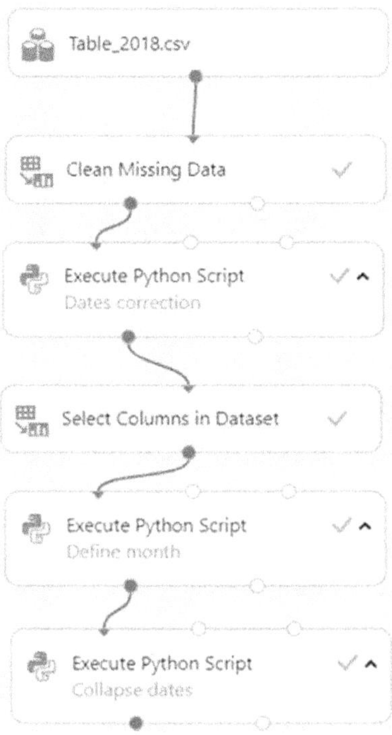

Python script

```
import pandas as pd
import matplotlib
import matplotlib.pyplot as plt

def azureml_main(dataframe1):
    matplotlib.use("agg")
    size = len(dataframe1)
    df2 = pd.DataFrame()
    n = []
    d = []
    m = []
    p = []
    count = 1
    num = 1
    temp = 0
    for x in range(size):
        temp += dataframe1.at[x, 'AES']
        date = dataframe1.at[x, 'Date']
        month = dataframe1.at[x, 'Month']
        if count == 24:
            s = temp / 24
            p.append(s)
            n.append(num)
            d.append(date)
            m.append(month)
            num += 1
            count = 0
            temp = 0
        count += 1
    data = {'Num': n, 'Date': d, 'Month': m, 'AES': p}
    df2 = pd.DataFrame(data)

    # plotting a line graph
    fig = plt.figure()
    plt.plot(df2["Num"], df2["AES"])
    plt.xlabel("Days")
    plt.ylabel("Power, kW")
    plt.grid()
    fig.savefig("chart.png")

    # Return value of pandas.DataFrame
    return df2,
```

To run a predictive experiment, the *Run* button at the bottom of the experiment area executes the corresponding procedure. After the experiment is completed, a new web service is created using the

Figure 22. The result of executing scripts for processing table data

Energy › Execute Python Script › Results dataset

rows 364 columns 4

AES	Date	Month	Num
9601	01.01.2018	January	1
9592	02.01.2018	January	2
8496	03.01.2018	January	3
8452	04.01.2018	January	4
8515	05.01.2018	January	5
8586	06.01.2018	January	6
8590	07.01.2018	January	7
8855	08.01.2018	January	8
9613	09.01.2018	January	9
9610	10.01.2018	January	10
9570	11.01.2018	January	11

Energy › Execute Python Script › Results dataset

rows 364 columns 4

10961	23.02.2018	February	54
10967	24.02.2018	February	55
10978	25.02.2018	February	56
10980	26.02.2018	February	57
10978	27.02.2018	February	58
10993	28.02.2018	February	59
11003	01.03.2018	March	60
11012	02.03.2018	March	61
10996	03.03.2018	March	62
10998	04.03.2018	March	63
10997	05.03.2018	March	64
10995	06.03.2018	March	65
10986	07.03.2018	March	66
10986	08.03.2018	March	67

Figure 23. AES energy production by measurement date – result of executing the Python script (see Figure 13)

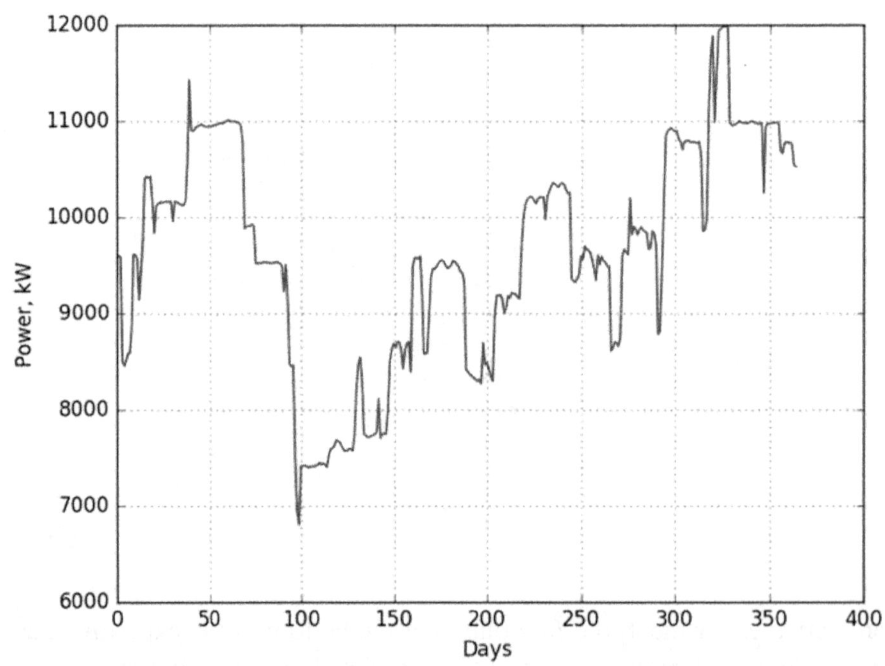

Figure 24. Setting up the data separation block

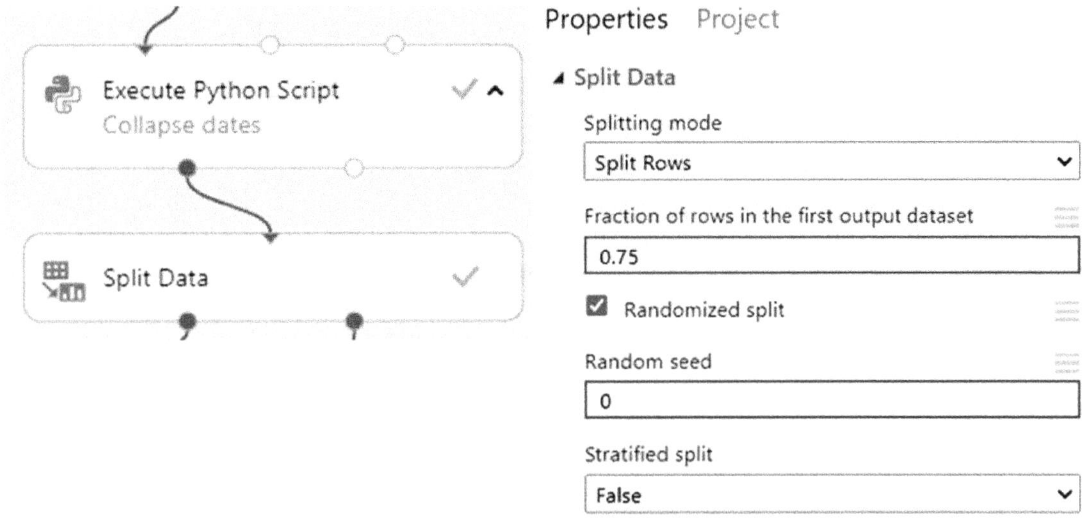

Figure 25. Sample target indicator for the Train Model block

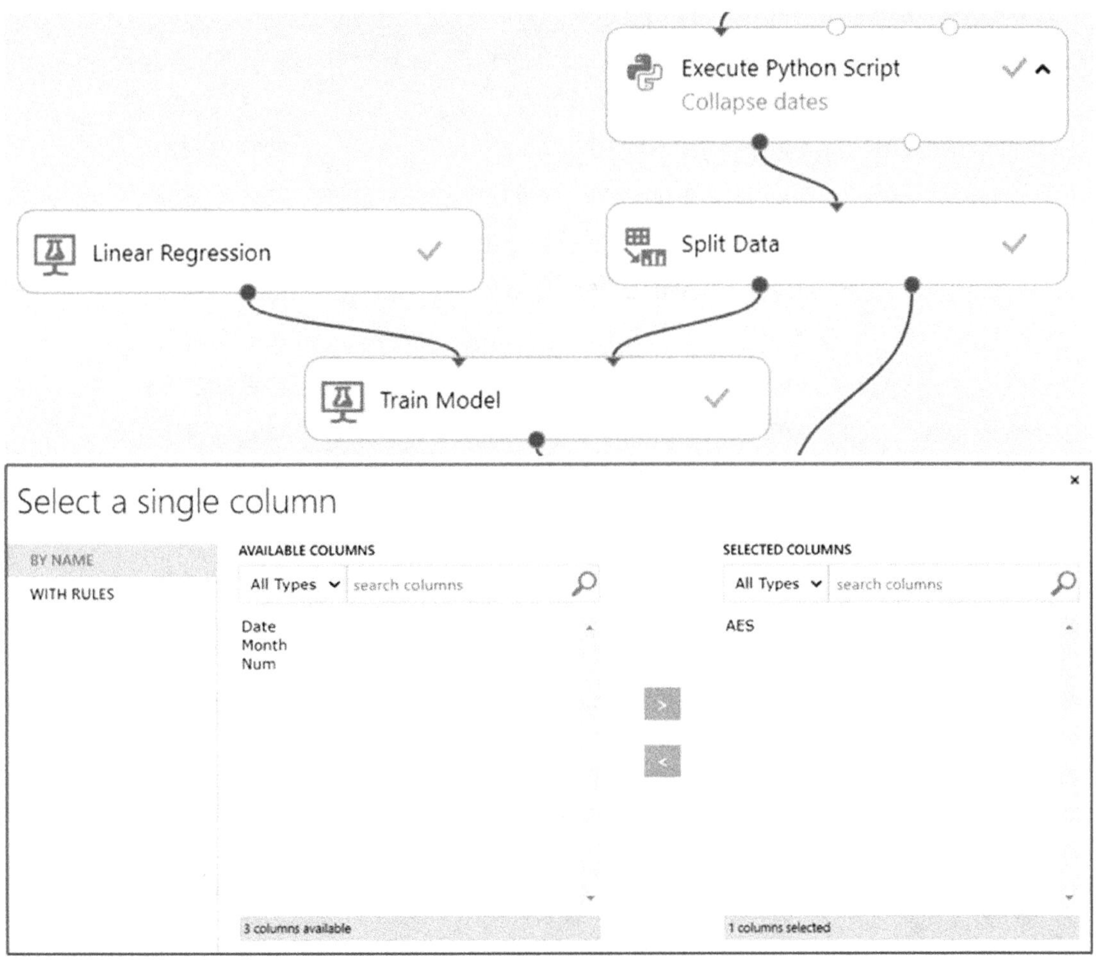

Figure 26. Prediction results

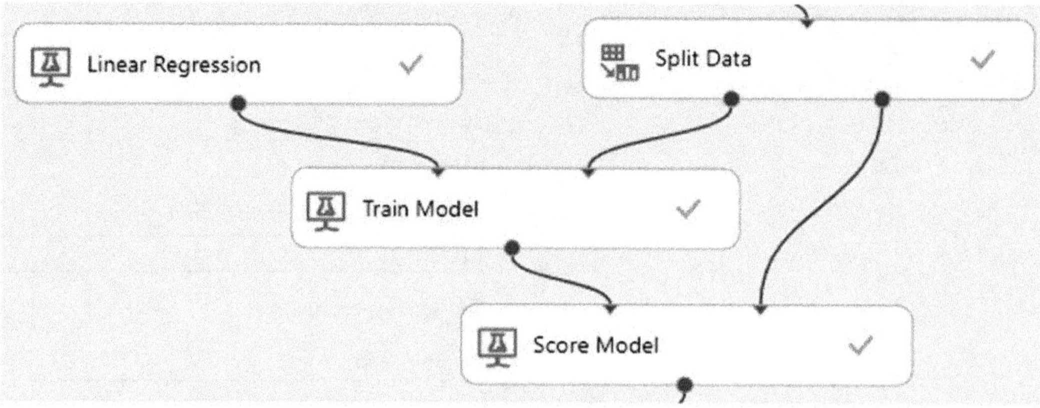

Energy › Score Model › Scored dataset

rows 91 columns 5

view as

AES	Date	Month	Num	Scored Labels
8003	27.05.2018	May	147	8103.390386
7445	20.04.2018	April	110	7877.756672
7404	15.04.2018	April	105	7815.999504
10209	18.08.2018	August	230	10107.121502
9523	20.03.2018	March	79	10076.394031
10111	21.01.2018	January	21	9629.012708
10942	12.02.2018	February	43	10598.821386
10787	01.11.2018	November	305	11028.024156
9685	14.10.2018	October	287	10016.193707
7732	19.05.2018	May	139	8004.578917
10961	27.11.2018	November	331	11349.161431

Deploy Web Service command. At the same time, a window will open with the parameters of the web service and tools for testing it (Figure 32).

The trained model for forecasting electricity production can already be used directly in the browser by clicking on the *Test* button (Figure 33). Additionally, the web service can be used in Excel. For this, it is enough to download the *Excel 2019 or later.xlsx* file from the web service settings window (see Figure 32). In the Excel file, it is needed to specify the row with the source data and the cell for displaying the

Figure 27. Complete machine learning model for forecasting AES energy production

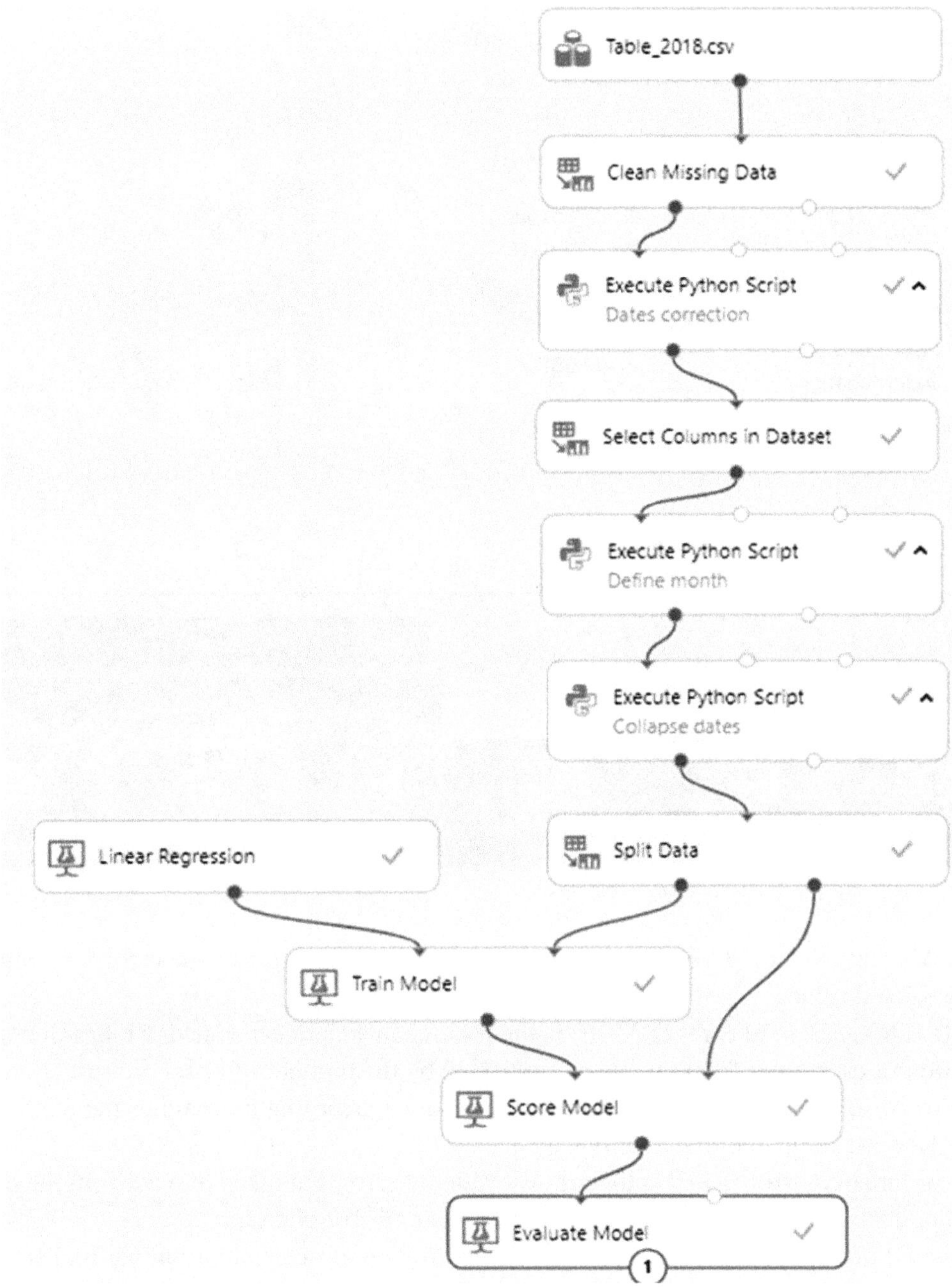

results in the web service settings. After click on the *Predict* button, the prediction result will be shown in the Excel field (Figure 34).

Based on the results of the performed work, forecasting models were developed for other types of power stations as well (see Figure 9). The obtained metrics are the result of the estimation of the forecasting model:

Figure 28. Statistical indicators of the predictive model

Figure 29. Converting a model to a web service

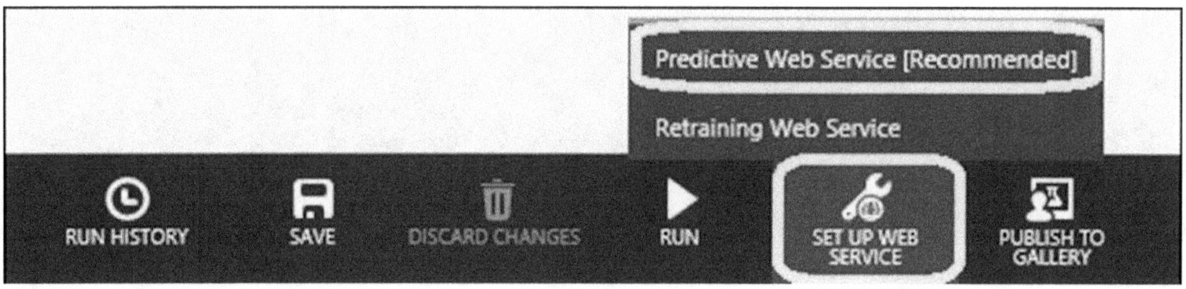

- Mean Absolute Error (MAE) – defines the average absolute value of the error between the actual and predicted values.
- The Root Mean Square Error (RMSE) is the root mean square error, which takes into account the deviation of each point from the mean multiplied by the number of observations.
- Relative Absolute Error (RAE) is the relative absolute error that determines the relative accuracy of the forecast compared to the output.
- Relative Squared Error (RSE) is the relative squared error that takes into account the deviation of each point from the mean multiplied by the number of observations.
- Coefficient of determination (R-squared) is a coefficient of determination that indicates how well the model explains the variance of the original data. The overall rating for each source of electricity is presented in the Table. 2.

Analyzing the obtained data, the following conclusions can be made. AES has a high coefficient of determination (R^2) indicating a high degree of correspondence between actual and predicted values (approximately 95.5%). Low values of MAE and RMSE indicate sufficient accuracy of the model in predicting energy values. TES has a high coefficient of determination and low values of MAE and RMSE, which indicate the high accuracy of the model in predicting energy parameters. Block TES shows good

Figure 30. Model of the predictive experiment

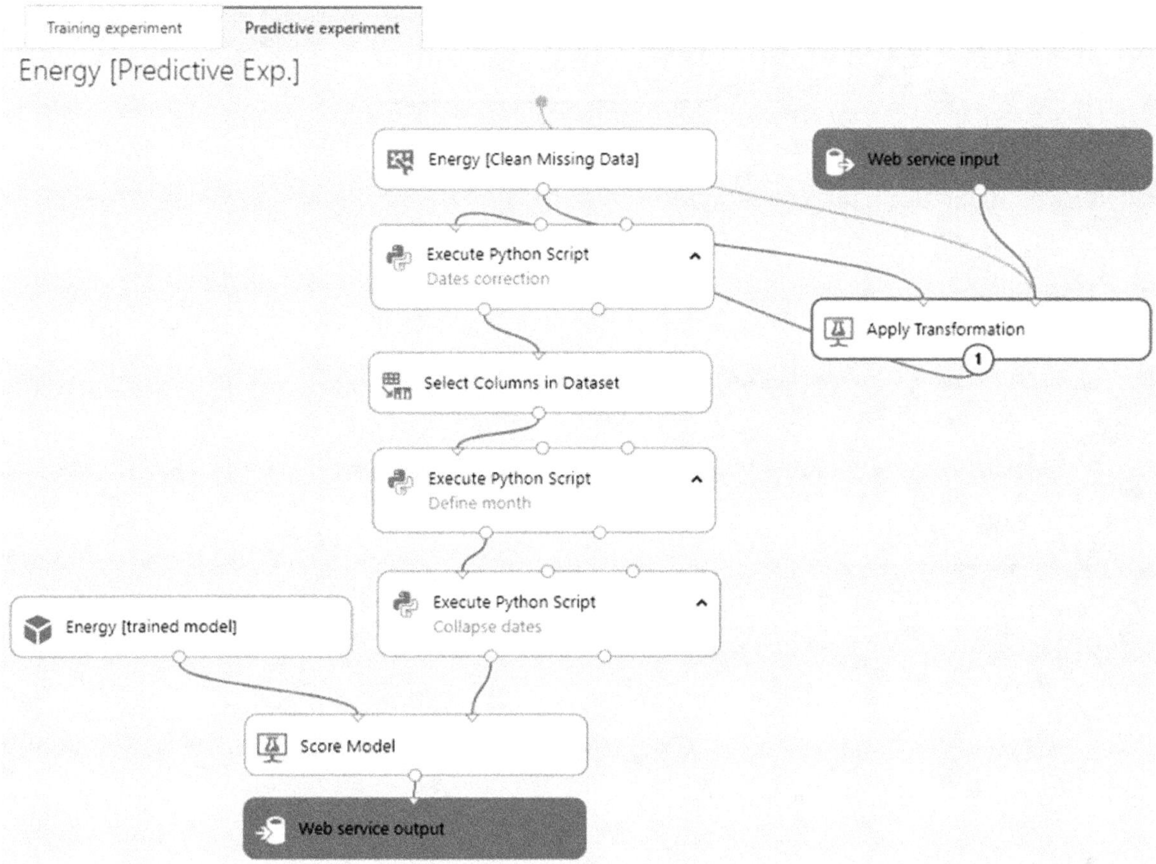

Figure 31. Final model of the predictive experiment

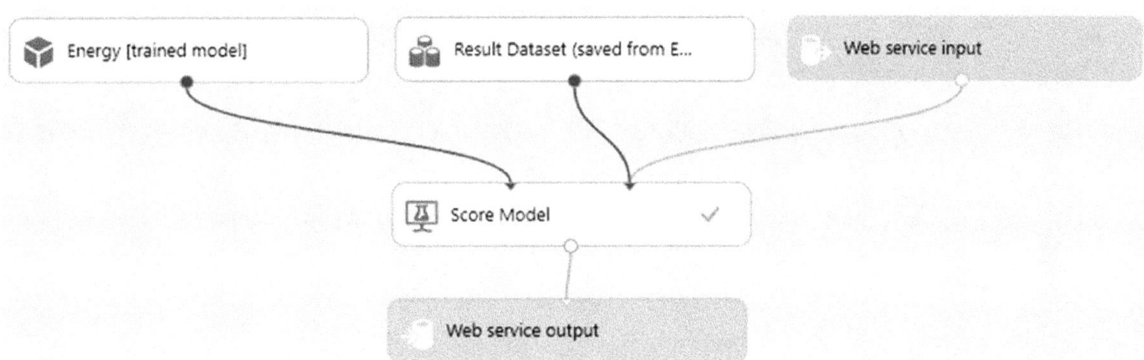

results with prediction accuracy expressed by high R-squared and low MAE and RMSE. As for GES, the high coefficient of determination and low values of RAE and RSE indicate the accuracy of the model for solar power plants. And already in WPP, the model has limited accuracy, which is expressed by a low coefficient of determination and large values of RAE and RSE. The general trend is that the models

Figure 32. Predictive Web Service Options window

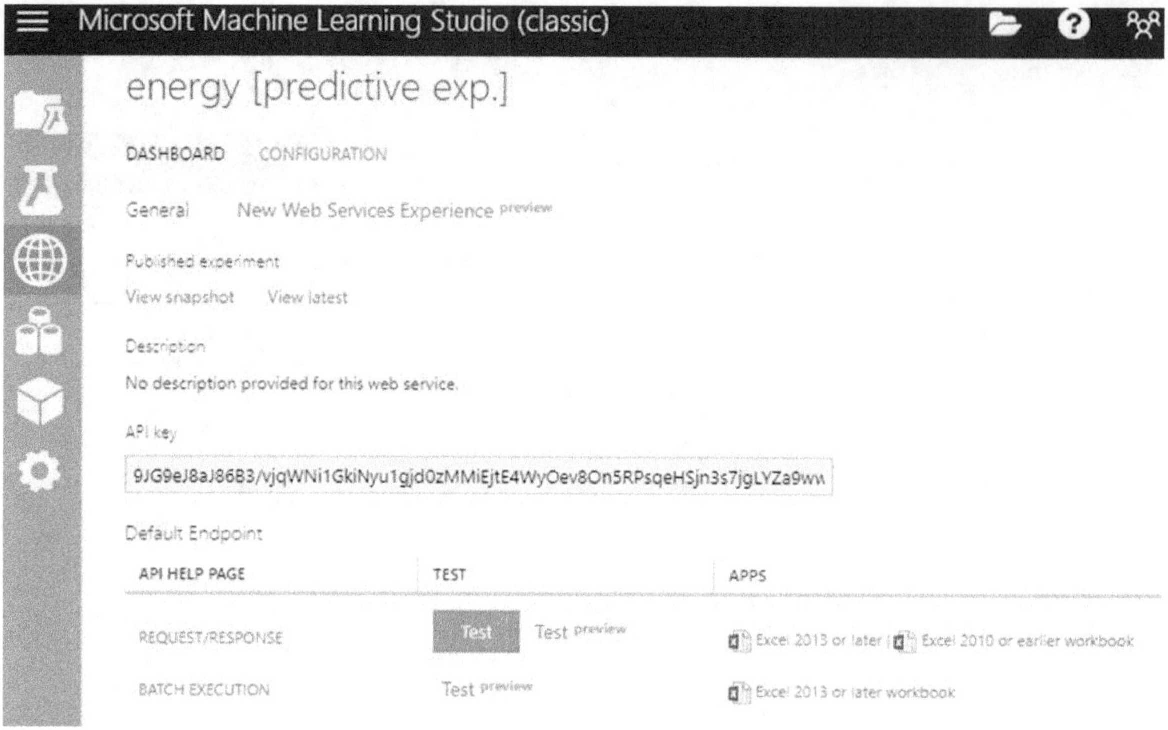

Figure 33. Testing a predictive web service

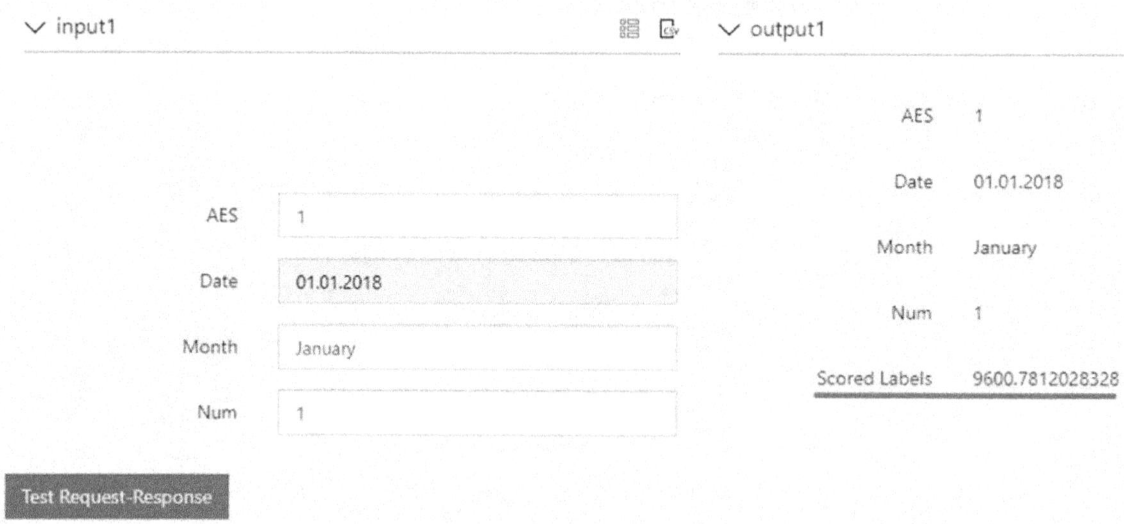

for AES, TES, and Block TES proved to be quite accurate and efficient in predicting energy parameters. In contrast, the model for GES appears to be less efficient, possibly due to a greater dependence on variational wind conditions.

Figure 34. Using the forecast service in Microsoft Excel

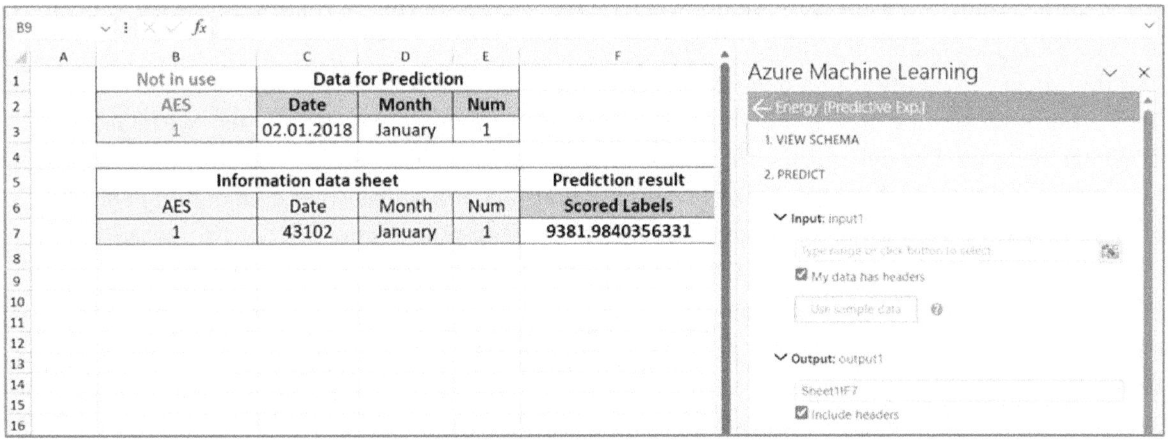

Table 2. Evaluation of accuracy of models for various energy objects

Model	MAE	RMSE	RAE	RSE	R-Squared
AES	97.2	118.55	0.19	0.045	0.96
TES	109.94	137.11	0.11	0.012	099
Block TES	54.94	70.19	0.18	0.042	0.96
GES	17.08	20.42	0.37	0.152	0.85

While modeling electricity consumption and production, it was possible to create a unique model that considers complex factors affecting energy processes. The obtained results consider the impact of various factors, which were divided into two categories: 1) original: such as time, date and types of power plants, on the consumption and production of electricity, providing valuable insights for optimizing energy processes; 2) generated by software inside MS Azure ML Studio using Python scripts: month names, date format, data counter, average electrical load etc. The modeling results presented in this paper should be considered in the context of the particularities of the energy domain and the accuracy requirements to ensure realistic predictions. The developed model is an important tool for improving management and planning in the power industry. Its application allows not only to effectively manage energy resources, but also contributes to the introduction of sustainable and rational consumption of electricity, meeting the challenges of the present and the future.

CONCLUSION

In the course of the work, an attempt was made to apply ML technology to forecasting the consumption of electrical energy produced within the country by various types of power plants. The obtained results showed a high convergence of real historical data and those found through predictive analytics of the Microsoft Azure Machine Learning service. This provides grounds for further research development and improvement of the developed model. The set of standard blocks of the Microsoft Azure Machine

Learning Studio is not fully able to meet the needs for a large variety of problems that are solved. This paper demonstrated how to implement user blocks with Python scripts for initial, intermediate and final data processing, including graphing.

In this paper, in addition to a detailed description of the method of building a model based on the machine learning service, it was shown how to deploy the project for its further use without the Microsoft Azure web service. The API service has been integrated into a Microsoft Excel spreadsheet and allows to perform predictive calculations even in the absence of an Internet connection. The results of the forecast show a very high convergence with real data, and the accuracy of the forecast (coefficient of determination) does not fall below 0.8.

The authors' next research will consist in improving the developed model and adding to the dataset data of various categories of electricity consumers, considering periodic shutdowns and startups of power units, emergency modes, etc. Another related line of work is the development of VBA scripts for Microsoft Excel to improve the interactivity for the end user, the machine learning unit and the results of the forecast execution and their interpretation.

ACKNOWLEDGMENT

This research is supported by the Ministry of Education and Science of Ukraine as a part of the scientific research projects No. 0123U102775, and 0124U000202.

REFERENCES

Akella, A. P., Alhoori, H., Kondamudi, P. R., Freeman, C., & Zhou, H. (2021). Early indicators of scientific impact: Predicting citations with altmetrics. *Journal of Informetrics*, *15*(2), 101128. doi:10.1016/j.joi.2020.101128

Akhtar, S., Adeel, M., Iqbal, M., Namoun, A., Tufail, A., & Kim, K. H. (2023). Deep learning methods utilization in electric power systems. *Energy Reports*, *10*, 2138–2151. doi:10.1016/j.egyr.2023.09.028

Aksonov, Y., Kombarov, V., Tsegelnyk, Y., Plankovskyy, S., Fojtù, O., & Piddubna, L. (2021). Visualization and analysis of technological systems experimental operating results. In *2021 IEEE 16th International Conference on Computer Sciences and Information Technologies (CSIT)* (Vol. 2, pp. 141–146). IEEE. 10.1109/CSIT52700.2021.9648592

Alchokr, R., Haider, R., Shakeel, Y., Leich, T., Saake, G., & Krüger, J. (2023). Forecasting Publication's Success Using Machine Learning. *CEUR Workshop Proceedings*, *3617*, 77–89. https://ceur-ws.org/Vol-3617/paper-08.pdf

Ampomah, E. K., Qin, Z., & Nyame, G. (2020). Evaluation of tree-based ensemble machine learning models in predicting stock price direction of movement. *Information (Basel)*, *11*(6), 332. doi:10.3390/info11060332

Balaji, P., Alelyani, S., Qahmash, A., & Mohana, M. (2021). Contributions of machine learning models towards student academic performance prediction: A systematic review. *Applied Sciences (Basel, Switzerland)*, *11*(21), 10007. doi:10.3390/app112110007

Balal, A., Jafarabadi, Y. P., Demir, A., Igene, M., Giesselmann, M., & Bayne, S. (2023). Forecasting solar power generation utilizing machine learning models in Lubbock. *Emerging Science Journal*, *7*(4), 1052–1062. doi:10.28991/ESJ-2023-07-04-02

Bojer, C. S. (2022). Understanding machine learning-based forecasting methods: A decomposition framework and research opportunities. *International Journal of Forecasting*, *38*(4), 1555–1561. doi:10.1016/j.ijforecast.2021.11.003

Bomfim, T. S. (2020). Evolution of machine learning in smart grids. In *2020 IEEE 8th International Conference on Smart Energy Grid Engineering (SEGE)* (pp. 82–87). IEEE. 10.1109/SEGE49949.2020.9182023

Chen, Y., Wang, H., Zhang, B., & Zhang, W. (2022). A method of measuring the article discriminative capacity and its distribution. *Scientometrics*, *127*(6), 3317–3341. doi:10.1007/s11192-022-04371-0

Dayupay, J.P., Ajibade, S.S., Villagonzalo, M.C., Baird, J.C., Antecristo, C.P., & Adediran, A.O. (2024). Application of machine learning techniques in energy power production: A publication trend and bibliometrics analysis (2012-2023). *E3S Web of Conferences*, *491*, 01003. doi:10.1051/e3sconf/202449101003

Duchesne, L., Karangelos, E., & Wehenkel, L. (2020). Recent developments in machine learning for energy systems reliability management. *Proceedings of the IEEE*, *108*(9), 1656–1676. doi:10.1109/JPROC.2020.2988715

Entezari, A., Aslani, A., Zahedi, R., & Noorollahi, Y. (2023). Artificial intelligence and machine learning in energy systems: A bibliographic perspective. *Energy Strategy Reviews*, *45*, 101017. doi:10.1016/j.esr.2022.101017

Eskandarpour, R., & Khodaei, A. (2016). Machine learning based power grid outage prediction in response to extreme events. *IEEE Transactions on Power Systems*, *32*(4), 3315–3316. doi:10.1109/TPWRS.2016.2631895

Forootan, M. M., Larki, I., Zahedi, R., & Ahmadi, A. (2022). Machine learning and deep learning in energy systems: A review. *Sustainability (Basel)*, *14*(8), 4832. doi:10.3390/su14084832

Franki, V., Majnarić, D., & Višković, A. (2023). A comprehensive review of artificial intelligence (AI) companies in the power sector. *Energies*, *16*(3), 1077. doi:10.3390/en16031077

Hofmann, E., & Rutschmann, E. (2018). Big data analytics and demand forecasting in supply chains: A conceptual analysis. *International Journal of Logistics Management*, *29*(2), 739–766. doi:10.1108/IJLM-04-2017-0088

Houdou, A., El Badisy, I., Khomsi, K., Abdala, S. A., Abdulla, F., Najmi, H., Obtel, M., Belyamani, L., Ibrahimi, A., & Khalis, M. (2024). Interpretable machine learning approaches for forecasting and predicting air pollution: A systematic review. *Aerosol and Air Quality Research*, *24*(1), 230151. doi:10.4209/aaqr.230151

Htun, H. H., Biehl, M., & Petkov, N. (2023). Survey of feature selection and extraction techniques for stock market prediction. *Financial Innovation*, *9*(1), 26. doi:10.1186/s40854-022-00441-7 PMID:36687795

Huang, G., Wu, F., & Guo, C. (2022). Smart grid dispatch powered by deep learning: A survey. *Frontiers of Information Technology & Electronic Engineering*, *23*(5), 763–776. doi:10.1631/FITEE.2000719

Huotari, M., Malhi, A., & Främling, K. (2024). Machine learning applications for smart building energy utilization: A survey. *Archives of Computational Methods in Engineering*, •••, 1–20. doi:10.1007/s11831-023-10054-7

Ke, J., Zheng, H., Yang, H., & Chen, X. M. (2017). Short-term forecasting of passenger demand under on-demand ride services: A spatio-temporal deep learning approach. *Transportation Research Part C, Emerging Technologies*, *85*, 591–608. doi:10.1016/j.trc.2017.10.016

Kirubakaran, S., Nagarajan, V., Dhayapulley, S. C., Soubache, I. D., Pasupuleti, S. R., Kumar, A., Rastogi, R., & Vasudevan, S. (2023). Hybrid deep learning-based grid-supportive renewable energy systems for maximizing power generation using optimum sizing. *Electric Power Components and Systems*, *51*(15), 1597–1611. doi:10.1080/15325008.2023.2201249

Kombarov, V., Sorokin, V., Tsegelnyk, Y., Plankovskyy, S., Aksonov, Y., & Fojtů, O. (2021). Numerical control of machining parts from aluminum alloys with sticking minimization. *International Journal of Mechatronics and Applied Mechanics*, *I*(9), 209–216. doi:10.17683/ijomam/issue9.30

Kumbhar, A., Dhawale, P. G., Kumbhar, S., Patil, U., & Magdum, P. (2021). A comprehensive review: Machine learning and its application in integrated power system. *Energy Reports*, *7*, 5467–5474. doi:10.1016/j.egyr.2021.08.133

Kureljusic, M., & Karger, E. (2023). Forecasting in financial accounting with artificial intelligence – A systematic literature review and future research agenda. *Journal of Applied Accounting Research*, *25*(1), 81–104. doi:10.1108/JAAR-06-2022-0146

Martins, E., & Galegale, N. V. (2023). Sales forecasting using machine learning algorithms. *Revista de Gestão e Secretariado*, *14*(7), 11294–11308. doi:10.7769/gesec.v14i7.1670

Microsoft. (2020). *The Next AI Guardians*. https://www.microsoft.com/en-cy/aibootcamps

Mira, K., Bugiotti, F., & Morosuk, T. (2023). Artificial intelligence and machine learning in energy conversion and management. *Energies*, *16*(23), 7773. doi:10.3390/en16237773

Moradzadeh, A., Mansour-Saatloo, A., Nazari-Heris, M., Mohammadi-Ivatloo, B., & Asadi, S. (2021). Introduction and literature review of the application of machine learning/deep learning to load forecasting in power system. In M. Nazari-Heris & ... (Eds.), *Application of Machine Learning and Deep Learning Methods to Power System Problems* (pp. 119–135). Springer., doi:10.1007/978-3-030-77696-1_6

Pliuhin, V., Korobka, V., Karyuk, A., Pan, M., & Sukhonos, M. (2019a). Using Azure Machine Learning Studio with python scripts for induction motors optimization web-deploy project. In *2019 IEEE International Scientific-Practical Conference Problems of Infocommunications, Science and Technology (PIC S&T)* (pp. 631–634). IEEE. 10.1109/PICST47496.2019.9061447

Pliuhin, V., Pan, M., Yesina, V., & Sukhonos, M. (2018). Using Azure Machine Learning cloud technology for electric machines optimization. In *2018 International Scientific-Practical Conference Problems of Infocommunications. Science and Technology (PIC S&T)* (pp. 55–58). IEEE. 10.1109/INFOCOMMST.2018.8632093

Pliuhin, V., Sukhonos, M., Pan, M., Petrenko, O., & Petrenko, M. (2019b). Implementing of Microsoft Azure machine learning technology for electric machines optimization. *Electrical Engineering & Electromechanics*, *1*(1), 23–28. doi:10.20998/2074-272X.2019.1.04

Pliuhin, V., Teterev, V., & Lapko, A. (2021). Smart Grid technologies as a concept of innovative energy development: Initial proposals for the development of Ukraine. *Lighting Engineering & Power Engineering*, *60*(2), 47–65. doi:10.33042/2079-424X.2021.60.2.02

Rahimi, I., Chen, F., & Gandomi, A. H. (2023). A review on COVID-19 forecasting models. *Neural Computing & Applications*, *35*(33), 23671–23681. doi:10.1007/s00521-020-05626-8 PMID:33564213

Rohaan, D., Topan, E., & Groothuis-Oudshoorn, C. G. (2022). Using supervised machine learning for B2B sales forecasting: A case study of spare parts sales forecasting at an after-sales service provider. *Expert Systems with Applications*, *188*, 115925. doi:10.1016/j.eswa.2021.115925

Srinivasan, S., Kumarasamy, S., Andreadakis, Z. E., & Lind, P. G. (2023). Artificial intelligence and mathematical models of power grids driven by renewable energy sources: A survey. *Energies*, *16*(14), 5383. doi:10.3390/en16145383

Sukhonos, M., Babaiev, V., Pliuhin, V., Teterev, V., & Khudiakov, I. (2023). Load forecasting and electricity consumption by regression model. In *Smart Technologies in Urban Engineering* (pp. 302–314). Springer. doi:10.1007/978-3-031-20141-7_28

Tarallo, E., Akabane, G. K., Shimabukuro, C. I., Mello, J., & Amancio, D. (2019). Machine learning in predicting demand for fast-moving consumer goods: An exploratory research. *IFAC-PapersOnLine*, *52*(13), 737–742. doi:10.1016/j.ifacol.2019.11.203

Thelwall, M., Kousha, K., Wilson, P., Makita, M., Abdoli, M., Stuart, E., Levitt, J., Knoth, P., & Cancellieri, M. (2023). Predicting article quality scores with machine learning: The UK Research Excellence Framework. *Quantitative Science Studies*, *4*(2), 547–573. doi:10.1162/qss_a_00258

Theoharidis, A. F., Guillén, D. A., & Lopes, H. (2023). Deep learning models for inflation forecasting. *Applied Stochastic Models in Business and Industry*, *39*(3), 447–470. doi:10.1002/asmb.2757

Tsegelnyk, Y., Kombarov, V., Plankovskyy, S., Aksonov, Y., Pliuhin, V., & Aksonov, O. (2022). Investigation of the portal-type machine tool gear-belt gearbox. *International Journal of Mechatronics and Applied Mechanics*, *2022*(11), 295–302. doi:10.17683/ijomam/issue11.41

Wang, D., Gryshova, I., Balian, A., Kyzym, M., Salashenko, T., Khaustova, V., & Davidyuk, O. (2022a). Assessment of power system sustainability and compromises between the development goals. *Sustainability (Basel)*, *14*(4), 2236. doi:10.3390/su14042236

Wang, D., Gryshova, I., Kyzym, M., Salashenko, T., Khaustova, V., & Shcherbata, M. (2022b). Electricity price instability over time: Time series analysis and forecasting. *Sustainability (Basel)*, *14*(15), 9081. doi:10.3390/su14159081

Wen, M., Zhou, C., & Konstantin, M. (2023). Deep neural network for predicting changing market demands in the energy sector for a sustainable economy. *Energies*, *16*(5), 2407. doi:10.3390/en16052407

Xu, C., Liao, Z., Li, C., Zhou, X., & Xie, R. (2022). Review on interpretable machine learning in smart grid. *Energies*, *15*(12), 4427. doi:10.3390/en15124427

Yang, W. (2022). *Machine learning based wind power forecasting for operational decision support* [Doctoral thesis, University of Vaasa]. OSUVA. https://urn.fi/URN:NBN:fi-fe2022091559103

Zhou, L., Pan, S., Wang, J., & Vasilakos, A. V. (2017). Machine learning on big data: Opportunities and challenges. *Neurocomputing*, *237*, 350–361. doi:10.1016/j.neucom.2017.01.026

ADDITIONAL READING

Barnes, J. (2015). *Azure Machine Learning: Microsoft Azure Essentials*. Microsoft Press.

Campesato, O. (2020). *Artificial Intelligence, Machine Learning, and Deep Learning*. Mercury Learning and Information. doi:10.1515/9781683924654

Elston, S. F. (2015). *Data Science in the Cloud with Microsoft Azure Machine Learning and R*. O'Reilly Media Inc.

Etaati, L. (2019). *Machine Learning with Microsoft Technologies*. Apress., doi:10.1007/978-1-4842-3658-1

Hwang, K. (2017). *Cloud Computing for Machine Learning and Cognitive Applications*. MIT Press.

Khanuja, M., Sabir, F., Subramanian, S., & Potgieter, T. (2022). *Applied Machine Learning and High-Performance Computing on AWS*. Packt Publishing.

Nielsen, F. (2016). *Introduction to HPC with MPI for Data Science*. Springer., doi:10.1007/978-3-319-21903-5

Robey, R., & Zamora, Y. (2021). *Parallel and High Performance Computing*. Manning.

Terzo, O., & Martinovič, J. (Eds.). (2022). *HPC, Big Data, and AI Convergence Towards Exascale: Challenge and Vision*. CRC Press. doi:10.1201/9781003176664

Vetter, J. S. (Ed.). (2017). *Contemporary High Performance Computing*. CRC Press.

KEY TERMS AND DEFINITIONS

Artificial Intelligence (AI): The intelligence of machines or software, as opposed to the intelligence of humans or animals. It is a field of study in computer science which develops and studies intelligent machines. Such machines may be called AIs.

Big Data: Refers to data sets that are too large or complex to be dealt with by traditional data-processing application software. Data with many entries (rows) offer greater statistical power, while data with higher complexity (more attributes or columns) may lead to a higher false discovery rate.

Data Science: Is an interdisciplinary academic field that uses statistics, scientific computing, scientific methods, processes, algorithms and systems to extract or extrapolate knowledge and insights from potentially noisy, structured, or unstructured data. Data science also integrates domain knowledge from the underlying application domain (e.g., natural sciences, information technology, and medicine). Data science is multifaceted and can be described as a science, a research paradigm, a research method, a discipline, a workflow, and a profession.

Data Set (or Dataset): Is a collection of data. In the case of tabular data, a data set corresponds to one or more database tables, where every column of a table represents a particular variable, and each row corresponds to a given record of the data set in question. The data set lists values for each of the variables, such as for example height and weight of an object, for each member of the data set. Data sets can also consist of a collection of documents or files.

Deep Learning: Is a class of machine learning algorithms that uses multiple layers to progressively extract higher-level features from the raw input. For example, in image processing, lower layers may identify edges, while higher layers may identify the concepts relevant to a human such as digits or letters or faces.

Machine Learning: Is a field of study in AI concerned with the development and study of statistical algorithms that can effectively generalize and thus perform tasks without explicit instructions. Recently, generative artificial neural networks have been able to surpass many previous approaches in performance. Machine learning approaches have been applied to large language models, computer vision, speech recognition, email filtering, agriculture, and medicine, where it is too costly to develop algorithms to perform the needed tasks.

Microsoft Azure (or Azure): Is a cloud computing platform run by Microsoft. It offers access, management, and the development of applications and services through global data centers. It also provides a range of capabilities, including software as a service (SaaS), platform as a service (PaaS), and infrastructure as a service (IaaS). Microsoft Azure supports many programming languages, tools, and frameworks, including Microsoft-specific and third-party software and systems.

ENDNOTE

1. Due to the invasion of russian federation on the territory of Ukraine and the start of war actions, data from the electricity supply is closed for open use. In this paper, as an example of building a machine learning service, historical data of the electrical energy production for 2018 was used.

Chapter 14
Probabilistic Forecasting of Hourly Wind Power Load in South Africa

Katleho Makatjane
https://orcid.org/0000-0002-3687-4098
University of Botswana, Botswana

Claris Shoko
https://orcid.org/0000-0002-0896-4387
University of Botswana, Botswana

Ntebogang Moroke
https://orcid.org/0000-0001-8545-1860
North West University, South Africa

ABSTRACT

Forecasters must make some educated assumptions about future electricity demand and effectively explain them because real electricity demand and forecasted electricity demand always differ. In order to forecast future wind power load base stations in South Africa, the authors evaluate hourly wind power generation in this chapter. The predicted time series can be used to show the flow of the load demand trend for electricity. Because of the noise in the original time series, they provide the enhanced gradient-boosted decision tree algorithm based on Kalman filter (GBDT-KF). They compare how well the proposed GBDT-KF method performs with varying numbers of decision trees. Re-sampling is cross-validated three times at a 10 fold interval. The employment of MSE, RMSE, MAE, and MAPE allowed for the selection of the best model. As a result, the total findings showed that 90% of the test data and 92% of the training data were successful. The findings of this study will be helpful to the energy sector and decision-makers for planning and future use in the hourly electricity forecasting domain.

DOI: 10.4018/979-8-3693-2355-7.ch014

INTRODUCTION

The central function of any organization is decision-making at the operational, tactical, and strategic levels. Such decisions are supported by forecasts that are fundamentally different. For instance, long-term, aggregate predictions are necessary for strategic decisions, but in short-term, extremely precise forecasts are necessary for highly dynamic operational decisions (Athanasopoulos et al., 2017). The way these projections are produced is impacted by variations in time granularity. Strategic projections for the long term are typically produced by taking into account high-level unstructured data from the business environment. These mostly rely on the judgement and experience of senior management, as reliable long-term statistical forecasts that capture the market dynamics can be quite tough to make (Kolsarici and Vakratsas, 2015). Short-term operational forecasts, on the other hand, are typically produced utilising organised but constrained data sources, including historical power demand load. Most of this is based on statistical techniques. Furthermore, the forecasting literature is well aware that different forecast horizons necessitate different approaches because these forecasts are generated by different methodologies and are based on different information sets, and it is expected that they may not agree, even within the same decision-making level (Mir et al., 2020).

Decisions resulting from this disagreement may not be in line. For instance, daily electricity projections that assist with production and inventory decisions can offer a steady overall picture of the market. On the other hand, market growth may be anticipated by the strategic level predictions, which might have a variety of effects on the investment plan and financial choices (Château, 2022). Decisions must be made together if a corporation such as ESKOM (the South African Power utility supplier) is to be able to satisfy its growth in all possible operational, tactical, and strategic projections. If not, this misalignment may result in waste, missed opportunities, and extra expenses. There are several difficulties in forecasting a hierarchy of time series. First, there may be noticeable differences in the properties displayed by the series at various hierarchical levels, which correspond to varying degrees of aggregation yet, hourly electricity demand, load, and supply data are marked by skewness, high volatility, and intermittency, according to Mircetic et al. (2022). As a result, aggregated series will inherently evolve more smoothly, be less skewed, and exhibit more obvious signs of seasonality and weather dependence. The forecasting method that should be used at each level of the hierarchy is affected by these differences (Athanasopoulos et al., 2017).

Summing the forecasts of the respective series at the lower levels would yield the aggregates' forecasts. But the results of this bottom-up strategy are usually not very good. This relates to a second major challenge: to provide coherent decision-making at various hierarchical levels, the prediction of each aggregated series must equal the sum of the forecasts of the corresponding disaggregated series. If the forecasts are generated independently for every series in the hierarchy, this aggregation condition is unlikely to be met. Finally, in many applications, such as the hourly electricity demand load, the hierarchy can consist of thousands, or even millions, of time series at the bottom level, which has implications in the choice of forecasting method in terms of computational load. The absence of probabilistic forecasting is a prominent characteristic in the literature of hierarchical forecasting. This is a serious drawback since forecast uncertainty evaluation is necessary for optimum decision making in many different applications. Therefore, Gabalawy et al. (2021) highlighted that the stochastic unit commitment models, power supply planning, operating reserve setting, price forecasting, and electricity market trading are all based on probabilistic projections of the entire system load in the context of electricity demand. However, probabilistic demand projections in a hierarchy with several degrees of aggregation—substations, cities, regions—are necessary for optimal decision-making. The forecasts ought to be consistent, meaning that the total forecasts for related disaggregated series and the forecast of aggregated series should match (Gabalawy et al., 2021).

Since the distribution of power at one rung of the hierarchy depends on the proper quantity being supplied from the rung above, coherency is therefore crucial. Consequently, we produce probabilistic forecasts in wind power (i.e., electricity) generating situations in this chapter. By looking at how the hierarchy of wind power generation might be used for forecasting, an alternative to the conventional methods of bottom-up versus top-down forecasting is presented. We investigate the possibility of using all the series for forecasting as opposed to picking series from certain areas of the hierarchy. Furthermore, we view the hierarchical structure as a distinguishing characteristic of the data generating process and use it to instantly construct forecasts for every level of the hierarchy, rather than using it after the initial forecasts are generated. This integrated technique uses the Kalman filter to explicitly address hourly demand load dependencies, such as complementarity of products and energy substitution which are normally ignored.

We anticipate that our suggested approach provides reliable and accurate estimations of trends in energy consumption. We give crucial information to all parties involved, including Eskom, by doing this. These insights enable the development of strategic infrastructure, resource allocation, and the smooth integration of environmentally friendly power suppliers into the national grid. To guarantee that South Africa has a future dependable and sustainable energy, this cooperative effort is crucial. There are four benefits to our method. Firstly, a density forecast is generated for each series in the hierarchy, and this is taken together to meet the requirements of coherency. Second, all estimated densities and dependence structures are the products of an integration of data from various hierarchy levels. Third, the assumptions of a distribution are not required. Fourth, the problem is decomposed into lower-dimension sub problems, enabling the approach to be scalable to large hierarchies.

System Architecture

This is a conceptual model that defines the structure, behaviour, and more views of a system. The description of an architecture is a formal description and representation of a system which is organized in a way that it supports reasoning about the structures and behaviours of the system.

Problem Definition

Assuming that original time series is denoted as $\{r_1,r_2,\ldots,r_n\}$, a Kalman filter is used to reduce the noise of this series, thus obtaining a smoothed time series as $\{x_1,\ldots,x_n\}$ (Gupta et al., 2018). Given this smoothed time series, all its sliding windows is computed by $SW_i^L = \{x_1, x_{i+1}, \cdots, x_{i+L-1}\}$, for i=1,2,…,n–L and the training dataset along with the training labels as $\{\ell_1,\ldots,\ell_{n-L}\}$, where L is the length of each time series. The algorithm needs to learn a function f which is defined as

$$\ell_i = f\left(SW_i^L\right). \tag{1}$$

So, the solution is formally defined as

$$\arg\min_{\Omega} \sum_{i=1}^{N} \left(\ell_i - f\left(SW_i^L \right) \right)^2 \tag{2}$$

where Ω is the set of parameters for model *f* and *N* is the size of the training set. In this chapter, we use gradient boosting decision tree (GBDT) to interpret model *f*. It outperforms other models such as deep learning techniques in terms of time complexity, while the accuracy of them remains almost the same.

Proposed Gradient Boosting Decision Tree-Kalman Filter Algorithm

While gradient boosting decision tree performs remarkably well in time series forecasting, overfitting is a real issue with its applications in real-world. Noisy data in the training set is typically the reason for this decline in its performance. Thus, by pre-processing the data with the use of a Kalman filter and thereby reducing the noise in the initial training set, overfitting issue of GBDT is resolved. Hence, the Kalman filter addresses the general problem by trying to estimate the state $X \in \Re^n$ of a discrete-time stochastic process that is governed by a linear stochastic difference equation that is given by

$$X_t = F_t X_{t-1} + \beta_t u_t + e_t \tag{3}$$

with a measurement $Z \in \Re^m$ that is computed by

$$Z_t = H_t X_t + \upsilon_t . \tag{4}$$

On the other hand, Li et al. (2020) declared that Z_t is the vector of measurements and H_t is the transformation matrix that maps the state vector into the measurement space. Here, W_t and *υt* are random variables that represent the process and measurement noise respectively. According to Kumar (2017), the state space representations and the associated Kalman filter have a profound impact on many applications. A state space model is generally represented in two equations: the first equation is called the state equation which finds the state Xt_{+1} in time (t+1) using the previous state X*t* and a noise term as shown in model (5) as

$$X_{t+1} = \alpha_t X_t + \varpi_t, t = 1, 2 \cdots . \tag{5}$$

At this juncture, *αt* is a sequence of B×B matrix and *ϖt* $_i$s the error term that follows a normal distribution with parameters (0,{Qt*}*$_)$ respectively. The second equation is the observation equation which expresses a w—*d*imensional observation Zt $_a$s a function of a v-*d*imensional state variable Xt $_a$nd the noise. Thus,

$$Z_t = G_t X_t + V_t, t = 1, 2, \cdots \tag{6}$$

where V_t is the measurement noise and G_t is a sequence of *w*×*v* matrix and, the correlation between W_t and v_t is zero. For that reason, this chapter employs previous hourly wind power generation from

1st April 2019 to 29th December 2023 as input for predicting the next 61 days generation load values (30th December 2023 to 29 February 2024) using the prediction scheme developed using KFT; and, the standard Kalman filter equations for the prediction stage are given as

$$\widehat{X}_{t|t-1} = F_t \widehat{X}_{t-1|t-1} + B_t u_t \tag{7}$$

$$P_{t|t-1} = F_t P_{t-1|t-1} F_t^T + Q_t \tag{8}$$

where $\widehat{X}_t$ is an estimate of X_t and u_t is the vector that contains any control input. F_t is the transition matrix which applies the effect of each state parameter at time t–1 on system state at time t. Additionally, B_t is the control input matrix that applies the effect of each control input parameter in the vector u_t for the state vector; and, the terms of P_t in this case provide a covariance of the estimated state vector X_t. Finally, Q_t is the covariance matrix of the noise process that is associated with a noisy control input variable. The measurement update equations are

$$\widehat{X}_{t|t} = \widehat{X}_{t|t-1} + K_t \left(Z_t - H_t \widehat{X}_{t|t-1} \right) \tag{9}$$

$$P_{t|t} = P_{t|t-1} - K_t H_t P_{t|t-1} \tag{10}$$

$$K_t = P_{t|t-1} - H_t^T \left(H_t P_{t|t-1} H_t^T + R_t \right)^{-1}. \tag{11}$$

The parameters of a Kalman filter can be obtained by means of expectation–maximization (EM) algorithm or can be decided according to physical models in advance (Carey and Ramsay, 2021).

A gradient boosting decision tree is a kind of ensemble learning method. By creating multiple weak prediction models, the decisions made by these weak learners are combined and a single stronger model is generated iteratively. The GBDT uses classification and regression trees (CART) as base learners. Assuming that there are K trees, the estimation with respect to the input feature x_i is defined in additive form as

$$\hat{y}_i = \sum_{t=1}^{K} f_t(x_i), f_t \in \mathcal{F} \tag{12}$$

where K denotes the number of base functions, $\mathcal{F}$ is the space of functions containing all regression trees. Thus, the learning stage of GBDT algorithm is to find a set of functions in the following space $\mathcal{F}$; while at the same time minimizing the loss function which is formally computed by

$$\arg\min_{f\in\mathcal{F}} \sum_{i=1}^{N} L\left(y_i, \hat{y}_i\right). \tag{13}$$

In equation (13), *N*denotes the total training sample. If the CART algorithm is used as base learner, then the base function is now given by

$$f_t\left(x_i\right) = w_{c(x_i)} \tag{14}$$

where $w \in R^T$, c: $R^d \mapsto \{1,2,\ldots,T\}$. Here, T is the number of leaves. For an input $x_i \in R^d$, the outputs $c(x_i)$ are the corresponding estimation values of the leaf node for a decision tree and, Moroke and Makatjane (2022) declared that gradient boosting aims to minimize the loss function $L(y,f(x))$ in every iteration. When the loss function is estimated by $L(y,f(x))= {}^1/_2(y-f(x))^2$, the gradient itself become a residue. Consequently, the decision trees fit this residue of all previous stages each time, and from this time,

$$\tilde{y} = \frac{\partial L\left(y, f\left(x\right)\right)}{\partial f\left(x\right)} = \frac{1}{2}\frac{\partial\left(y - f\left(x\right)\right)^2}{\partial f\left(x\right)} = y - f\left(x\right). \tag{15}$$

In summary, the pseudo-code of GBDT algorithm is presented in Algorithm 1.

Algorithm 1: *Gradient Boosting Decision tree*
Initialize: $f_0(X)=0,\eta,d0\_0$
for $j=1,\ldots,M$ do
$\omega_i = \exp\left(-y_i f_{j-1}\left(x_i\right)\right), i = 1,\cdots,N$
$r_i= -y,\ i=1,\ldots,N$
$S = \left\{\left(x_i, \omega_i, r_i\right)\right\}_{1,\cdots,N}$ and $v = \left\{x \mid x \in \Re\right\}$ here, v is a shrinkage factor to training data and the classifier.
$R= GROWTREE(s,v,\eta 0\ d0)$
$f_j\left(x\right) = f_{j-1}\left(x\right) - v \cdot \sum_{k=1}^{|R|} \eta_k \delta\left(x \in \Re_k\right),$
where $\Re_k, \eta_k \in R$
end
Procedure SPLITLEARNING(S)
$\left(\eta^{\ell_*}, \eta^{r_*}\tau^*\right) = \arg\min_{\eta^{\ell},\eta^{r},\tau} \varepsilon\left(\tau\right)$ in equation 13
Return $\left(\eta^{\ell_*}, \eta^{r_*}\tau^*\right)$

There is a trade-off between time complexity and precision when using a Kalman filter. The GBDT algorithm is trained in batch mode, which eliminates multiple hidden layers in deep learning. It turned out to be rather simple to train for the decision trees because this method no longer has any deep hidden layers. However, the accuracy of such a method slightly decreases due to lack of deep nonlinear mappings. But, given its time-saving feature, we contend that the suggested method is useful as long as the accuracy decline is acceptable.

Application to Hourly Wind Power Generation

The data used in this chapter is hourly wind power generation by ESKOM for the period of 2019-04-01 to 2023-12-30. This makes a total of 39960 observations. The data is split into two sets: training and testing sets. Figure 1 presents a univariate data analysis for the variable of interest: the hourly wind power generation for South Africa. The normal histogram with kernel density, normal Q-Q, and box and whisker plot and time series are presented in this Figure.

Preparation of Training and Validation Datasets

For the purpose of this chapter, classical decomposition is performed to deseasonalize the series. Thereafter, we develop hierarchies from: hour, day, week, month, quarter, half-year and year from this deseasonalized series and Figure 2 presents trend analysis of each hierarchy. It is clearly seen that

Figure 1. Hourly wind power generation

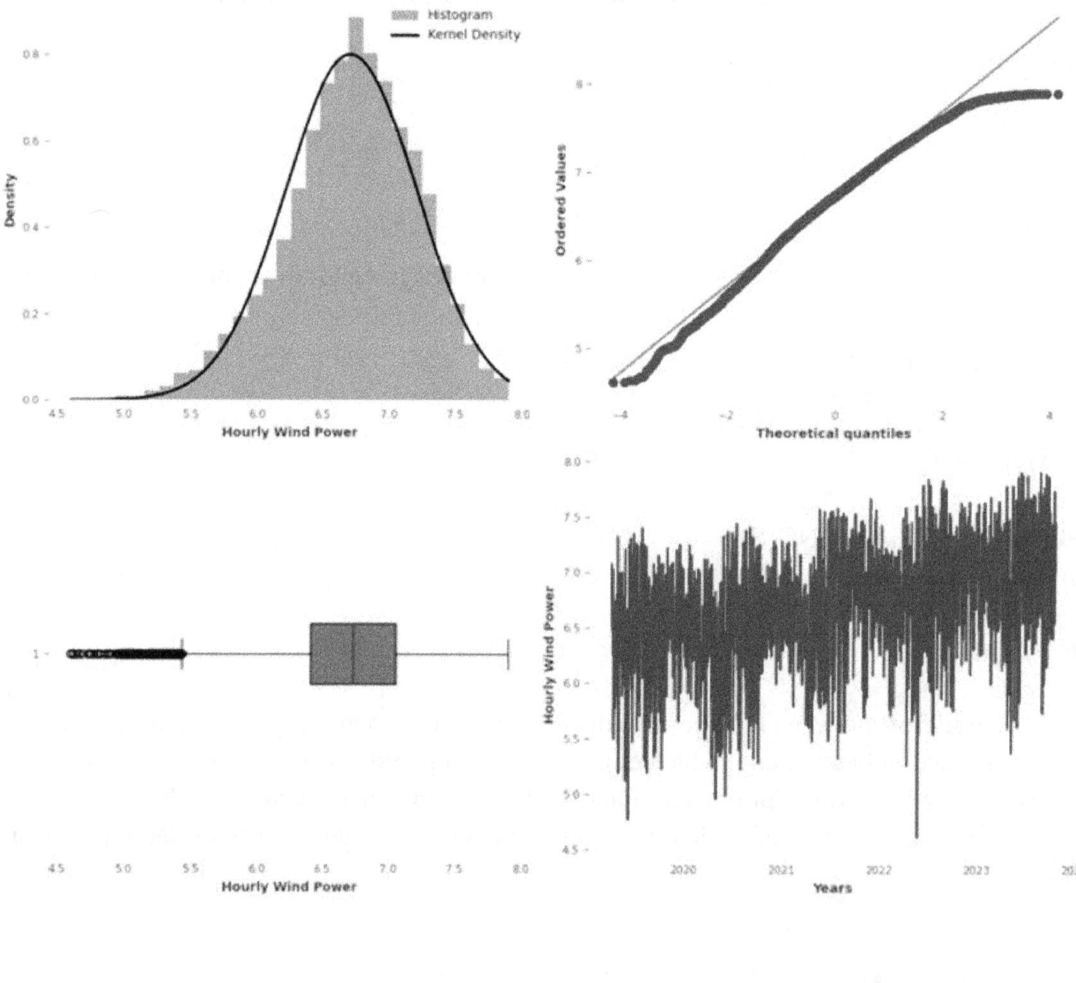

Table 1. Descriptive statistics

Min	25%	Mean	sd	50%	75%	Max	Kurtosis	Skewness
19.8030	564.463	973.772	535.968	895.60450	1294.1072	3102.2250	0.3239	0.7569

the trend has been increasing from time to time. For instance, a half-year trend in the second bottom of the Figure 2 clear shows an upward trend for wind power generation in South Africa. The cause of this increase in wind power generation in South Africa according to Karamanski and Erfort (2023) is because, independent power producers (IPPs) have been introduced to assist with sustaining South Africa's electricity demand. The IPPs that have come online are all renewable energy sources, as this technology has become more financially competitive with conventional energy sources around the world (Kåberger, 2018). The renewable energy sector in South Africa, specifically wind energy, has seen a rapid growth in recent years, starting in 2011 with 633.99 MW (Peters, 2011) and now with 3357 MW of installed capacity, which is expected to increase to 17742 MW by 2030. There are a total of 36 wind farms in operation in South Africa. The renewable energy detractors maintain that this type of energy is unreliable and requires significant backup to be viable. In the UK, discussions around wind's ability to provide power when needed highlight the benefits of offshore wind as a more secure supply source Thornton et al. (2017). A highly interconnected grid allows for the rapid adoption of renewable energy due to the grid flexibility Sijm et al. (2019), but this is not always replicable. A study was conducted in 2022, where the placement of renewable energy sources was found to be very important for an efficient implementation Kockel et al. (2022).

Adefarati and Obikoya (2019), also indicated that this increase is the result of South Africa's Department of Energy implementing an integrated resource plan (IRP). The Western Cape, Free State, and Eastern Cape provinces are well-suited to large-scale wind farm development. The abundance of renewable energy resources in the country has made several provinces suitable for large-scale grid-connected and freestanding power systems. The country has a reasonable potential for wind energy applications ranging from modest to large-scale microgrid power plants.

Beaufort West in Western Cape Province is the windiest area in South Africa; the town has tranquil conditions for only 2% of the time during the year. The average wind speed in the town is about 6.9 m/s, with 42.1% of wind speeds greater than 8 m/s. The most appealing locations for wind turbine (WT) installation in South Africa include coastal areas like as Cape Town, as well as mountainous areas in the Eastern Cape, Free State, and Northern Cape provinces. Lesotho's mountainous environment, which is only a few miles from Bloemfontein and Kimberley, has contributed significantly to the area's wind energy output. The inland hilly areas of the Eastern Cape such as Bhisho is another attractive location for installation of the WT in South Africa (Nobela et al., 2019). A total of 2078 MW has been established as the capacity of wind farms in South Africa. Mostafaeipour et al. (2020) state that an additional 840.9 MW of wind farms are currently under construction, with the largest wind farm in South Africa, with a capacity of 148 MW, being built at Nxuba (Sagel et al., 2023). South Africa has around 40% more installed wind power capacity than solar electricity. Hence the significance increase seen in Figure 1. Another concern is that the use of wind and solar electricity has expanded by approximately 8.5 times, from 467 MW to 3957 MW, in the last six years. Of course, the fact that the expansion of installed

Figure 2. Trend plot for each hierarchies

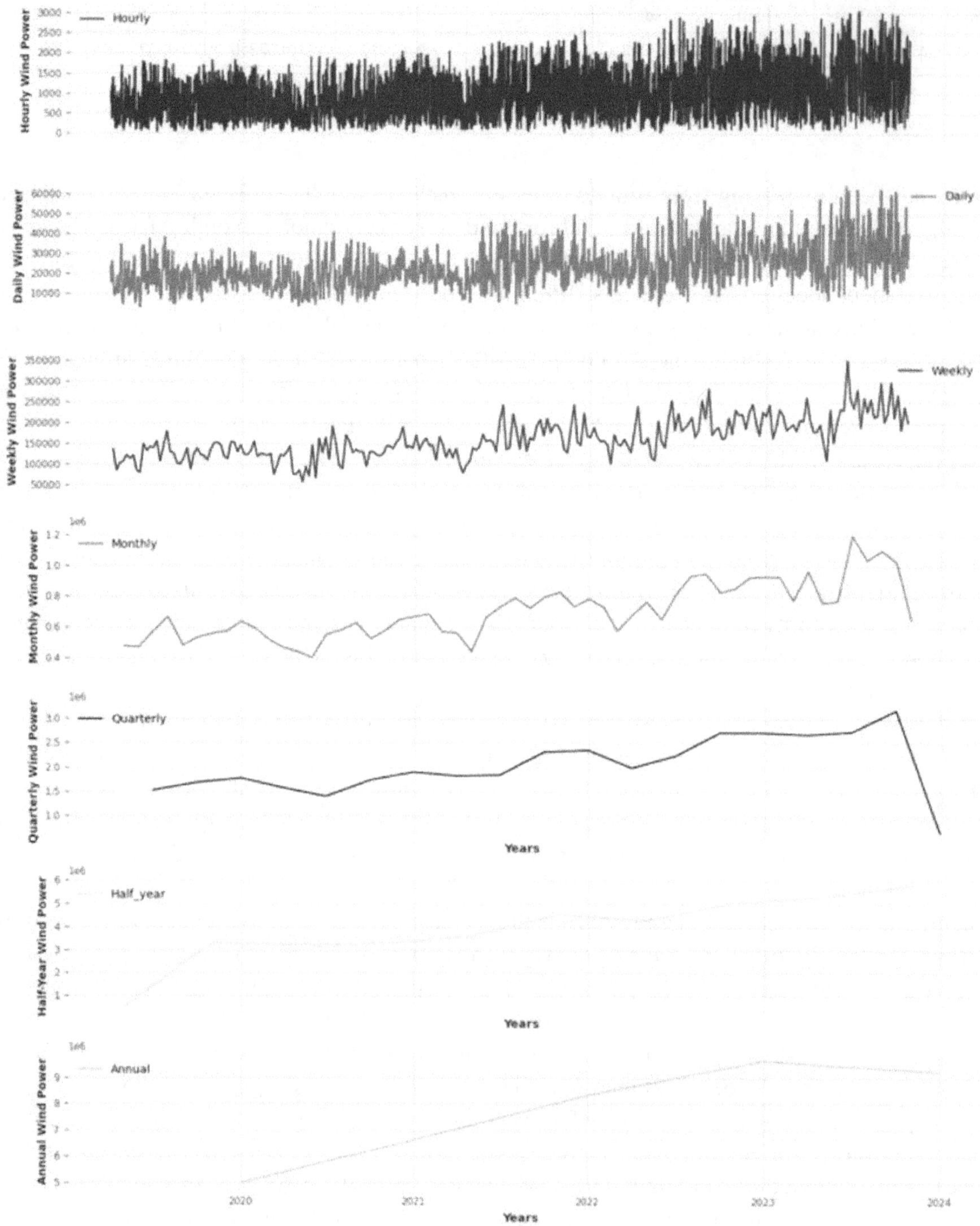

renewable electricity capacity has nearly ceased in 2018 is a negative. It is significant because South Africa, although having the world's third-largest GDP, meets the majority of its demands through coal.

Moreover, we fitted a linear regression model and our results showed that there is an increasing trend in hourly wind power generation in South Africa. The fitted regression model is given as $\widehat{X}_t =$ $-1096000+0.4459_t$ where $\widehat{X}_t$ presents the estimated hourly wind power from deseasonalized series.

Forecasting Performance of Hourly Wind Power Generation

Firstly, the performance of our proposed GBDT-KF algorithm under different number of decision trees is compared. The processing scale and centre of the algorithm are set to 450 and re-sampling is cross-validated at 10 fold which is repeated 3 times. Tuning parameter 'shrinkage' is held constant at a value of 0.01; while, tuning parameter 'n.minobsinnode' is held constant at a value of 50. The performance is measured using mean square error (MSE), root mean square error (RMSE), mean absolute error (MAE) and mean absolute percentage error (MAPE) of GBDT-KF with different number of CART regressors. Grid search is used to determine the initial parameters for Kalman filter, and the maximum depth of a single regression tree is set to one, two and three. The final values used for the model are n.trees = 450, interaction.depth = 3, shrinkage = 0.01 and n.minobsinnode = 10. As recommended by Tien et al. (2016) and Subramani et al. (2023), all the condition factors are removed with null predictive values. Experimental results show that the error matrices used decreases as the number of trees are growing. In general, the more the base learners (CART), the better the prediction precision in the test dataset. However, as the number of regression trees exceeds 2000, we find that the performance drops due to overfitting of the model.

An alternative perspective holds that the GBDT-KF ensemble's performance and that of each base leaners are influenced by the depth of the regression tree. Regression trees with more depth according to Hu et al. (2023) are typically predicted to better suit training sets. In this chapter, we keep hourly wind power generation to 1500 gigawatts and altered the depth of regression tree from 1 to 10. Thus, we assess the performance of GBDT-KF model. Figure 3 outlines this procedure and shows that, for an hourly wind power generation dataset, performance peaks when the depth equals to 6 and thereafter declines beyond depth of 7. Similar results are found in the work of Li et al. (2020) who found that at depth 7, the model begins to produce worse results.

Table 2. Re-sampling results across tuning parameters

interaction.depth	n.trees	MSE	RMSE	MAE	MAPE
1	150	0.4003459	0.632729	0.4946109	7.77475579
1	300	0.3944174	0.6280266	0.4910403	7.72107925
1	450	0.3898782	0.6244023	0.4883816	7.68115497
2	150	0.3979669	0.6308462	0.4928004	7.74737532
2	300	0.3900066	0.6245051	0.4878643	7.67279107
2	450	0.3837741	0.619495	0.4839715	7.61391489
3	150	0.3954349	0.6288361	0.4911103	7.72149552
3	300	0.3856379	0.6209975	0.4848759	7.62695738
3	450	**0.3782103**	**0.6149881**	**0.4800127**	**7.55311836**

Figure 3. The performance measured in MSE with different maximum depth of regression trees

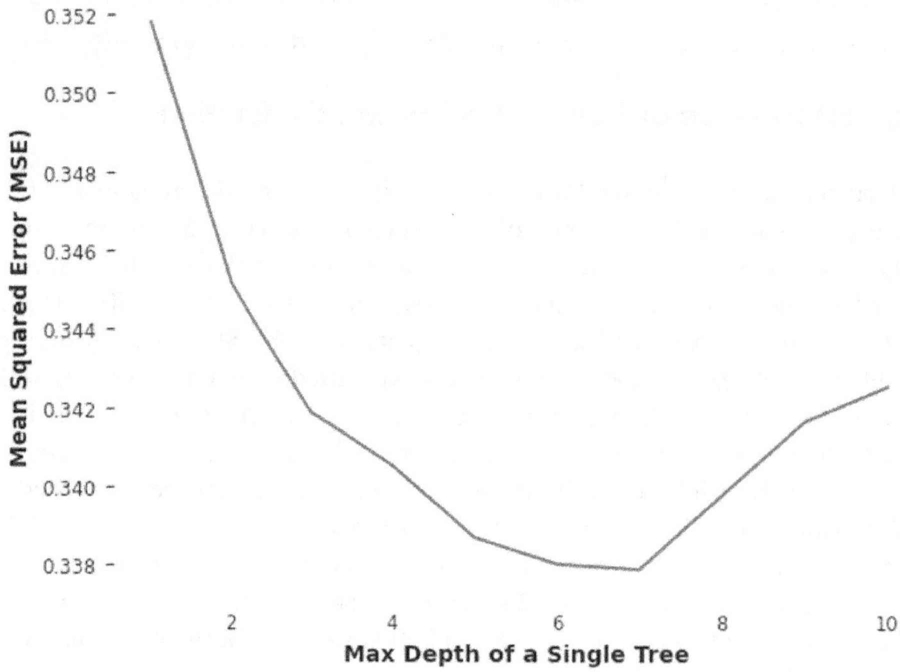

Next, we examine how the Kalman filter affects the performance of the suggested GBDT-KF model. The ability of the estimated GBDT-KF model to generalise is enhanced by using the Kalman filter prior to data entry into the regression tree (Ye, 2024). The counts for regression tree ranges from 500 to 1500, with 500 serving as the initial step with a maximum depth of 10. Figure 4 compares the results of GBDT with various error matrices and shows that a lower MSE, RMSE, MAE, and MAPE can be obtained by applying a Kalman filter before the general GBDT technique.

Lastly, model power and Wilcoxon signed-rank test are employed to compare the significance of the model. The former is seen in the work of Chen et al. (2017) and Bui et al. (2016) while the latter is found in the empirical analysis of Makatjane and Moroke (2016). The null hypothesis for this approach is that the model is statistically insignificant, whereas the alternative is in contradiction. A 5% significance level is used in the calculation of the results. The model is further evaluated considerably using the z and p-values. The null hypothesis is rejected and conclude that the performance of the model is clearly different when the z values are greater than the critical values and if the calculated probability values are less than the significant value of 5%. Table 3 displays the outcomes of the Wilcoxon signed-rank and power test. As may be observed, there have been a notable differences in the GBDT-KF model performances since the calculated probability value is 0.01% which his less than 5% level of significance. Additionally, the model shows a high prediction power of 92% in the training data and 90% on the test data.

Figure 4. Model performance

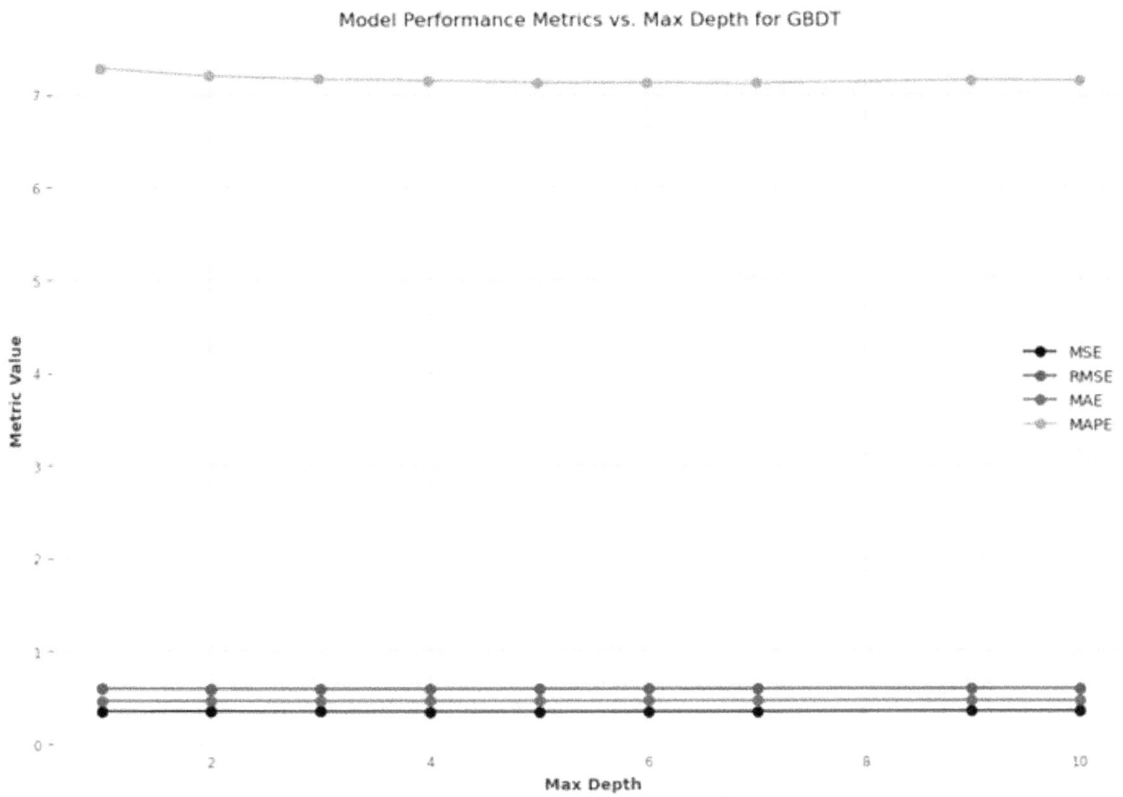

Table 3. Pair-wise comparison and power test

Wilcoxon Signed-Rank Test		Model Power Test		
Parameters	**GBDT-KF**	**Data**	**Mean Difference**	**Actual Power**
z-value	10.798	Training-set	0.0716	0.9198
p-value	0.0001	Testing-set	0.0348	0.8956

Discussion, Recommendation, and Conclusion

Precise long-term predictions for electricity demand are vital for strategic planning and decision-making within power utility companies and governmental bodies that oversees energy security. These forecasts impact key aspects such as determining the appropriate generation capacity, maintenance of scheduling, selection of an optimal energy mix, and planning future expansions of infrastructure. According to Conteh et al. (2023), forecasts ensure that resources are efficiently allocated, preventing either excess capacity or shortages. Moreover, the forecasts guide economic planning, influence policy formulation, and facilitate the implementation efficiency initiatives for energy. Overall, these predictions play a fundamental role in fostering a reliable, sustainable, and economically sound energy supply (Aziz et al., 2023). Decisions in the realm of electricity demand planning are inherently influenced by uncertainties; and, utilizing probabilistic modelling is a strategy to account for these uncertainties, therefore, empowering decision-makers to enhance quality of their choices when faced with such unpredictabilities. The consequences of these decisions are significant, as opting for infrastructure expansion might lead to the construction of unnecessary power generation facilities, while choosing not to expand the existing infrastructure could result in a failure to meet future electricity demand.

In the context of planning, Rangapuram et al. (2023) emphasised that, integrating probabilistic forecasts for long-term electricity demand enables managers to explore a spectrum of potential future values for electricity demand. This approach facilitates more informed decision-making regarding the necessity for additional (or reduced) power stations, mitigating risks associated with such strategic choices. Within this chapter, we employ the enhanced GBDT-KF algorithm designed for time series forecasting. Our application focuses on predicting future wind power generation in South Africa, providing insights into the 24-hour wind power generation cycle. By incorporating a Kalman filter to eliminate the noise from the initial time series, we elevate the performance of the conventional GBDT. Notably, the GBDT-KF algorithm boasts a relatively low level of complexity when compared to neural network approaches like LSTM. This methodological choice underscores the efficiency and effectiveness of the algorithm in handling wind power generation time series forecasting.

In 2019, the South African government announced intentions to split Eskom into three entities: generating, transmission, and distribution. The decision was spurred by an urgent necessity to address the utility's substantial debt levels. South Africa's National Treasury is formulating a measure to handle the USD23 billion debt load of the troubled national utility (Mohlakoana et al., 2023). The National Energy Regulator of South Africa (NERSA) has awarded the National Transmission Company South Africa (NTC) authorization to operate a transmission system inside South Africa's territorial limitations beginning in July 2023. Eskom, a South African electrical utility provider, has welcomed the NERSA action, which constitutes a critical step in the process of legally separating the transmission sector. With rising demand for embedded generation, the South African energy storage industry is predicted to reach ZAR14.5 billion by 2035, becoming a cornerstone of the future energy services market (Jahns, 2023). This is evidenced by the considerable increasing trends observed in the empirical analysis of this chapter. This will provide opportunities for investors, manufacturers, suppliers, and energy end users along the energy storage value chain. Because of the warning of Eskom death spiral by the national energy regulator of South Africa (NERSA), price elasticity of industrial demand is found to be an emerging primary driver of the lack of demand. This has therefore, served to exacerbate a "vicious cycle" in which increasing electricity prices drive declining sales, thereby resulting in the utility having to recover the same cost base from a shrinking customer base. Hence the results of this chapter are more precise to predict precise hourly load demand.

Future Research Directions

The practical application of our findings enables service providers to anticipate network congestion locations and times, allowing for proactive measures. This capability empowers them to address potential issues before they occur. Subsequent efforts should focus on incorporating additional factors into the time series prediction process for a more comprehensive analysis. Additionally, our future endeavours involve exploring alternative nonlinear filters with more intricate structures, supplementing the Kalman filter in the pre-processing stage. This approach aims to enhance the versatility and effectiveness of our predictive modelling framework.

CONCLUSION

This chapter used the enhanced gradient-boosted decision tree algorithm, based on the Kalman filter technique, to predict and forecast short-term hourly load wind power time series in South Africa. We study the potential of using all of the series for prediction rather than selecting the series from certain parts of the hierarchy. Furthermore, we see the hierarchical structure as a defining feature of the data generation process, and use it to instantaneously produce forecasts for each level of the hierarchy. This integrated technique employs the Kalman filter to directly handle hourly demand load dependencies such as products of complementarity and energy substitution, which are typically neglected. The results of this chapter showed our enhanced GBDT-KF algorithm is the perfect algorithm to predict short-term hourly wind power load in South Africa with prediction power of 92% in the training data and 90% on the test data.

ACKNOWLEDGMENT

The authors are grateful for the pleasant and helpful remarks and ideas made by the chapter's anonymous reviewers.

REFERENCES

Adefarati, T., & Obikoya, G. (2019). Evaluation of wind resources potential and economic analysis of wind power generation in South Africa. *International Journal of Engineering Research in Africa*, *44*, 150–181. doi:10.4028/www.scientific.net/JERA.44.150

Athanasopoulos, G., Hyndman, R. J., Kourentzes, N., & Petropoulos, F. (2017). Forecasting with temporal hierarchies. *European Journal of Operational Research*, *262*(1), 60–74. doi:10.1016/j.ejor.2017.02.046

Aziz, S., Ahmed, I., Khan, K., & Khalid, M. (2023). Emerging trends and approaches for designing net-zero low-carbon integrated energy networks: A review of current practices. Arabian Journal for Science and Engineering, 1-23.

Carey, M., & Ramsay, J. O. (2021). Fast stable parameter estimation for linear dynamical systems. *Computational Statistics & Data Analysis*, *156*, 107124. doi:10.1016/j.csda.2020.107124

Château, B. (2022). Energy Demand Drivers. In M. Hafner & G. Luciani (Eds.), *The Palgrave Handbook of International Energy Economics* (pp. 511–543). Springer International Publishing. doi:10.1007/978-3-030-86884-0_26

Chen, W., Xie, X., Wang, J., Pradhan, B., Hong, H., Bui, D. T., & Ma, J. (2017). A comparative study of logistic model tree, random forest, and classification and regression tree models for spatial prediction of landslide susceptibility. *Catena*, *151*, 147–160. doi:10.1016/j.catena.2016.11.032

Conteh, F., Furukakoi, M., Rangarajan, S. S., Collins, E. R., Conteh, M. A., Rashwan, A., & Senjyu, T. (2023). Long-Term Forecast of Sierra Leone's Energy Supply and Demand (2019–2040): A LEAP Model Application for Sustainable Power Generation System. *Sustainability (Basel), 15*(15), 11838. doi:10.3390/su151511838

Gabalawy, M., Hosny, N. S., & Adly, A. R. (2021). Probabilistic forecasting for energy time series considering uncertainties based on deep learning algorithms. *Electric Power Systems Research, 196*, 107216. doi:10.1016/j.epsr.2021.107216

Gupta, S., Pawar, S., Ramrakhiyani, N., Palshikar, G. K., & Varma, V. (2018). Semi-supervised recurrent neural network for adverse drug reaction mention extraction. *BMC Bioinformatics, 19*(S8), 1–7. doi:10.1186/s12859-018-2192-4 PMID:29897321

Hu, M., Guo, L., Liu, J., & Song, Y. (2023). Classification and Recognition of Building Appearance Based on Optimized Gradient-Boosted Decision Tree Algorithm. *Sensors (Basel), 23*(11), 5353. doi:10.3390/s23115353 PMID:37300080

Jahns, S. (2023). How Has the Current Electricity Crisis in South Africa Affected the Development of Renewable Energy within the Independent Power Producer Procurement Programmes? (Bachelor), Lund University, Kåberger, T. (2018). Progress of renewable electricity replacing fossil fuels. *Global Energy Interconnection, 1*(1), 48–52.

Karamanski, S., & Erfort, G. (2023). Wind Energy Supply Profiling and Offshore Potential in South Africa. *Energies, 16*(9), 3668. doi:10.3390/en16093668

Kockel, C., Nolting, L., Priesmann, J., & Praktiknjo, A. (2022). Does renewable electricity supply match with energy demand? – A spatio-temporal analysis for the German case. *Applied Energy, 308*, 118226. doi:10.1016/j.apenergy.2021.118226

Kolsarici, C., & Vakratsas, D. (2015). Correcting for misspecification in parameter dynamics to improve forecast accuracy with adaptively estimated models. *Management Science, 61*(10), 2495–2513. doi:10.1287/mnsc.2014.2027

Kumar, S. V. (2017). Traffic Flow Prediction using Kalman Filtering Technique. *Procedia Engineering, 187*, 582–587. doi:10.1016/j.proeng.2017.04.417

Li, L., Dai, S., Cao, Z., Hong, J., Jiang, S., & Yang, K. (2020). Using improved gradient-boosted decision tree algorithm based on Kalman filter (GBDT-KF) in time series prediction. *The Journal of Supercomputing, 76*(9), 6887–6900. doi:10.1007/s11227-019-03130-y

Makatjane, K. D., & Moroke, N. D. (2016). Comparative study of Holt-Winters triple exponential smoothing and seasonal Arima: forecasting short term seasonal car sales in South Africa. Risk governance and control: financial markets and institutions, 6(1), 71-82.

Mir, A. A., Alghassab, M., Ullah, K., Khan, Z. A., Lu, Y., & Imran, M. (2020). A Review of Electricity Demand Forecasting in Low and Middle Income Countries: The Demand Determinants and Horizons. *Sustainability (Basel), 12*(15), 5931. doi:10.3390/su12155931

Mircetic, D., Rostami-Tabar, B., Nikolicic, S., & Maslaric, M. (2022). Forecasting hierarchical time series in supply chains: An empirical investigation. *International Journal of Production Research*, *60*(8), 2514–2533. doi:10.1080/00207543.2021.1896817

Mohlakoana, N., Wolpe, P., & David, A. (2023). Unbundling Eskom: How would a new distribution system impact on energy poverty. *AFD Research Papers*, (269), 1–46.

Moroke, N. D., & Makatjane, K. (2022). Predictive Modelling for Financial Fraud Detection Using Data Analytics: A Gradient-Boosting Decision Tree. In Applications of Machine Learning and Deep Learning for Privacy and Cybersecurity (pp. 25-45): IGI Global. doi:10.4018/978-1-7998-9430-8.ch002

Mostafaeipour, A., Jahangiri, M., Haghani, A., Dehshiri, S. J. H., Dehshiri, S. S. H., Issakhov, A., & Techato, K. (2020). Statistical evaluation of using the new generation of wind turbines in South Africa. *Energy Reports*, *6*, 2816–2827. doi:10.1016/j.egyr.2020.09.035

Nobela, O., Bansal, R., & Justo, J. (2019). A review of power quality compatibility of wind energy conversion systems with the South African utility grid. *Renewable Energy Focus*, *31*, 63–72. doi:10.1016/j.ref.2019.10.001

Peters, H. D. (2011). Statement by the Minister of Energy, Hon. Dipuo Peters, MP, on the Occasion of the Announcement of Preferred Bidders in Respect of the First Window of 3 725 MW under the Renewable Energy IPP Bidding Programme. 2011. Retrieved from https://www.energy.gov.za/files/media/pr/2011/MinisterRemarks_IPP_BidsAnnouncement_7Dec2011.pdf

Rangapuram, S. S., Kapoor, S., Nirwan, R. S., Mercado, P., Januschowski, T., Wang, Y., & Bohlke-Schneider, M. (2023). Coherent probabilistic forecasting of temporal hierarchies. Paper presented at the International Conference on Artificial Intelligence and Statistics.

Sagel, V. N., Rouwenhorst, K. H., & Faria, J. A. (2023). Leveraging Green Ammonia for Resilient and Cost-Competitive Islanded Electricity Generation from Hybrid Solar Photovoltaic–Wind Farms: A Case Study in South Africa. *Energy & Fuels*, *37*(18), 14383–14392. doi:10.1021/acs.energyfuels.3c01950 PMID:37753452

Sijm, J., Koutstaal, P., Özdemir, Ö., & van Hout, M. (2019). Energy Transition Implications for Demand and Supply of Power System Flexibility: A Case Study of the Netherlands Within an EU Electricity Market and Trading Context. In E. Gawel, S. Strunz, P. Lehmann, & A. Purkus (Eds.), *The European Dimension of Germany's Energy Transition: Opportunities and Conflicts* (pp. 363–394). Springer International Publishing. doi:10.1007/978-3-030-03374-3_21

Subramani, N., Veerappampalayam Easwaramoorthy, S., Mohan, P., Subramanian, M., & Sambath, V. J. B. D. (2023). A Gradient Boosted Decision Tree-Based Influencer Prediction in Social Network Analysis. *Big Data and Cognitive Computing*, *7*(1), 6. doi:10.3390/bdcc7010006

Thornton, H. E., Scaife, A. A., Hoskins, B. J., & Brayshaw, D. J. (2017). The relationship between wind power, electricity demand and winter weather patterns in Great Britain. *Environmental Research Letters*, *12*(6), 064017. doi:10.1088/1748-9326/aa69c6

Tien, B. D., Tuan, T. A., Klempe, H., Pradhan, B., & Revhaug, I. (2016). Spatial prediction models for shallow landslide hazards: A comparative assessment of the efficacy of support vector machines, artificial neural networks, kernel logistic regression, and logistic model tree. *Landslides*, *13*(2), 361–378. doi:10.1007/s10346-015-0557-6

Ye, C. (2024). Using Time Series Analysis to Study the Changing Trend of Business English Teaching Effectiveness in Colleges and Universities. *Applied Mathematics and Nonlinear Sciences*, *9*(9), 1–15. doi:10.2478/amns.2023.2.01108

ADDITIONAL READING

Das, S., Alshehri, Y. M., Stenchikov, G. L., & Genton, M. G. J. S. (2023). A space-time model with temporal cyclostationarity for probabilistic forecasting and simulation of solar irradiance data. *Stat (International Statistical Institute)*, *12*(1), e583. doi:10.1002/sta4.583

Fan, H., Zhen, Z., Liu, N., Sun, Y., Chang, X., Li, Y., & Mi, Z. (2023). Fluctuation pattern recognition based ultra-short-term wind power probabilistic forecasting method. *Energy*, *266*, 126420. doi:10.1016/j.energy.2022.126420

Finamore, A. R., Calderaro, V., Galdi, V., Graber, G., Ippolito, L., & Conio, G. (2023). Improving Wind Power Generation Forecasts: A Hybrid ANN-Clustering-PSO Approach. *Energies*, *16*(22), 7522. doi:10.3390/en16227522

Islam, M., Hassan, N., Rasul, M., Emami, K., & Chowdhury, A. A. (2023). Forecasting of Solar and Wind Resources for Power Generation. *Energies*, *16*(17), 6247. doi:10.3390/en16176247

Kamani, D., & Ardehali, M. M. (2023). Long-term forecast of electrical energy consumption with considerations for solar and wind energy sources. *Energy*, *268*, 126617. doi:10.1016/j.energy.2023.126617

Konstantinou, T., & Hatziargyriou, N. J. I. T. S. E. (2023). Day-Ahead Parametric Probabilistic Forecasting of Wind and Solar Power Generation using Bounded Probability Distributions and Hybrid Neural Networks. *IEEE Transactions on Sustainable Energy*, *14*(4), 2109–2120. Advance online publication. doi:10.1109/TSTE.2023.3270968

Powanga, L., & Kwakwa, P. A. (2023). Determinants of Electricity Transmission and Distribution Losses in South Africa. *Journal of Renewable Energy*, *2023*, 1–9. Advance online publication. doi:10.1155/2023/2376449

Shonhiwa, C., Makaka, G., Mukumba, P., & Shambira, N. (2023). Investigation of Wind Power Potential in Mthatha, Eastern Cape Province, South Africa. *Applied Sciences (Basel, Switzerland)*, *13*(22), 12237. doi:10.3390/app132212237

Sivhugwana, K. S., & Ranganai, E. (2024). An Ensemble Approach to Short-Term Wind Speed Predictions Using Stochastic Methods, Wavelets and Gradient Boosting Decision Trees. *Wind*, *4*(1), 3. doi:10.3390/wind4010003

Uniejewski, B. (2023). Smoothing Quantile Regression Averaging: A new approach to probabilistic forecasting of electricity prices. arXiv preprint arXiv:00411. https://doi.org//arXiv.2302.00411 doi:10.48550

Wen, K., Wu, W., & Wu, X. (2023). Electricity demand forecasting and risk management using Gaussian process model with error propagation. *Journal of Forecasting*, *42*(4), 957–969. doi:10.1002/for.2925

Compilation of References

. Reedman, L., Chew, M. S., Gordon, J., Sue, W., Brinsmead, T., Hayward, J., & Havas, L. (2021). Multi-sector energy modelling.

. White, D., Winchester, N., Atkins, M., Ballingall, J., Coates, S., de Miguel Mercader, F., & Suomalainen, K. (2018). Energy-and multi-sector modelling of climate change mitigation in New Zealand: current practice and future needs. Academic Press.

Abadi, M., Chu, A., Goodfellow, I., McMahan, H. B., Mironov, I., Talwar, K., & Zhang, L. (2016). Deep learning with differential privacy. In *Proceedings of the 2016 ACM SIGSAC Conference on Computer and Communications Security* (pp. 308-318). 10.1145/2976749.2978318

Abbas, F., Habib, S., Feng, D., & Yan, Z. (2018). Optimizing Generation Capacities Incorporating Renewable Energy with Storage Systems Using Genetic Algorithms. *Electronics (Basel)*, *7*(7), 100. doi:10.3390/electronics7070100

Abdel-Nasser & Mahmoud. (2019). Accurate photovoltaic power forecasting models using deep LSTM-RNN. Neural Computing and Applications, 31, 2727–2740.

Abdul Majid, M. H., & Ibrahim, K. (2021). On Bayesian approach to composite Pareto models. *PLoS One*, *16*(9), e0257762.

Abedini, F., Bahaghighat, M., & S'hoyan, M. (2019). Wind turbine tower detection using feature descriptors and deep learning. Facta Universitatis, Series. *Electronics and Energetics*, *33*(1), 133–153.

Abualigah, L., Zitar, R. A., Almotairi, K. H., Hussein, A. M., Abd Elaziz, M., Nikoo, M. R., & Gandomi, A. H. (2022). Wind, solar, and photovoltaic renewable energy systems with and without energy storage optimization: A survey of advanced machine learning and deep learning techniques. *Energies*, *15*(2), 578. doi:10.3390/en15020578

Ackermann, T., & Söder, L. (2000). Wind energy technology and current status: A review. *Renewable & Sustainable Energy Reviews*, *4*(4), 315–374. doi:10.1016/S1364-0321(00)00004-6

Adefarati, T., & Obikoya, G. (2019). Evaluation of wind resources potential and economic analysis of wind power generation in South Africa. *International Journal of Engineering Research in Africa*, *44*, 150–181. doi:10.4028/www.scientific.net/JERA.44.150

Adhikari, R. J. N. (2015). A neural network based linear ensemble framework for time series forecasting. Academic Press.

Aggarwal, C. C., & Aggarwal, C. C. (2017). *Time series and multidimensional streaming outlier detection. Outlier analysis*. Springer International Publishing. doi:10.1007/978-3-319-47578-3_9

Ahmad, S. T., Samarth, B., & Shivkumar, K. (2019). A deep learning approach to solar-irradiance forecasting in sky-videos. In IEEE winter conference on applications of computer vision (WACV). IEEE.

Ahmadi, A., Daccache, A., Sadegh, M., & Snyder, R. L. (2023), Statistical and deep learning models for reference evapotranspiration time series forecasting: A comparison of accuracy, complexity, and data efficiency. Computers and Electronics in Agriculture, 215. doi:10.1016/j.compag.2023.108424

Ahmad, T., & Chen, H. (2020). A review on machine learning forecasting growth trends and their real-time applications in different energy systems. *Sustainable Cities and Society*, *54*, 102010.

Ahmad, T., Madonski, R., Zhang, D., Huang, C., & Mujeeb, A. (2022). Data-driven probabilistic machine learning in sustainable smart energy/smart energy systems: Key developments, challenges, and future research opportunities in the context of smart grid paradigm. *Renewable & Sustainable Energy Reviews*, *160*, 112128. doi:10.1016/j.rser.2022.112128

Ahmad, T., Zhang, D., Huang, C., Zhang, H., Dai, N., Song, Y., & Chen, H. (2021). Artificial intelligence in sustainable energy industry: Status Quo, challenges and opportunities. *Journal of Cleaner Production*, *289*, 125834. doi:10.1016/j.jclepro.2021.125834

Ahmed, M. A., Eltamaly, A. M., Alotaibi, M. A., Alolah, A. I., & Kim, Y.-C. (2020). Wireless Network Architecture for Cyber Physical Wind Energy System. *IEEE Access : Practical Innovations, Open Solutions*, *8*, 40180–40197. doi:10.1109/ACCESS.2020.2976742

Ahmed, S. F., Mofijur, M., Parisa, T. A., Islam, N., Kusumo, F., Inayat, A., Lee, V. G., Badruddin, I. A., Khan, T. M. Y., & Ong, H. C. (2022). Progress and Challenges of Contaminate Removal from Wastewater Using Microalgae Biomass. *Chemosphere*, *286*, 131656. doi:10.1016/j.chemosphere.2021.131656 PMID:34325255

Ajibade, S. S. M., Bekun, F. V., Adedoyin, F. F., Gyamfi, B. A., & Adediran, A. O. (2023). Machine learning applications in renewable energy (MLARE) research: A publication trend and bibliometric analysis study (2012–2021). *Cleanroom Technology*, *5*(2), 497–517. doi:10.3390/cleantechnol5020026

Ajit, A., Acharya, K., & Samanta, A. (2020). A Review of Convolutional Neural Networks. Proceedings of the 2020 International Conference on Emerging Trends in Information Technology and Engineering (ic-ETITE 2020), 1–5. 10.1109/ic-ETITE47903.2020.049

Akella, A. P., Alhoori, H., Kondamudi, P. R., Freeman, C., & Zhou, H. (2021). Early indicators of scientific impact: Predicting citations with altmetrics. *Journal of Informetrics*, *15*(2), 101128. doi:10.1016/j.joi.2020.101128

Akhtar, S., Adeel, M., Iqbal, M., Namoun, A., Tufail, A., & Kim, K. H. (2023). Deep learning methods utilization in electric power systems. *Energy Reports*, *10*, 2138–2151. doi:10.1016/j.egyr.2023.09.028

Akram, M. W., Li, G., Jin, Y., Chen, X., Zhu, C., Zhao, X., Khaliq, A., Faheem, M., & Ahmad, A. (2019). CNN based automatic detection of photovoltaic cell defects in electroluminescence images. *Energy*, *189*, 116319. doi:10.1016/j.energy.2019.116319

Aksonov, Y., Kombarov, V., Tsegelnyk, Y., Plankovskyy, S., Fojtù, O., & Piddubna, L. (2021). Visualization and analysis of technological systems experimental operating results. In *2021 IEEE 16th International Conference on Computer Sciences and Information Technologies (CSIT)* (Vol. 2, pp. 141–146). IEEE. https://doi.org/10.1109/CSIT52700.2021.9648592

Alchokr, R., Haider, R., Shakeel, Y., Leich, T., Saake, G., & Krüger, J. (2023). Forecasting Publication's Success Using Machine Learning. *CEUR Workshop Proceedings*, *3617*, 77–89. https://ceur-ws.org/Vol-3617/paper-08.pdf

Aldhshan, S. R. S., Abdul Maulud, K. N., Wan Mohd Jaafar, W. S., Karim, O. A., & Pradhan, B. (2021). Energy Consumption and Spatial Assessment of Renewable Energy Penetration and Building Energy Efficiency in Malaysia: A Review. *Sustainability (Basel)*, *13*(16), 9244. doi:10.3390/su13169244

Aldieri, L., Grafström, J., & Vinci, C. P. (2021). The effect of Marshallian and Jacobian knowledge spillovers on jobs in the solar, wind and energy efficiency sector. *Energies*, *14*(14), 4269. doi:10.3390/en14144269

Alduailij, M. A., Petri, I., Rana, O., Alduailij, M. A., & Aldawood, A. S. (2021). Forecasting peak energy demand for smart buildings. *The Journal of Supercomputing*, *77*, 6356–6380.

Alharkan, H., Habib, S., & Islam, M. (2023). Solar power prediction using dual stream cnn-lstm architecture. *Sensors (Basel)*, *23*(2), 945.

Ali, O. M. (2022). Solar energy technology and its roles in sustainable development. *Clean Energy*, *6*(3), 476–483. doi:10.1093/ce/zkac023

Alkabbani, H., Ahmadian, A., Zhu, Q., & Elkamel, A. (2021). Machine Learning and Metaheuristic Methods for Renewable Power Forecasting: A Recent Review. *Frontiers in Chemical Engineering*, *3*, 665415. Advance online publication. doi:10.3389/fceng.2021.665415

AlKandari & Ahmad. (2019). Solar power generation forecasting using ensemble approach based on deep learning and statistical methods. Appl. Comput. Inform.

AlRashidi, M. R., & El-Naggar, K. M. (2010). Long-term electric load forecasting based on particle swarm optimization. *Applied Energy*, *87*(1), 320–326. doi:10.1016/j.apenergy.2009.04.024

Al-Saedi, W., Lachowicz, S., Habibi, D., & Bass, O. (2018). A review of algorithms for control and optimization for energy systems. *Journal of Renewable and Sustainable Energy*, *10*(5), 053502. doi:10.1063/1.5032146

Alsayed, A. O., Rahim, M. S. M., AlBidewi, I., Hussain, M., Jabeen, S. H., Alromema, N., & Jibril, M. L. (2021). Selection of the Right Undergraduate Major by Students Using Supervised Learning Techniques. *Applied Sciences (Basel, Switzerland)*, *11*(22), 10639.

Alsharif, M. H., Younes, M. K., & Kim, J. (2019). Time series ARIMA model for prediction of daily and monthly average global solar radiation: The case study of Seoul, South Korea. *Symmetry*, *11*(2), 240. doi:10.3390/sym11020240

Al-Sumaiti, A., Salama, M., & Al-Hinai, A. (2018). A review of optimal planning and operation of hybrid renewable energy systems. *International Journal of Energy Research*, *42*(10), 3179–3204.

Al-tabatabaie, K. F., Hossain, M. B., Islam, M. K., Awual, M. R., TowfiqulIslam, A. R. M., Hossain, M. A., Esraz-Ul-Zannat, M., & Islam, A. (2022). Taking strides towards decarbonization: The viewpoint of Bangladesh. *Energy Strategy Reviews*, *44*, 100948. doi:10.1016/j.esr.2022.100948

Altuwaijri, M., & Bayoumi, M. A. (1994). Arabic Text Recognition Using Neural Network. *ISCAS 94. IEEE International Symposium on Circuits and systems*, *6*(30). 10.1109/ISCAS.1994.409614

Al-Yousefi, H., & Udpa, S. S. (1988). Recognition of handwritten Arabic characters. *Proc. SPIE 32nd Ann. Int. Tech. Symp. Opt. Optoelectric Applied Sci. Eng.*

Amer, M., Goldstein, M., & Abdennadher, S. 2013, August. Enhancing one-class support vector machines for unsupervised anomaly detection. In *Proceedings of the ACM SIGKDD workshop on outlier detection and description* (pp. 8-15). 10.1145/2500853.2500857

Ampomah, E. K., Qin, Z., & Nyame, G. (2020). Evaluation of tree-based ensemble machine learning models in predicting stock price direction of movement. *Information (Basel)*, *11*(6), 332. doi:10.3390/info11060332

Amran, Y. A., Amran, Y. M., Alyousef, R., & Alabduljabbar, H. (2020). Renewable and sustainable energy production in Saudi Arabia according to Saudi Vision 2030; Current status and future prospects. *Journal of Cleaner Production*, *247*, 119602. doi:10.1016/j.jclepro.2019.119602

Anand, B., Shankar, R., Murugavelh, S., Rivera, W., Midhun Prasad, K., & Nagarajan, R. (2021). A review on solar photovoltaic thermal integrated desalination technologies. *Renewable & Sustainable Energy Reviews*, *141*, 110787. doi:10.1016/j.rser.2021.110787

Ananthi, J., Sengottaiyan, N., Anbukaruppusamy, S., Upreti, K., & Dubey, A. K. (2022). Forest fire prediction using IoT and deep learning. *International Journal of Advanced Technology and Engineering Exploration*, *9*(87), 246–256. doi:10.19101/IJATEE.2021.87464

Anderson, R. N., Boulanger, A., Powell, W. B., & Scott, W. (2011). Adaptive stochastic control for the smart grid. *Proceedings of the IEEE*, *99*(6), 1098–1115. doi:10.1109/JPROC.2011.2109671

Angulo, A., van der Linde, P., Gardeniers, H., Modestino, M., & Rivas, D. F. (2020). Influence of Bubbles on the Energy Conversion Efficiency of Electrochemical Reactors. [Google Scholar] [CrossRef] [Green Version]. *Joule*, *4*(3), 555–579. doi:10.1016/j.joule.2020.01.005

Arbabzadeh, M., Gençer, E., Morris, J. F., Paltsev, S., & Armstrong, R. C. (2019). *Plausible Energy Futures: A Framework for Evaluating Options*. Impacts, and National Energy Choices.

Arent, D. J., Green, P., Abdullah, Z., Barnes, T., Bauer, S., Bernstein, A., & Turchi, C. (2022). Challenges and opportunities in decarbonizing the US energy system. *Renewable & Sustainable Energy Reviews*, *169*, 112939. doi:10.1016/j.rser.2022.112939

Aslam, M., Lee, J.-M., Kim, H.-S., Lee, S.-J., & Hong, S. (2019). Deep learning models for long-term solar radiation forecasting considering microgrid installation: A comparative study. *Energies*, *13*(1), 147. doi:10.3390/en13010147

Aslam, S., Herodotou, H., Mohsin, S. M., Javaid, N., Ashraf, N., & Aslam, S. (2021). A survey on deep learning methods for power load and renewable energy forecasting in smart microgrids. *Renewable & Sustainable Energy Reviews*, *144*, 110992. doi:10.1016/j.rser.2021.110992

Atalay, Y., Biermann, F., & Kalfagianni, A. (2016). Adoption of renewable energy technologies in oil-rich countries: Explaining policy variation in the Gulf Cooperation Council states. *Renewable Energy*, *85*, 206–214. doi:10.1016/j.renene.2015.06.045

Athanasopoulos, G., Hyndman, R. J., Kourentzes, N., & Petropoulos, F. (2017). Forecasting with temporal hierarchies. *European Journal of Operational Research*, *262*(1), 60–74. doi:10.1016/j.ejor.2017.02.046

Atique, S., Noureen, S., Roy, V., Subburaj, V., Bayne, S., & Macfie, J. (2019). Forecasting of total daily solar energy generation using ARIMA: A case study. *Proceedings of the 2019 IEEE 9th Annual Computing and Communication Workshop and Conference (CCWC)*, 114–119. 10.1109/CCWC.2019.8666481

Aung, S. (2018). Electric power is the main driving force for industrialization. Academic Press.

Aziz, S., Ahmed, I., Khan, K., & Khalid, M. (2023). Emerging trends and approaches for designing net-zero low-carbon integrated energy networks: A review of current practices. Arabian Journal for Science and Engineering, 1-23.

Babaie, M., Shiri, M. E., & Bahaghighat, M. (2018, April). A new descriptor for UAV images mapping by applying discrete local radon. In 2018 8th conference of AI & robotics and 10th RoboCup Iranopen international symposium (IRANOPEN) (pp. 52-56). IEEE. 10.1109/RIOS.2018.8406631

Bacher, P., Madsen, H., & Nielsen, H. A. (2009). Online short-term solar power forecasting. *Solar Energy*, *83*(10), 1772–1783. doi:10.1016/j.solener.2009.05.016

Badi, I., Muhammad, L., Abubakar, M., & Bakır, M. (2022). Measuring sustainability performance indicators using FUCOM-MARCOS methods. Operational Research in Engineering Sciences: Theory and Applications, 5(2), 99-116. doi:10.31181/oresta040722060b

Badi, K., & Shimura, M. (1982). Machine recognition of Arabic cursive script. *Trans. Inst. Electron. Commun. Eng.*, *E65*(2), 107–114.

Baeyens, J., Zhang, H., Nie, J., Appels, L., Dewil, R., Ansart, R., & Deng, Y. (2020). Reviewing the Potential of Bio-Hydrogen Production by Fermentation. *Renewable & Sustainable Energy Reviews*, *131*, 110023. doi:10.1016/j.rser.2020.110023

Bagnall, A., Lines, J., Hills, J., & Bostrom, A. (2015). Time-series classification with COTE: The collective of transformation-based ensembles. *IEEE Transactions on Knowledge and Data Engineering*, *27*(9), 2522–2535. doi:10.1109/TKDE.2015.2416723

Bahaghighat, M., & Motamedi, S. A. (2018). Vision inspection and monitoring of wind turbine farms in emerging smart grids. Facta Universitatis, Series. *Electronics and Energetics*, *31*(2), 287–301.

Bahlmann, C., Haasdonk, B., & Burkhardt, H. (2002). Online Handwriting Recognition with Support Vector Machine – A Kernel Approach. *Proceeding of the 8th Int. Workshop in Handwriting Recognition (IWHFR)*, 49-54. 10.1109/IWFHR.2002.1030883

Balaji, P., Alelyani, S., Qahmash, A., & Mohana, M. (2021). Contributions of machine learning models towards student academic performance prediction: A systematic review. *Applied Sciences (Basel, Switzerland)*, *11*(21), 10007. doi:10.3390/app112110007

Balal, A., Jafarabadi, Y. P., Demir, A., Igene, M., Giesselmann, M., & Bayne, S. (2023). Forecasting solar power generation utilizing machine learning models in Lubbock. *Emerging Science Journal*, *7*(4), 1052–1062. doi:10.28991/ESJ-2023-07-04-02

Baljit, S. S. S., Chan, H.-Y., & Sopian, K. (2016). Review of building integrated applications of photovoltaic and solar thermal systems. *Journal of Cleaner Production*, *137*, 677–689. doi:10.1016/j.jclepro.2016.07.150

Balle, B., Bell, J., Gascon, A., & Nissim, K. (2018). The privacy blanket of the shuffle model. In *Proceedings of the 2018 ACM SIGSAC Conference on Computer and Communications Security* (pp. 638-650). ACM.

Bandara, K., Bergmeir, C., & Smyl, S. (2020). Forecasting across time series databases using recurrent neural networks on groups of similar series: A clustering approach. *Expert Systems with Applications*, *140*, 112896. doi:10.1016/j.eswa.2019.112896

Bapai, A., & Duchon, M. (2019). A hybrid approach of solar power forecasting using machine learning. *Proceedings of the 3rd International Conference on Smart Grid and Smart Cities.*

Barbier, E. (2002). Geothermal energy technology and current status: an overview. *Renewable and Sustainable Energy Reviews, 6*(1-2), 3-65.

Bartels, J. R., Pate, M. B., & Olson, N. K. (2010). An economic survey of hydrogen production from conventional and alternative energy sources. *International Journal of Hydrogen Energy*, *35*(16), 8371–8384. doi:10.1016/j.ijhydene.2010.04.035

Bartle, A. (2002). Hydropower potential and development activities. *Energy Policy*, *30*(14), 1231–1239. doi:10.1016/S0301-4215(02)00084-8

Basheer, M., Nechifor, V., Calzadilla, A., Ringler, C., Hulme, D., & Harou, J. J. (2022). Balancing national economic policy outcomes for sustainable development. *Nature Communications*, *13*(1), 5041. doi:10.1038/s41467-022-32415-9 PMID:36028486

Beigi, H. S. M. (1993). An Overview of Handwriting Recognition. *Proceedings of the 1st Annual Conference on Technological Advancements in Developing Countries*, 30-46.

Benghanem, M., Mellit, A., & Alamri, S. N. (2009). A. Mellit, S. Alamri, ANN-based modelling and estimation of daily global solar radiation data: A case study. *Energy Conversion and Management*, *50*(7), 1644–1655. doi:10.1016/j.enconman.2009.03.035

Berry, Holland, & Watkiss. (1994). Externalities of fuel cycles' ExternE' project. Coal fuel cycle. Estimation of physical impacts and monetary valuation for priority impact pathways. Academic Press.

Beshr, E. H., Abdelghany, H., & Eteiba, M. (2018). Novel optimization technique of isolated microgrid with hydrogen energy storage. *PLoS One*, *13*(2), e0193224. doi:10.1371/journal.pone.0193224 PMID:29466433

Bharath, H., Pablo, A., Ross, G., & Jitendra, M. (2015). Hypercolumns for object segmentation and fine-grained localization. *Proceedings of the IEEE conference on computer vision and pattern recognition*, 447–56.

Blanco. (2009). The economics of wind energy. *Renewable and sustainable energy reviews,* 13(6-7), 1372-1382.

Bogdanov, D., Ram, M., Aghahosseini, A., Gulagi, A., Oyewo, A. S., Child, M., & Breyer, C. (2021). Low-cost renewable electricity as the key driver of the global energy transition towards sustainability. *Energy*, *227*, 120467. doi:10.1016/j.energy.2021.120467

Bojer, C. S. (2022). Understanding machine learning-based forecasting methods: A decomposition framework and research opportunities. *International Journal of Forecasting*, *38*(4), 1555–1561. doi:10.1016/j.ijforecast.2021.11.003

Bolatkhan, K., Kossalbayev, B. D., Zayadan, B. K., Tomo, T., Veziroglu, T. N., & Allakhverdiev, S. I. (2019). Hydrogen Production from Phototrophic Microorganisms: Reality and Perspectives. *International Journal of Hydrogen Energy*, *44*(12), 5799–5811. doi:10.1016/j.ijhydene.2019.01.092

Boltasseva, A., & Shalaev, V. M. (2008). Fabrication of optical negative-index metamaterials: Recent advances and outlook. *Metamaterials (Amsterdam)*, *2*(1), 1–17. doi:10.1016/j.metmat.2008.03.004

Bomfim, T. S. (2020). Evolution of machine learning in smart grids. In *2020 IEEE 8th International Conference on Smart Energy Grid Engineering (SEGE)* (pp. 82–87). IEEE. https://doi.org/10.1109/SEGE49949.2020.9182023

Bouktif, S. (2005). Improving software Quality prediction by combining and adapting predictive models. Montreal University Montreal. doi:10.1198/jasa.2011.tm09771

Bouzerdoum, Mellit, & Massi. (2013). A hybrid model (SARIMA-SVM) for short-term power forecasting of a small-scale grid-connected photovoltaic plant. Sol. Energy, 98, 226–235.

Boza, P., & Evgeniou, T. (2021). Artificial intelligence to support the integration of variable renewable energy sources to the power system. *Applied Energy*, *290*, 116754. doi:10.1016/j.apenergy.2021.116754

BP Energy. (2018). *Outlook country and regional insights-India*. https://www.bp.com/ content/dam/ bp/en/corporate/pdf/energy-economics/energy-outlook/bp-energy-outlook-2018-country-insight-india.pdf

Breeze, P. (2018). *Hydropower*. Academic Press.

Brophy, P. (1997). Environmental advantages to the utilization of geothermal energy. Renewable Energy, 10(2-3), 367-377.

Budak, G., Budak, E., & Kaya, I. (2019). A systematic approach to assess renewable energy and identify the energy alternatives for a city. *Energy*, *189*, 116233.

Budiman, I. (2018). Enabling community participation for social innovation in the energy sector. *Indonesian Journal of Energy*, *1*(2), 21–31. doi:10.33116/ije.v1i2.23

Bujang, A. S., Bern, C. J., & Brumm, T. J. (2016). Summary of energy demand and renewable energy policies in Malaysia. *Renewable & Sustainable Energy Reviews*, *53*, 1459–1467. doi:10.1016/j.rser.2015.09.047

Cabal. (2017). Solar-thermal and hybrid photovoltaic-thermal systems for renewable heating. Academic Press.

Cali, U., & Sharma, V. (2019). Short-term wind power forecasting using long-short term memory based recurrent neural network model and variable selection. *Int. J. Smart Grid Clean Energy*, *8*, 103–110. doi:10.12720/sgce.8.2.103-110

Camila, C. J., Cardemil, J. M., Droguett, E. L., & Behzad, M. (2020). Assessment of deep learning techniques for prognosis of solar thermal systems. *Renewable Energy*, *145*, 2178–2191. doi:10.1016/j.renene.2019.07.100

Cardona, J. L., Howland, M. F., & Dabiri, J. O. (2019). Seeing the wind: Visual wind speed prediction with a coupled convolutional and recurrent neural network. arXiv 2019, arXiv:1905.13290

Carey, M., & Ramsay, J. O. (2021). Fast stable parameter estimation for linear dynamical systems. *Computational Statistics & Data Analysis*, *156*, 107124. doi:10.1016/j.csda.2020.107124

Carolin Mabel, M., & Fernandez, E. (2008). Analysis of wind power generation and prediction using ANN: A case study. *Renewable Energy*, *33*(5), 986–992. doi:10.1016/j.renene.2007.06.013

Carrillo, A. J., González-Aguilar, J., Romero, M., & Coronado, J. M. (2019). Solar Energy on Demand: A Review on High Temperature Thermochemical Heat Storage Systems and Materials. *Chemical Reviews*, *119*(7), 4777–4816. doi:10.1021/acs.chemrev.8b00315 PMID:30869873

Cellura, M., Cirrincione, G., Marvuglia, A., & Miraoui, A. (2008). Wind speed spatial estimation for energy planning in {Sicily}: {A} neural kriging application. *Renewable Energy*, *33*(6), 1251–1266. doi:10.1016/j.renene.2007.08.013

Chae, K. J., Choi, M. J., Kim, K. Y., Ajayi, F. F., Chang, I. S., & Kim, I. S. (2010). Selective inhibition of methanogens for the improvement of biohydrogen production in microbial electrolysis cells. *International Journal of Hydrogen Energy*, *35*(24), 13379–13386. doi:10.1016/j.ijhydene.2009.11.114

Chahla, C., Snoussi, H., Merghem, L., & Esseghir, M. (2020). A deep learning approach for anomaly detection and prediction in power consumption data. *Energy Efficiency*, *13*(8), 1633–1651. doi:10.1007/s12053-020-09884-2

Chambers, D. (2013). Social media and personal relationships: Online intimacies and networked friendship. Springer.

Chandrasekhar, K., Kumar, S., Lee, B. D., & Kim, S. H. (2020). Waste based hydrogen production for circular bioeconomy: Current status and future directions. *Bioresource Technology*, *302*, 122920. doi:10.1016/j.biortech.2020.122920 PMID:32029301

Chang, Y., Lee, J., & Yoon, H. (2010). Alternative projection of the world energy consumption-in comparison with the 2010 international energy outlook. *Energy Policy*, *50*, 154–160. doi:10.1016/j.enpol.2012.07.059

Château, B. (2022). Energy Demand Drivers. In M. Hafner & G. Luciani (Eds.), *The Palgrave Handbook of International Energy Economics* (pp. 511–543). Springer International Publishing. doi:10.1007/978-3-030-86884-0_26

Chen, C. J. (2011) Physics of solar energy. John Wiley & Sons. doi:10.1002/9781118172841

Chen, Q., & Folly, K. (2019). Effect of Input Features on the Performance of the ANN-based Wind Power Forecasting. Proceedings of the Southern African Universities Power Engineering Conference/Robotics and Mechatronics/Pattern Recognition Association of South Africa (SAUPEC/RobMech/PRASA), 673–678.

Chen, C., Chang, H., & Chang, J. (2016). Producing carbohydrate-rich microalgal biomass grown under mixotrophic conditions as feedstock for biohydrogen production. *International Journal of Hydrogen Energy*, *41*(7), 4413–4420. doi:10.1016/j.ijhydene.2015.05.163

Chen, J., Yu, J., Song, M., & Valdmanis, V. (2019). Factor decomposition and prediction of solar energy consumption in the United States. *Journal of Cleaner Production*, *234*, 1210–1220. doi:10.1016/j.jclepro.2019.06.173

Chen, W., Xie, X., Wang, J., Pradhan, B., Hong, H., Bui, D. T., & Ma, J. (2017). A comparative study of logistic model tree, random forest, and classification and regression tree models for spatial prediction of landslide susceptibility. *Catena*, *151*, 147–160. doi:10.1016/j.catena.2016.11.032

Chen, Y., Wang, H., Zhang, B., & Zhang, W. (2022). A method of measuring the article discriminative capacity and its distribution. *Scientometrics*, *127*(6), 3317–3341. doi:10.1007/s11192-022-04371-0

Chen, Y., & Wang, L. (2020). Application of One-Class Support Vector Machine Model in Anomaly Detection of Renewable Energy Generation Systems. *Renewable Energy*, *145*, 155–163.

Cheung, K., Yeung, D., & Chin, R. T. (1998). A Bayesian framework for deformable pattern recognition with application to handwritten character recognition. *IEEE Transactions on Pattern Analysis and Machine Intelligence*, *20*(12), 382–1388.

Chookaew, T., O-Thong, S., & Prasertsan, P. (2015). Biohydrogen production from crude glycerol by two stage of dark and photo fermentation. *International Journal of Hydrogen Energy*, *40*(24), 7433–7438. doi:10.1016/j.ijhydene.2015.02.133

Chowdhury, M. S., Rahman, K. S., Selvanathan, V., Nuthammachot, N., Suklueng, M., Mostafaeipour, A., Habib, A., Akhtaruzzaman, M., Amin, N., & Techato, K. (2021). Current trends and prospects of tidal energy technology. *Environment, Development and Sustainability*, *23*(6), 8179–8194. doi:10.1007/s10668-020-01013-4 PMID:33041645

Chow, T. T. (2018). A review on photovoltaic/thermal hybrid solar technology. *Renewable Energy*, *4*(88), 119.

Cohen, S. M., Iyer, G. C., Brown, M., Macknick, J., Wise, M., Binsted, M., & Hejazi, M. (2021). How structural differences influence cross-model consistency: An electric sector case study. *Renewable & Sustainable Energy Reviews*, *144*, 111009. doi:10.1016/j.rser.2021.111009

Connolly, D., Lund, H., Mathiesen, B. V., & Leahy, M. (2010). A review of computer tools for analysing the integration of renewable energy into various energy systems. *Applied Energy*, *87*(4), 1059–1082. doi:10.1016/j.apenergy.2009.09.026

Conteh, F., Furukakoi, M., Rangarajan, S. S., Collins, E. R., Conteh, M. A., Rashwan, A., & Senjyu, T. (2023). Long-Term Forecast of Sierra Leone's Energy Supply and Demand (2019–2040): A LEAP Model Application for Sustainable Power Generation System. *Sustainability (Basel)*, *15*(15), 11838. doi:10.3390/su151511838

Cordeiro-Costas, M., Villanueva, D., Eguía-Oller, P., Martínez-Comesaña, M., & Ramos, S. (2023). Load Forecasting with Machine Learning and Deep Learning Methods. *Applied Sciences (Basel, Switzerland)*, *13*(13), 7933. doi:10.3390/app13137933

Cozzi, L., Gould, T., & Bouckart, S. (2020). *World Energy Outlook 2020*. International Energy Agency.

Croese, E., Pereira, M. A., Euverink, G. J. W., Stams, A. J. M., & Geelhoed, J. S. (2011). Analysis of the microbial community of the biocathode of a hydrogen-producing microbial electrolysis cell. *Applied Microbiology and Biotechnology*, *92*(5), 1083–1093. doi:10.1007/s00253-011-3583-x PMID:21983651

Croisier, J.-L., Ganteaume, S., Binet, J., Genty, M., & Ferret, J.-M. (2008). Strength imbalances and prevention of hamstring injury in professional soccer players: A prospective study. *The American Journal of Sports Medicine*, *36*(8), 1469–1475. doi:10.1177/0363546508316764 PMID:18448578

Cruz-Rojas, T., Franco, J. A., Hernandez-Escobedo, Q., Ruiz-Robles, D., & Juarez-Lopez, J. M. (2023). A novel comparison of image semantic segmentation techniques for detecting dust in photovoltaic panels using machine learning and deep learning. *Renewable Energy*, *217*, 119126. doi:10.1016/j.renene.2023.119126

Dai, H., Xie, X., Xie, Y., Liu, J., & Masui, T. (2016). Green growth: The economic impacts of large-scale renewable energy development in China. *Applied Energy*, *162*, 435–449. doi:10.1016/j.apenergy.2015.10.049

Dalal, N., & Triggs, B. (2005). Histograms of oriented gradients for human detection. *Proceedings of the IEEE Computer Society Conference on Computer Vision And Pattern Recognition. CVPR 2005*, 886–893. 10.1109/CVPR.2005.177

Damman, S., Sandberg, E., Rosenberg, E., Pisciella, P., & Graabak, I. (2021). A hybrid perspective on energy transition pathways: Is hydrogen the key for Norway? *Energy Research & Social Science*, *78*, 102116. doi:10.1016/j.erss.2021.102116

Damousis, I. G., Alexiadis, M. C., Theocharis, J. B., & Dokopoulos, P. S. (2004). A fuzzy model for wind speed prediction and power generation in wind parks using spatial correlation. *IEEE Transactions on Energy Conversion*, *19*(2), 352–361. doi:10.1109/TEC.2003.821865

Danfei, X., Dragomir, A., & Jain, P. A. (2018). Deep sensor fusion for 3d bounding box estimation. *Proceedings of the IEEE conference on computer vision and pattern recognition*, 244–53.

Danish, M. S. S., & Senjyu, T. (2023). AI-Enabled Energy Policy for a Sustainable Future. *Sustainability (Basel)*, *15*(9), 7643. doi:10.3390/su15097643

Dash, P. K., Samantaray, S. R., & Panda, G. (2006). Fault classification and section identification of an advanced series-compensated transmission line using support vector machine. *IEEE Transactions on Power Delivery*, *22*(1), 67–73. doi:10.1109/TPWRD.2006.876695

Dawood, F., Anda, M., & Shafiullah, G. M. (2020). Hydrogen Production for Energy: An Overview. *International Journal of Hydrogen Energy*, *45*(7), 3847–3869. doi:10.1016/j.ijhydene.2019.12.059

Daxini, N., Sharma, S., & Patel, R. (2015). Real Time Animal Detection System using HAAR Like Feature. *International Journal of Innovative Research in Computer and Communication Engineering*, *3*(6), 5177–5182. doi:10.15680/ijircce.2015.0306032

Dayupay, J.P., Ajibade, S.S., Villagonzalo, M.C., Baird, J.C., Antecristo, C.P., & Adediran, A.O. (2024). Application of machine learning techniques in energy power production: A publication trend and bibliometrics analysis (2012-2023). *E3S Web of Conferences, 491*, 01003. https://doi.org/ doi:10.1051/e3sconf/202449101003

De Meyer, A., Cattrysse, D., Rasinmäki, J., & Van Orshoven, J. (2014). Methods to optimise the design and management of biomass-for-bioenergy supply chains: A review. *Renewable & Sustainable Energy Reviews*, *31*, 657–670. doi:10.1016/j.rser.2013.12.036

Dehghani, M., Riahi-Madvar, H., Hooshyaripor, F., Mosavi, A., Shamshirband, S., Zavadskas, E. K., & Chau, K.-W. (2019). Prediction of hydropower generation using Grey wolf optimization adaptive neuro-fuzzy inference system. *Energies*, *12*(2), 289. doi:10.3390/en12020289

del Real, A. J., Dorado, F., & Durán, J. (2020). Energy demand forecasting using deep learning: Applications for the French grid. *Energies*, *13*(9), 2242.

Del Ser, J., Casillas-Perez, D., Cornejo-Bueno, L., Prieto-Godino, L., Sanz-Justo, J., Casanova-Mateo, C., & Salcedo-Sanz, S. (2022). Randomization-based machine learning in renewable energy prediction problems: Critical literature review, new results and perspectives. *Applied Soft Computing*, *118*, 108526. doi:10.1016/j.asoc.2022.108526

Dent, C. M. (2012). Renewable energy and East Asia's new developmentalism: Towards a low carbon future? *The Pacific Review*, *25*(5), 561–587. doi:10.1080/09512748.2012.728240

Devabhaktuni, V., Alam, M., Depuru, S. S. S. R., Green, R. C. II, Nims, D., & Near, C. (2013). Solar energy: Trends and enabling technologies. *Renewable & Sustainable Energy Reviews*, *19*, 555–564. doi:10.1016/j.rser.2012.11.024

Dicharry, S. Y. (2012). "Wind Energy Production Compensation Scheme: Oil-Like Royalties or Oyster-Like Rent." *Loy. L. REv.*, *58*, 179.

Dickson & Fanelli. (2013). Geothermal energy: utilization and technology. Academic Press.

Dincer, I., & Acar, C. (2015). Review and Evaluation of Hydrogen Production Methods for Better Sustainability. *International Journal of Hydrogen Energy*, *40*(34), 11094–11111. doi:10.1016/j.ijhydene.2014.12.035

Ding, L., Xia, A., Lin, R., Li, Y., Zhou, J., & Cen, K. (2015). Hydrogen production using amino acids obtained by protein degradation in waste biomass by combined dark- and photo-fermentation. *Bioresource Technology*, *179*, 13–19. doi:10.1016/j.biortech.2014.11.109 PMID:25514397

Doneus, M., Verhoeven, G., Fera, M., Briese, C., Kucera, M., & Neubauer, W. (2011). From deposit to point cloud–a study of low-cost computer vision approaches for the straightforward documentation of archaeological excavations. *Geoinformatics FCE CTU*, *6*, 81–88. doi:10.14311/gi.6.11

Doost Mohammadian, H., & Rezaie, F. (2020). Blue-green smart mobility technologies as readiness for facing tomorrow's urban shock toward the world as a better place for living (Case studies: Songdo and Copenhagen). *Technologies*, *8*(3), 39. doi:10.3390/technologies8030039

Dorado-Moreno, M., Navarin, N., Gutierrez, P. A., Prieto, L., Sperduti, A., Salcedo-Sanz, S., & Hervas-Martinez, C. (2020a). Multi-task learning for the prediction of wind power ramp events with deep neural networks. *Neural Networks*, *123*, 401–411. doi:10.1016/j.neunet.2019.12.017 PMID:31926464

Downton, A. C., Tregidgo, R. W. S., & Kabir, E. (1991). Recognition and Verification of Handwritten and Hand- Printed British Postal Addresses. *International Journal of Pattern Recognition and Artificial Intelligence*, *5*(01n02), 265–292. doi:10.1142/S0218001491000168

Duchesne, L., Karangelos, E., & Wehenkel, L. (2020). Recent developments in machine learning for energy systems reliability management. *Proceedings of the IEEE*, *108*(9), 1656–1676. doi:10.1109/JPROC.2020.2988715

Durairaj, D. M., & Mohan, B. K. (2022). A convolutional neural network-based approach to financial time series prediction. *Neural Computing & Applications*, *34*(16), 13319–13337.

EIA. (2018). *International energy outlook 2018 (IEO2018).* https://www.eia.gov / pressroom/presentations/capuano_07242018.pdf

EIA. (2019). *Energy outlook 2019 with projections to 2050.* https://www.eia.gov/outlooks/aeo/pdf/ aeo2019.pdf

Eicker, U., Pietruschka, D., Schmitt, A., & Haag, M. (2015). Comparison of photovoltaic and solar thermal cooling systems for office buildings in different climates. *Solar Energy*, *118*, 243–255. doi:10.1016/j.solener.2015.05.018

Ekanayake, P., Peiris, A. T., Jayasinghe, J. M. J. W., & Rathnayake, U. (2021). Development of Wind Power Prediction Models for Pawan Danavi Wind Farm in Sri Lanka. *Mathematical Problems in Engineering*, *2021*, 4893713. doi:10.1155/2021/4893713

El Bassam, N. (2010). *Handbook of bioenergy crops: a complete reference to species, development and applications*. Routledge. doi:10.4324/9781849774789

Elavarasan, R. M., Shafiullah, G. M., Padmanaban, S., Kumar, N. M., Annam, A., Vetrichelvan, A. M., & Holm-Nielsen, J. B. (2020). A comprehensive review on renewable energy development, challenges, and policies of leading Indian states with an international perspective. *IEEE Access : Practical Innovations, Open Solutions*, *8*, 74432–74457. doi:10.1109/ACCESS.2020.2988011

El-Dalatony, M. M., Zheng, Y., Ji, M.-K., Li, X., & Salama, E.-S. (2020). Metabolic Pathways for MicroalgalBiohydrogen Production: Current Progress and Future Prospectives. *Bioresource Technology*, *318*, 124253. doi:10.1016/j.biortech.2020.124253 PMID:33129070

ElFar, O. A., Chang, C. K., Leong, H. Y., Peter, A. P., Chew, K. W., & Show, P. L. (2021). Prospects of Industry 5.0 in algae: Customization of production and new advance technology for clean bioenergy generation. *Energy Conversion and Management: X*, *10*, 100048.

Eljack, F., & Kazi, M. K. (2021). Prospects and challenges of green hydrogen economy via multi-sector global symbiosis in Qatar. *Frontiers in Sustainability*, *1*, 612762. doi:10.3389/frsus.2020.612762

Elrahmani, A., Hannun, J., Eljack, F., & Kazi, M. K. (2021). Status of renewable energy in the GCC region and future opportunities. *Current Opinion in Chemical Engineering*, *31*, 100664. doi:10.1016/j.coche.2020.100664

Energy Information Administration. (n.d.). https://www.eia.gov/energyexplained/?page=biomass_home

Ensafi, Y., Amin, S. H., Zhang, G., & Shah, B. (2022). Time-series forecasting of seasonal items sales using machine learning–a comparative analysis. *International Journal of Information Management Data Insights*, *2*(1), 100058. doi:10.1016/j.jjimei.2022.100058

Entezari, A., Aslani, A., Zahedi, R., & Noorollahi, Y. (2023). Artificial intelligence and machine learning in energy systems: A bibliographic perspective. *Energy Strategy Reviews*, *45*, 101017. doi:10.1016/j.esr.2022.101017

Eriksson, E. L. V., & Gray, E. M. (2019). Optimization of renewable hybrid energy systems – A multi-objective approach. *Renewable Energy*, *133*, 971–999. doi:10.1016/j.renene.2018.10.053

Ernst, B., Oakleaf, B., Ahlstrom, M. L., Lange, M., Moehrlen, C., Lange, B., Focken, U., & Rohrig, K. (2007). Predicting the wind. *IEEE Power & Energy Magazine*, *5*(6), 78–89. doi:10.1109/MPE.2007.906306

Eroglu, E., & Melis, A. (2016). Microalgal Hydrogen Production Research. *International Journal of Hydrogen Energy*, *41*(30), 12772–12798. doi:10.1016/j.ijhydene.2016.05.115

Erol-Kantarci, M., & Mouftah, H. T. (2011). Wireless multimedia sensor and actor networks for the next generation power grid. *Ad Hoc Networks*, *9*(4), 542–551. doi:10.1016/j.adhoc.2010.08.005

Erten, M. Y., & Aydilek, H. (2022). Solar Power Prediction using Regression Models. *Int. J. Eng. Res. Dev.*, *14*, s333–s342.

Eskandarpour, R., & Khodaei, A. (2016). Machine learning based power grid outage prediction in response to extreme events. *IEEE Transactions on Power Systems*, *32*(4), 3315–3316. doi:10.1109/TPWRS.2016.2631895

Eze, E. C., & Chatwin, C. R. (2019). Enhanced recurrent neural network for short-term wind farm power output prediction. *J. Appl. Sci.*, *5*, 28–35.

Factsheet-China-FC-EN. (n.d.). https://www.now-gmbh.de/wp-content/uploads/2020/09/Factsheet-China-FC-EN.pdf

Fadel, E., Gungor, V. C., Nassef, L., Akkari, N., Malik, M. A., Almasri, S., & Akyildiz, I. F. (2015). A survey on wireless sensor networks for smart grid. *Computer Communications*, *71*, 22–33. doi:10.1016/j.comcom.2015.09.006

Falcao, A. F. (2010). Wave energy utilization: A review of the technologies. *Renewable & Sustainable Energy Reviews*, *14*(3), 899–918. doi:10.1016/j.rser.2009.11.003

Fang, X., & Li, D. (2013). Solar photovoltaic and thermal technology and applications in China. *Renewable & Sustainable Energy Reviews*, *23*, 330–340. doi:10.1016/j.rser.2013.03.010

Faroudja & Izeboudjen. (2020) Decision tree based system on chip for forest fires prediction. In *International Conference on Electrical Engineering (ICEE)*. IEEE.

Feng, C., & Zhang, J. (2018). Hourly-similarity based solar forecasting using multi-model machine learning blending. arXiv 2018, arXiv:1803.03623 doi:10.1109/PESGM.2018.8586091

Feng, C., Cui, M., Hodge, B. M., & Zhang, J. (2017). A data-driven multi-model methodology with deep feature selection for short-term wind forecasting. *Applied Energy*, *190*, 1245–1257. doi:10.1016/j.apenergy.2017.01.043

Filik, Ü. B., & Filik, T. (2017). Wind speed prediction using artificial neural networks based on multiple local measurements in Eskisehir. *Energy Procedia*, *107*, 264–269. doi:10.1016/j.egypro.2016.12.147

Fischer, C., & Preonas, L. (2010). Combining policies for renewable energy: Is the whole less than the sum of its parts? *Resource for the future discussion paper*, (10-19).

Fischetti, M., & Fraccaro, M. (2019). Machine learning meets mathematical optimization to predict the optimal production of offshore wind parks. *Computers & Operations Research*, *106*, 289–297. doi:10.1016/j.cor.2018.04.006

Forootan, M. M., Larki, I., Zahedi, R., & Ahmadi, A. (2022). Machine learning and deep learning in energy systems: A review. *Sustainability (Basel)*, *14*(8), 4832. doi:10.3390/su14084832

Fosket, J., & Mamo, L. (2009). Living green: Communities that sustain. New Society Publishers.

Franki, V., Majnarić, D., & Višković, A. (2023). A comprehensive review of artificial intelligence (AI) companies in the power sector. *Energies*, *16*(3), 1077. doi:10.3390/en16031077

Fridleifsson, I. B. (2001). Geothermal energy for the benefit of the people. *Renewable & Sustainable Energy Reviews*, *5*(3), 299–312. doi:10.1016/S1364-0321(01)00002-8

Gabalawy, M., Hosny, N. S., & Adly, A. R. (2021). Probabilistic forecasting for energy time series considering uncertainties based on deep learning algorithms. *Electric Power Systems Research*, *196*, 107216. doi:10.1016/j.epsr.2021.107216

Gaétan, B., Lionel, D., Matthieu, M., & Florent, L. (2019). Low-power neural networks for semantic segmentation of satellite images. *Proceedings of the IEEE/CVF international conference on computer vision workshops*.

García-Hinde, O., Terrén-Serrano, G., Hombrados-Herrera, M. A., Gómez-Verdejo, V., Jiménez-Fernández, S., Casanova-Mateo, C., Sanz-Justo, J., Martínez-Ramón, M., & Salcedo-Sanz, S. (2018). Evaluation of dimensionality reduction methods applied to numerical weather models for solar radiation forecasting. *Engineering Applications of Artificial Intelligence*, *69*, 157–167. doi:10.1016/j.engappai.2017.12.003

Garrido, S., Sequeira, T., & Santos, M. (2020). Renewable energy and sustainability from the supply side: A critical review and analysis. *Applied Sciences (Basel, Switzerland)*, *10*(17), 5755. doi:10.3390/app10175755

Gasparin, A., Lukovic, S., & Alippi, C. (2022). Deep learning for time series forecasting: The electric load case. Transactions on Intelligence Technology, 7(1), 1-25. doi:10.1049/cit2.12060

Gellert, A., Fiore, U., Florea, A., Chis, R., & Palmieri, F. (2022). Forecasting electricity consumption and production in smart homes through statistical methods. *Sustainable Cities and Society*, *76*, 103426.

Ge, S., Foong, S. Y., Ma, N. L., Liew, R. K., Wan Mahari, W. A., Xia, C., Yek, P. N. Y., Peng, W., Nam, W. L., Lim, X. Y., Liew, C. M., Chong, C. C., Sonne, C., & Lam, S. S. (2020). Vacuum Pyrolysis Incorporating Microwave Heating and Base Mixture Modification: An Integrated Approach to Transform Biowaste into Eco-Friendly Bioenergy Products. *Renewable & Sustainable Energy Reviews*, *127*, 109871. doi:10.1016/j.rser.2020.109871

Ghai, D., Tripathi, S. L., Saxena, S., Chanda, M., & Alazab, M. (Eds.). (2022). *Machine Learning Algorithms for Signal and Image Processing*. John Wiley & Sons. doi:10.1002/9781119861850

Ghimire, A., Valentino, S., Frunzo, L., Trably, E., Escudié, R., Pirozzi, F., Lens, P. N. L., & Esposito, G. (2015). Biohydrogen production from food waste by coupling semi-continuous dark-photofermentation and residue post-treatment to anaerobic digestion: A synergy for energy recovery. *International Journal of Hydrogen Energy*, *40*(46), 16045–16055. doi:10.1016/j.ijhydene.2015.09.117

Gholkar, P., Shastri, Y., & Tanksale, A. (2021). Renewable Hydrogen and Methane Production from Microalgae: A Techno-Economic and Life Cycle Assessment Study. *Journal of Cleaner Production*, *279*, 123726. doi:10.1016/j.jclepro.2020.123726

Gielen, D. (2019). *The role of renewable energy in the global energy transformation*. Elsevier.

Goswami, R. K., Mehariya, S., Obulisamy, P. K., & Verma, P. (2021). Advanced Microalgae-Based Renewable Biohydrogen Production Systems: A Review. *Bioresource Technology*, *320*, 124301. doi:10.1016/j.biortech.2020.124301 PMID:33152683

Grubb, M., Drummond, P., Poncia, A., McDowall, W., Popp, D., Samadi, S., & Pavan, G. (2021). Induced innovation in energy technologies and systems: A review of evidence and potential implications for CO2 mitigation. *Environmental Research Letters*, *16*(4), 043007. doi:10.1088/1748-9326/abde07

Gupta, H. K., & Roy, S. (2006). Geothermal energy: an alternative resource for the 21st century. Elsevier.

Gupta, S., Pawar, S., Ramrakhiyani, N., Palshikar, G. K., & Varma, V. (2018). Semi-supervised recurrent neural network for adverse drug reaction mention extraction. *BMC Bioinformatics*, *19*(S8), 1–7. doi:10.1186/s12859-018-2192-4 PMID:29897321

Hagenau, M., Liebmann, M., & Neumann, D. (2013). Automated news reading: Stock price prediction based on financial news using context-capturing features. *Decision Support Systems*, *55*(3), 685–697. doi:10.1016/j.dss.2013.02.006

Hainsch, K., Löffler, K., Burandt, T., Auer, H., del Granado, P. C., Pisciella, P., & Zwickl-Bernhard, S. (2022). Energy transition scenarios: What policies, societal attitudes, and technology developments will realize the EU Green Deal? *Energy*, *239*, 122067. doi:10.1016/j.energy.2021.122067

Hakeem, K. R., Jawaid, M., & Rashid, U. (2016). *Biomass and bioenergy*. Springer International Pu.

Hakeem, K. R., Jawaid, M., & Rashid, U. (Eds.). (2014). *Biomass and bioenergy: Applications*. Springer.

Hamed, M. M., Ali, H., & Abdelal, Q. (2022). Forecasting annual electric power consumption using a random parameters model with heterogeneity in means and variances. *Energy*, *255*, 124510. doi:10.1016/j.energy.2022.124510

Han, S., Li, J., & Liu, Y. (2011). Tabu search algorithm optimized ANN model for wind power prediction with NWP. *Energy Procedia, 12*, 733–740. doi:10.1016/j.egypro.2011.10.099

Han, T., Liu, C., Yang, W., & Jiang, D. (2019). A novel adversarial learning framework in deep convolutional neural network for intelligent diagnosis of mechanical faults. *Knowledge-Based Systems, 165*, 474–487. doi:10.1016/j.knosys.2018.12.019

Harbola, S., & Coors, V. (2019). One dimensional convolutional neural network architectures for wind prediction. *Energy Conversion and Management, 195*, 70–75. doi:10.1016/j.enconman.2019.05.007

Harrou, F., Dairi, A., Taghezouit, B., & Sun, Y. J. S. E. (2019). An unsupervised monitoring procedure for detecting anomalies in photovoltaic systems using a one-class support vector machine. Academic Press.

Hassan, Q., Viktor, P., Al-Musawi, T. J., Ali, B. M., Algburi, S., Alzoubi, H. M., & Jaszczur, M. (2024). The renewable energy role in the global energy Transformations. *Renewable Energy Focus, 48*, 100545. doi:10.1016/j.ref.2024.100545

HassanzadehFard, H., Tooryan, F., Collins, E. R., Jin, S., & Ramezani, B. (2020). Design and optimum energy management of a hybrid renewable energy system based on efficient various hydrogen production. *International Journal of Hydrogen Energy, 45*(55), 30113–30128. doi:10.1016/j.ijhydene.2020.08.040

Hemavathi, B., Vidya, G., Anantharaju, K. S., & Pai, R. K. (2024). Machine Learning in the Era of Smart Automation for Renewable Energy Materials. e-Prime-Advances in Electrical Engineering, Electronics and Energy, 100458.

Herbert, G. M. (2007). A review of wind energy technologies. *Renewable & Sustainable Energy Reviews, 11*(6), 1117–1145. doi:10.1016/j.rser.2005.08.004

Hochreiter S, & Schmidhuber J (1997). Long Short-Term Memory. *Neural Comput,* 9(8), 1735–1780. doi:10.1162/neco.1997.9.8.1735

Hoffmann, M. W., Wildermuth, S., Gitzel, R., Boyaci, A., Gebhardt, J., Kaul, H., & Tornede, T. (2020). Integration of novel sensors and machine learning for predictive maintenance in medium voltage switchgear to enable the energy and mobility revolutions. *Sensors (Basel), 20*(7), 2099. doi:10.3390/s20072099 PMID:32276442

Hofmann, E., & Rutschmann, E. (2018). Big data analytics and demand forecasting in supply chains: A conceptual analysis. *International Journal of Logistics Management, 29*(2), 739–766. doi:10.1108/IJLM-04-2017-0088

Hollingsworth, J., Smith, A., & Johnson, R. (2018). Anomaly Detection in Renewable Energy Systems Using One-Class Support Vector Machines. *Renewable Energy, 123*, 743–752.

Hong, W. C. (2008). Rainfall forecasting by technological machine learning models. *Applied Mathematics and Computation, 200*(1), 41–57. doi:10.1016/j.amc.2007.10.046

Houdou, A., El Badisy, I., Khomsi, K., Abdala, S. A., Abdulla, F., Najmi, H., Obtel, M., Belyamani, L., Ibrahimi, A., & Khalis, M. (2024). Interpretable machine learning approaches for forecasting and predicting air pollution: A systematic review. *Aerosol and Air Quality Research, 24*(1), 230151. doi:10.4209/aaqr.230151

Hou, Y., Vidu, R., & Stroeve, P. (2011). Solar energy storage methods. *Industrial & Engineering Chemistry Research, 50*(15), 8954–8964. doi:10.1021/ie2003413

Htun, H. H., Biehl, M., & Petkov, N. (2023). Survey of feature selection and extraction techniques for stock market prediction. *Financial Innovation, 9*(1), 26. doi:10.1186/s40854-022-00441-7 PMID:36687795

Huang, G. B., Zhu, Q. Y., & Siew, C.-K. (2006). Extreme learning machine: Theory and applications. *Neurocomputing, 70*(1-3), 489–501. doi:10.1016/j.neucom.2005.12.126

Huang, G., Wu, F., & Guo, C. (2022). Smart grid dispatch powered by deep learning: A survey. *Frontiers of Information Technology & Electronic Engineering*, *23*(5), 763–776. doi:10.1631/FITEE.2000719

Huang, S.-H., & Pan, Y.-C. (2014). Ergonomic job rotation strategy based on an automated RGB-D anthropometric measuring system. *Journal of Manufacturing Systems*, *33*(4), 699–710. doi:10.1016/j.jmsy.2014.02.005

Huang, S.-H., & Pan, Y.-C. (2015). Automated visual inspection in the semiconductor industry: A survey. *Computers in Industry*, *66*, 1–10. doi:10.1016/j.compind.2014.10.006

Huide, F., Xuxin, Z., Lei, M., Tao, Z., Qixing, W., & Hongyuan, S. (2017). A comparative study on three types of solar utilization technologies for buildings: Photovoltaic, solar thermal and hybrid photovoltaic/thermal systems. *Energy Conversion and Management*, *140*, 1–13. doi:10.1016/j.enconman.2017.02.059

Huiyu, G., & Miaomiao, L. (2022). Short-term solar irradiance prediction from sky images with a clear sky model. *Proceedings of the IEEE/CVF winter conference on applications of computer vision*, 2475–83.

Hu, M., Guo, L., Liu, J., & Song, Y. (2023). Classification and Recognition of Building Appearance Based on Optimized Gradient-Boosted Decision Tree Algorithm. *Sensors (Basel)*, *23*(11), 5353. doi:10.3390/s23115353 PMID:37300080

Huotari, M., Malhi, A., & Främling, K. (2024). Machine learning applications for smart building energy utilization: A survey. *Archives of Computational Methods in Engineering*, •••, 1–20. doi:10.1007/s11831-023-10054-7

Hussain, S., Muhammad, L., Ishaq, F., Yakubu, A., & Mohammed, I. (2019). Performance evaluation of various data mining algorithms on road traffic accident datasets. Paper presented at the Information and Communication Technology for Intelligent Systems: Proceedings of ICTIS 2018, 1.

Hussain, A., Rahman, M., & Memon, J. A. (2016). Forecasting electricity consumption in Pakistan: The way forward. *Energy Policy*, *90*, 73–80.

Hu, Z., Hu, Z., & Du, X. (2019). One-class support vector machines with a bias constraint and its application in system reliability prediction. *Artificial Intelligence for Engineering Design, Analysis and Manufacturing*, *33*(3), 346–358. doi:10.1017/S0890060419000155

Hwangbo, S., Nam, K., Heo, S., & Yoo, C. (2019). Hydrogen-based self-sustaining integrated renewable electricity network (HySIREN) using a supply-demand forecasting model and deep-learning algorithms. *Energy Conversion and Management*, *185*, 353–367. doi:10.1016/j.enconman.2019.02.017

Hydrogeninsight. (2023). https://www.hydrogeninsight.com/policy/south-korea-to-create-six-hydrogen-cities-that-would-use-h2-in-buildings-and-transport-as-part-of-daily-life/2-1-1385821

Hydropower. (2018). https://www.hydropower.org/publications/2018-hydropower-status-report

Ibrahim, M., Imam, Y., & Ghanem, A. (2019). Optimal planning and design of run-of-river hydroelectric power projects. *Renewable Energy*, *141*, 858–873. doi:10.1016/j.renene.2019.04.009

Ibrahim, S., Daut, I., Irwan, Y., Irwanto, M., Gomesh, N., & Farhana, Z. (2012). Linear Regression Model in Estimating Solar Radiation in Perlis. *Energy Procedia*, *18*, 1402–1412. doi:10.1016/j.egypro.2012.05.156

Igba, J., Alemzadeh, K., Durugbo, C., & Henningsen, K. (2015). Performance assessment of wind turbine gearboxes using in-service data: Current approaches and future trends. *Renewable & Sustainable Energy Reviews*, *50*, 144–159. doi:10.1016/j.rser.2015.04.139

Ilyas, A., Parkinson, S., Vinca, A., Byers, E., Manzoor, T., Riahi, K., & Muhammad, A. (2022). Balancing smart irrigation and hydropower investments for sustainable water conservation in the Indus basin. *Environmental Science & Policy*, *135*, 147–161. doi:10.1016/j.envsci.2022.04.012

International Energy Agency. (2020). Power Systems in Transition. Retrieved from https://www.iea.org/reports/power-systems-in-transition

IRENA. (2018). Hydrogen from renewable power. IRENA.

IRENA. (2019). Hydrogen: a renewable energy perspective. IRENA.

Islam, M. Z., Islam, M. M., & Asraf, A. (2020). A combined deep CNN-LSTM network for the detection of novel coronavirus (COVID-19) using X-ray images. *Informatics in Medicine Unlocked*, *20*, 100412.

Jahns, S. (2023). How Has the Current Electricity Crisis in South Africa Affected the Development of Renewable Energy within the Independent Power Producer Procurement Programmes? (Bachelor), Lund University, Kåberger, T. (2018). Progress of renewable electricity replacing fossil fuels. *Global Energy Interconnection*, *1*(1), 48–52.

Jaiswal. (2022). *Renewable and sustainable clean energy development and impact on social, economic, and environmental health*. Elsevier.

Jamii, J., Mansouri, M., Trabelsi, M., Mimouni, M. F., & Shatanawi, W. (2022). Effective artificial neural network-based wind power generation and load demand forecasting for optimum energy management. *Frontiers in Energy Research*, *10*, 898413. doi:10.3389/fenrg.2022.898413

Jang, E., Kang, Y., Im, J., Lee, D.-W., Yoon, J., & Kim, S.-K. (2019). Detection and Monitoring of Forest Fires Using Himawari-8 Geostationary Satellite Data in South Korea. *Remote Sensing (Basel)*, *11*(3), 271. doi:10.3390/rs11030271

Japan Science and Technology Agency. (2021). https://sj.jst.go.jp/stories/2021/s0728-01a.html

Jevgenijs, S., deW, J., Kochnakyan, A., and Vivien, F. (2019). Forecasting electricity demand: an aid for practitioners. Academic Press.

Jiang, L., Zheng, S., & Lai, L. L. (2018). Application of One-Class Support Vector Machine in Anomaly Detection of Oil and Gas Data. *Journal of Petroleum Science Engineering*, *163*, 75–82.

Jiang, Y., & Huang, G. (2017). Short-term wind speed prediction: Hybrid of ensemble empirical mode decomposition, feature selection and error correction. *Energy Conversion and Management*, *144*, 340–350. doi:10.1016/j.enconman.2017.04.064

Jiménez, A. A., Márquez, F. P. G., Moraleda, V. B., & Muñoz, C. Q. G. (2019). Linear and nonlinear features and machine learning for wind turbine blade ice detection and diagnosis. *Renewable Energy*, *132*, 1034–1048. doi:10.1016/j.renene.2018.08.050

Jiménez-Llanos, J., Ramírez-Carmona, M., Rendón-Castrillón, L., & Ocampo-López, C. (2020). Sustainable Biohydrogen Production by Chlorella Sp. Microalgae: A Review. *International Journal of Hydrogen Energy*, *45*(15), 8310–8328. doi:10.1016/j.ijhydene.2020.01.059

Jin, L., & Dong, J. (2016). Ensemble deep learning for biomedical time series classification. *Computational Intelligence and Neuroscience*. doi:10.1155/2016/6212684 PMID:27725828

Jin, X., Sarkar, S., Ray, A., Gupta, S., & Damarla, T. (2012). Target Detection and Classification Using Seismic and PIR Sensors. *IEEE Sensors Journal*, *12*(6), 1709–1718. doi:10.1109/JSEN.2011.2177257

Jithish, J., Alangot, B., Mahalingam, N., & Yeo, K. S. (2023). Distributed Anomaly Detection in Smart Grids: A Federated Learning-Based Approach. *IEEE Access : Practical Innovations, Open Solutions*, *11*, 7157–7179. doi:10.1109/ACCESS.2023.3237554

Jogunola, O., Adebisi, B., Hoang, K. V., Tsado, Y., Popoola, S. I., Hammoudeh, M., & Nawaz, R. (2022). CBLSTM-AE: A Hybrid Deep Learning Framework for Predicting Energy Consumption. *Energies*, *15*(3), 810. doi:10.3390/en15030810

Jones, P. (2020). *City for Empire: An Anchorage History, 1919-1941*. University of Alaska Press.

Jordan, M. I. (1987). Attractor dynamics and parallelism in a connectionist sequential machine. *Proceeding of 8th Annual Conference of the Cognitive Science Society (CogSci)*, 112–127.

Jung, J., Lee, S. H., Cho, M. S., & Kim, J. H. (2011). Touch TT: Scene Text Extractor Using Touch Screen Interface. *ETRI Journal*, *33*(1), 78–88. doi:10.4218/etrij.11.1510.0029

Jursa, R. (2007). Wind power prediction with different artificial intelligence models. Proceedings of the 2007 European Wind Energy Conference and Exhibition (EWEC), 1–10.

Jursa, R., & Rohrig, K. (2008). Short-term wind power forecasting using evolutionary algorithms for the automated specification of artificial intelligence models. *International Journal of Forecasting*, *24*(4), 694–709. doi:10.1016/j.ijforecast.2008.08.007

Kabir, M. M., Roy, S. K., Alam, F., Nam, S. Y., Im, K. S., Tijing, L., & Shon, H. K. (2023). Machine learning-based prediction and optimization of green hydrogen production technologies from water industries for a circular economy. *Desalination*, *567*, 116992. doi:10.1016/j.desal.2023.116992

Kadier, A., Jiang, Y., Lai, B., Rai, P. K., Chandrasekhar, K., Mohamed, A., & Kalil, M. S. (2017). Biohydrogen production in microbial electrolysis cells from renewable resources. Bioenergy and Biofuels. Taylor & Francis. doi:10.1201/9781351228138-12

Kalogirou, S. A., Panteliou, S., & Dentsoras, A. (1999). Artificial neural networks used for the performance prediction of a thermosiphon solar water heater. *Renewable Energy*, *18*(1), 87–99. doi:10.1016/S0960-1481(98)00787-3

Kamadinata, O. J., Ken, T. L., & Suwa, T. (2019). Sky image-based solar irradiance prediction methodologies using artificial neural networks. *Renewable Energy*, *134*, 837–845. doi:10.1016/j.renene.2018.11.056

Kamal-Chaoui, L., & Robert, A. (2009). Competitive cities and climate change. Academic Press.

Kannan, N., & Vakeesan, D. (2016). Solar energy for future world:-A review. *Renewable & Sustainable Energy Reviews*, *62*, 1092–1105. doi:10.1016/j.rser.2016.05.022

Karamanski, S., & Erfort, G. (2023). Wind Energy Supply Profiling and Offshore Potential in South Africa. *Energies*, *16*(9), 3668. doi:10.3390/en16093668

Kartikasari, P., Warsito, B., Yasin, H., Utami, I. T., & Suhermi, N. (2023). Forecasting double seasonal electricity consumption with TBATS model. Paper presented at the AIP Conference Proceedings.

Kasinathan, P., Pugazhendhi, R., Elavarasan, R. M., Ramachandaramurthy, V. K., Ramanathan, V., Subramanian, S., & Alsharif, M. H. (2022). Realization of Sustainable Development Goals with Disruptive Technologies by Integrating Industry 5.0, Society 5.0, Smart Cities and Villages. *Sustainability (Basel)*, *14*(22), 15258. doi:10.3390/su142215258

Kaya, M., & Hajimirza, S. (2020). Neural network surrogate model for optical performance of thin film multi-layered amorphous-silicon-based solar cells. *Optics Express*, *28*(4), 4946–4957.

Kaygusuz & Bilgen. (2008). Energy related environmental policies in Turkey. *Energy Sources Part B, 3*, 396-410.

Kazem, H. A. (2016). Photovoltaic solar thermal (PV/T) collectors past, present and future: A review. *International Journal of Applied Engineering Research: IJAER*, *11*(22), 10757–10765.

Kechriotis, G., Zervas, E., & Manolakos, E. S. (1994). Using recurrent neural networks for adaptive communication channel equalization. *IEEE Transactions on Neural Networks*, *5*(2), 267–278. doi:10.1109/72.279190 PMID:18267796

Ke, J., Zheng, H., Yang, H., & Chen, X. M. (2017). Short-term forecasting of passenger demand under on-demand ride services: A spatio-temporal deep learning approach. *Transportation Research Part C, Emerging Technologies*, *85*, 591–608. doi:10.1016/j.trc.2017.10.016

Kelly-Richards, S., Silber-Coats, N., Crootof, A., Tecklin, D., & Bauer, C. (2017). Governing the transition to renewable energy: A review of impacts and policy issues in the small hydropower boom. *Energy Policy*, *101*, 251–264. doi:10.1016/j.enpol.2016.11.035

Khan, S. B., Ahmad, G., Ali, F., Faisal, F., Ahmed, I., & Elahi, S. (2018) Classification performance of linear binary pattern and histogram oriented features for Arabic characters images: A. Applied Physics of Condensed Matter (APCOM 2019), 5.

Khan, S., Javaid, N., Chand, A., Khan, A. B. M., Rashid, F., & Afridi, I. U. (2019). Electricity load forecasting for each day of week using deep CNN. Paper presented at the Web, Artificial Intelligence and Network Applications: Proceedings of the Workshops of the 33rd International Conference on Advanced Information Networking and Applications (WAINA-2019).

Khan, A. R., Mahmood, A., Safdar, A., Khan, Z. A., & Khan, N. A. (2016). Load forecasting, dynamic pricing and DSM in smart grid: A review. *Renewable & Sustainable Energy Reviews*, *54*, 1311–1322.

Khandakar, A., Chowdhury, M. E. H., Khoda Kazi, M., Benhmed, K., Touati, F., Al-Hitmi, M., & Gonzales, J. S. (2019). Machine learning based photovoltaics (PV) power prediction using different environmental parameters of Qatar. *Energies*, *12*, 2782. doi:10.3390/en12142782

Khetkorn, W., Khanna, N., Incharoensakdi, A., & Lindblad, P. (2014). Metabolic and genetic engineering of cyanobacteria for enhanced hydrogen production. *Biofuels*, *4*(5), 535–561. doi:10.4155/bfs.13.41

Kim, J., Rweyemamu, M., &Purevsuren, B. (2022). Machine Learning-Based Approach for Hydrogen Economic Evaluation of Small Modular Reactors. *Science and Technology of Nuclear Installations*. doi:10.1155/2022/9297122

Kim, G., & Gorivindaraju, V. (1997). A Lexicon Dnven Approach to Handwritten Word Recognition for Real-Time Applications. *IEEE Transactions on Pattern Analysis and Machine Intelligence*, *19*(4), 366–379. doi:10.1109/34.588017

Kim, H., Park, S., & Kim, S. (2023). Time-series clustering and forecasting household electricity demand using smart meter data. *Energy Reports*, *9*, 4111–4121.

Kim, S. G., Jung, J. Y., & Sim, M. K. (2019). A two-step approach to solar power generation prediction based on weather data using machine learning. *Sustainability (Basel)*, *11*(5), 1501. doi:10.3390/su11051501

Kim, Y., & Kim, S. (2021). Forecasting charging demand of electric vehicles using time-series models. *Energies*, *14*(5), 1487.

Kirubakaran, S., Nagarajan, V., Dhayapulley, S. C., Soubache, I. D., Pasupuleti, S. R., Kumar, A., Rastogi, R., & Vasudevan, S. (2023). Hybrid deep learning-based grid-supportive renewable energy systems for maximizing power generation using optimum sizing. *Electric Power Components and Systems*, *51*(15), 1597–1611. doi:10.1080/15325008.2023.2201249

Kittur, A. (2013). The future of crowd work. *Proceedings of the 2013 conference on Computer supported cooperative work*. 10.1145/2441776.2441923

Klaiber, J., & Van Dinther, C. (2023). Deep learning for variable renewable energy: A systematic review. *ACM Computing Surveys*, *56*(1), 1–37. doi:10.1145/3586006

Klemeš, J. J. (Ed.). (2015) Assessing and measuring environmental impact and sustainability. Butterworth-Heinemann.

Kockel, C., Nolting, L., Priesmann, J., & Praktiknjo, A. (2022). Does renewable electricity supply match with energy demand?–A spatio-temporal analysis for the German case. *Applied Energy*, *308*, 118226. doi:10.1016/j.apenergy.2021.118226

Kohzadi, N., Boyd, M. S., Kermanshahi, B., & Kaastra, I. (1996). A comparison of artificial neural network and time series models for forecasting commodity prices. *Neurocomputing*, *10*(2), 169–181. doi:10.1016/0925-2312(95)00020-8

Kolsarici, C., & Vakratsas, D. (2015). Correcting for misspecification in parameter dynamics to improve forecast accuracy with adaptively estimated models. *Management Science*, *61*(10), 2495–2513. doi:10.1287/mnsc.2014.2027

Kombarov, V., Sorokin, V., Tsegelnyk, Y., Plankovskyy, S., Aksonov, Y., & Fojtů, O. (2021). Numerical control of machining parts from aluminum alloys with sticking minimization. *International Journal of Mechatronics and Applied Mechanics*, *I*(9), 209–216. doi:10.17683/ijomam/issue9.30

Kong, X., Liu, X., Shi, R., & Lee, K. Y. (2015). Wind speed prediction using reduced support vector machines with feature selection. *Neurocomputing*, *169*, 449–456. doi:10.1016/j.neucom.2014.09.090

Kosourov, S., Murukesan, G., Seibert, M., & Allahverdiyeva, Y. (2017). Evaluation of light energy to H2 energy conversion efficiency in thin films of cyanobacteria and green alga under photoautotrophic conditions. *Algal Research*, *28*, 253–263. doi:10.1016/j.algal.2017.09.027

Kou, P., Liang, D., Gao, F., & Gao, L. (2014). Probabilistic wind power forecasting with online model selection and warped Gaussian process. *Energy Conversion and Management*, *84*, 649–663. doi:10.1016/j.enconman.2014.04.051

Kranjc, J., Smailović, J., Podpečan, V., Grčar, M., Žnidaršič, M., & Lavrač, N. (2015). Active learning for sentiment analysis on data streams: Methodology and workflow implementation in the ClowdFlows platform. *Information Processing & Management*, *51*(2), 187–203. doi:10.1016/j.ipm.2014.04.001

Krohn, S., Morthorst, P.-E., & Awerbuch, S. (2009). The economics of wind energy. European Wind Energy Association, 3.

Kuang, H., Guo, Q., Li, S., & Zhong, H. (2021). Short-term wind power forecasting model based on multi-feature extraction and CNN-LSTM. Paper presented at the IOP Conference Series: Earth and Environmental Science.

Kuang, M., Wang, L., & Song, Y. (2014). Controllable printing droplets for high-resolution patterns. *Advanced Materials*, *26*(40), 6950–6958. doi:10.1002/adma.201305416 PMID:24687946

Kulasekara, H., & Seynulabdeen, V. (2019). A review of geothermal energy for future power generation. *2019 5th International Conference on Advances in Electrical Engineering (ICAEE)*. IEEE. 10.1109/ICAEE48663.2019.8975470

Kumar, C. J., and Veerakumari, M. (2012). Load forecasting of Andhra Pradesh grid using PSO, DE algorithms. International Journal of Advanced Research in Computer Engineering and Technology, 1(9), 179-184.

Kumar, S., Sharma, S., Thakur, S., Mishra, T., Negi, P., Mishra, S., Hesham, A. E-L., Rastegari, A. A., Yadav, N., & Yadav, A. N. (2019). Bioprospecting of Microbes for Biohydrogen Production: Current Status and Future Challenges. Bioprocessing for Biomolecules Production. John Wiley & Sons Ltd. doi:10.1002/9781119434436.ch22

Kumar, A., Kumar, K., Kaushik, N., Sharma, S., & Mishra, S. (2010). Renewable energy in India: Current status and future potentials. *Renewable & Sustainable Energy Reviews*, *14*(8), 2434–2442. doi:10.1016/j.rser.2010.04.003

Kumar, M., Oyedun, A. O., & Kumar, A. (2019). A Comparative Analysis of Hydrogen Production from the Thermochemical Conversion of Algal Biomass. *International Journal of Hydrogen Energy*, *44*(21), 10384–10397. doi:10.1016/j.ijhydene.2019.02.220

Kumar, R., & Rosen, M. A. (2011). A critical review of photovoltaic–thermal solar collectors for air heating. *Applied Energy*, *88*(11), 3603–3614. doi:10.1016/j.apenergy.2011.04.044

Kumar, S. V. (2017). Traffic Flow Prediction using Kalman Filtering Technique. *Procedia Engineering*, *187*, 582–587. doi:10.1016/j.proeng.2017.04.417

Kumar, S., & Kaur, T. (2016). Development of ANN Based Model for Solar Potential Assessment Using Various Meteorological Parameters. *Energy Procedia*, *90*, 587–592. doi:10.1016/j.egypro.2016.11.227

Kumar, S., & Lopez, D. (2015). Feature selection used for wind speed forecasting with data driven approaches. *J. Eng. Sci. Technol. Rev.*, *8*(5), 124–127. doi:10.25103/jestr.085.17

Kumbhar, A., Dhawale, P. G., Kumbhar, S., Patil, U., & Magdum, P. (2021). A comprehensive review: Machine learning and its application in integrated power system. *Energy Reports*, *7*, 5467–5474. doi:10.1016/j.egyr.2021.08.133

Kureljusic, M., & Karger, E. (2023). Forecasting in financial accounting with artificial intelligence – A systematic literature review and future research agenda. *Journal of Applied Accounting Research*, *25*(1), 81–104. doi:10.1108/JAAR-06-2022-0146

Kusiak, A., & Li, W. (2011). The prediction and diagnosis of wind turbine faults. *Renewable Energy*, *36*(1), 16–23. doi:10.1016/j.renene.2010.05.014

Kusiak, A., Zheng, H., & Song, Z. (2009). Short-term prediction of wind farm power: A data mining approach. *IEEE Transactions on Energy Conversion*, *24*(1), 125–136. doi:10.1109/TEC.2008.2006552

Kusiak, A., Zheng, H., & Song, Z. (2010). Wind turbine performance assessment with data-mining models. *Renewable Energy*, *35*(12), 2734–2741.

Kuyumani, E., Hasan, A. N., & Shongwe, T. (2023). A Hybrid Model Based on CNN-LSTM to Detect and Forecast Harmonics: A Case Study of an Eskom Substation in South Africa. *Electric Power Components and Systems*, *51*(8), 746–760.

Kylili, A., Thabit, Q., Nassour, A., & Fokaides, P. A. (2021). Adoption of a holistic framework for innovative sustainable renewable energy development: A case study. *Energy Sources. Part A, Recovery, Utilization, and Environmental Effects*, 1–21. doi:10.1080/15567036.2021.1904058

Lagarrigue, C. (2021). How artificial intelligence can optimize renewable energy. *Forbes*.

Lahti. (2020). Multifactorial individualised programme for hamstring muscle injury risk reduction in professional football: protocol for a prospective cohort study. *BMJ Open Sport & Exercise Medicine,* 6(1).

Lansberry, J. E., Wozniak, L., & Goldberg, D. E. (1992). Optimal hydro generator governor tuning with a genetic algorithm. *IEEE Transactions on Energy Conversion*, *7*(4), 623–630. doi:10.1109/60.182643

Latin, H. A. (2012). Climate change policy failures: why conventional mitigation approaches cannot succeed. World Scientific. doi:10.1142/8194

Lauret, P., Voyant, C., Soubdhan, T., David, M., & Poggi, P. (2015). A benchmarking of machine learning techniques for solar radiation forecasting in an insular context. *Solar Energy*, *112*, 446–457. doi:10.1016/j.solener.2014.12.014

Lawrenz, L., Xiong, B., Lorenz, L., Krumm, A., Hosenfeld, H., Burandt, T., & Von Hirschhausen, C. (2018). Exploring energy pathways for the low-carbon transformation in India—A model-based analysis. *Energies*, *11*(11), 3001. doi:10.3390/en11113001

Lee, Cho, Jung, & Kim. (2010). *Scene Text Extraction with Edge Constraint and Text Collinearity Link. 20th International Conference on Pattern Recognition (ICPR)*, Istanbul, Turkey.

Leholo, S., Owolawi, P., & Akindeji, K. (2019). Solar energy potential forecasting and optimization using artificial neural network- South Africa case study. *Proceedings of the Amity International Conference on Artificial Intelligence*. 10.1109/AICAI.2019.8701372

Leron, J., & Sankaranarayanan, C. A. (2021). Precise forecasting of sky images using spatial warping. *Proceedings of the IEEE/CVF international conference on computer vision*, 1136–44.

Le, T., Vo, M. T., Vo, B., Hwang, E., Rho, S., & Baik, S. W. (2019). Improving electric energy consumption prediction using CNN and Bi-LSTM. *Applied Sciences (Basel, Switzerland)*, *9*(20), 4237. doi:10.3390/app9204237

Leva, S., Dolara, A., Grimaccia, F., Mussetta, M., & Sahin, E. (2017). Analysis and validation of 24 hours ahead neural network forecasting for photovoltaic output power. *Mathematics and Computers in Simulation*, *131*, 88–100. doi:10.1016/j.matcom.2015.05.010

Li. (2018). Hydropower development situation and prospects in China. *Renewable and Sustainable Energy Reviews,* 82, 232-239.

Li, C., Lin, S., Xu, F., Liu, D., & Liu, J. (2018). Short-term wind power prediction based on data mining technology and improved support vector machine method: A case study in Northwest China. *Journal of Cleaner Production*, *205*, 205. doi:10.1016/j.jclepro.2018.09.143

Li, L., Dai, S., Cao, Z., Hong, J., Jiang, S., & Yang, K. (2020). Using improved gradient-boosted decision tree algorithm based on Kalman filter (GBDT-KF) in time series prediction. *The Journal of Supercomputing*, *76*(9), 6887–6900. doi:10.1007/s11227-019-03130-y

Limongi, A. R., Viviano, E., De Luca, M., Radice, R. P., Bianco, G., & Martelli, G. (2021). Biohydrogen from Microalgae: Production and Applications. *Applied Sciences (Basel, Switzerland)*, *11*(4), 1616. doi:10.3390/app11041616

Li, S. (2003). Wind power prediction using recurrent multilayer perceptron neural networks. *Proceedings of the 2003 IEEE Power Engineering Society General Meeting*, 2325–2330. 10.1109/PES.2003.1270992

Li, S., Wunsch, D. C., O'Hair, E. A., & Giesselmann, M. G. (2001). Using neural networks to estimate wind turbine power generation. *IEEE Transactions on Energy Conversion*, *16*(3), 276–282. doi:10.1109/60.937208

Li, S., Zhang, Q., & Wei, L. (2020). Anomaly Detection in Renewable Energy Systems Using One-Class Support Vector Machines. *Journal of Renewable and Sustainable Energy*, *12*(4), 045301.

Liu, C. L., Nakashima, K., Sako, H., & Fujisawa, H. (2003). Benchmarking of state-of- the-art techniques. *Pattern Recognition*, *36*(10), 2271–2285. doi:10.1016/S0031-3203(03)00085-2

Liu, H., Mi, X., & Li, Y. (2018). Wind speed forecasting method based on deep learning strategy using empirical wavelet transform, long short term memory neural network and Elman neural network. *Energy Conversion and Management*, *156*, 498–514. doi:10.1016/j.enconman.2017.11.053

Liu, Q. (2013). *Electric car with solar and wind energy may change the environment and economy: A tool for utilizing the renewable energy resource*. AGU Earth's Future.

Liu, T., Tan, Z., Xu, C., Chen, H., & Li, Z. (2020). Study on deep reinforcement learning techniques for building energy consumption forecasting. *Energy and Building*, *208*, 109675. doi:10.1016/j.enbuild.2019.109675

Liu, Y., Guo, X., Gong, M., & Yang, X. (2020). Wind Turbine Anomaly Detection Using Vibration Signals Based on an Optimized One-Class SVM. *Energies*, *13*(15), 3998.

Liu, Z., Wang, J., Li, Y., & Wu, J. (2020). A deep reinforcement learning framework for optimal energy management of microgrids with renewable energy sources and energy storage systems. *Applied Energy*, *279*, 115836.

Li, W., Lu, C., & Zhang, Y. W. (2019). Prospective exploration of future renewable portfolio standard schemes in China via a multi-sector CGE model. *Energy Policy*, *128*, 45–56. doi:10.1016/j.enpol.2018.12.054

Li, X., Zhang, J., Wang, J., & Zhao, H. (2019). A machine learning method for solar power prediction using satellite images and numerical weather prediction data. *Applied Energy*, *253*, 113596.

Li, Y., Yang, P., & Wang, H. (2018). Short-term wind speed forecasting based on improved ant colony algorithm for LSSVM. *Cluster Computing*, *22*(S5), 11575–11581. doi:10.1007/s10586-017-1422-2

Lotfipoor, A., Patidar, S., & Jenkins, D. P. (2024). 'Deep neural network with empirical mode decomposition and Bayesian optimisation for residential load forecasting', Expert Systems with Applications, vol. 237, no. *Expert Systems with Applications*, *121355*. Advance online publication. doi:10.1016/j.eswa.2023.121355

Łukajtis, R., Hołowacz, I., Kucharska, K., Glinka, M., Rybarczyk, P., Przyjazny, A., & Kamiński, M. (2018). Hydrogen production from biomass using dark fermentation. *Renewable & Sustainable Energy Reviews*, *91*, 665–694. doi:10.1016/j.rser.2018.04.043

Lund, J. W., Freeston, D. H., & Boyd, T. L. (2005). Direct application of geothermal energy: 2005 worldwide review. *Geothermics*, *34*(6), 691–727. doi:10.1016/j.geothermics.2005.09.003

Machado, E., Pinto, T., Guedes, V., & Morais, H. (2021). Electrical Load Demand Forecasting Using Feed-Forward Neural Networks. *Energies*, *14*(22), 7644.

Madabhushi, S., & Dewri, R. A. (2023). A survey of anomaly detection methods for power grids. *International Journal of Information Security*, *22*(6), 1799–1832. doi:10.1007/s10207-023-00720-z

Mahmud, M. A., Huda, N., Farjana, S., & Lang, C. (2018). Environmental impacts of solar-photovoltaic and solar-thermal systems with life-cycle assessment. *Energies*, *11*(9), 2346. doi:10.3390/en11092346

Majidpour, M., Nazaripouya, H., Chu, P., Pota, H., & Gadh, R. (2018). Fast univariate time series prediction of solar power for real-time control of energy storage system. *Forecasting*, *1*(1), 8. doi:10.3390/forecast1010008

Makatjane, K. D., & Moroke, N. D. (2016). Comparative study of Holt-Winters triple exponential smoothing and seasonal Arima: forecasting short term seasonal car sales in South Africa. Risk governance and control: financial markets and institutions, 6(1), 71-82.

Makatjane, K., Moroke, N., & Ncube, B. (2021). Detecting Financial Fraud in South Africa: A Comparison of Logistic Model Tree and Gradient Boosting Decision Tree. Academic Press.

Makatjane, K. (2022). Deep Learning for Sentiment Analysis to Predict the Probability of Bank Loan Default. *American Journal of Data Mining and Knowledge Discovery*, *7*(2), 5–12.

Maleki, A. (2018). Modeling and optimum design of an off-grid PV/WT/FC/diesel hybrid system considering different fuel prices. *The International Journal of Low Carbon Technologies*, *13*(2), 140–147. doi:10.1093/ijlct/cty006

Maleki, A., & Pourfayaz, F. (2015). Sizing of stand-alone photovoltaic/wind/diesel system with battery and fuel cell storage devices by harmony search algorithm. *Journal of Energy Storage*, *2*, 30–42. doi:10.1016/j.est.2015.05.006

Maleki, A., Pourfayaz, F., & Ahmadi, M. H. (2016). Design of a cost-effective wind/photovoltaic/hydrogen energy system for supplying a desalination unit by a heuristic approach. *Solar Energy*, *139*, 666–675. doi:10.1016/j.solener.2016.09.028

Manoharan, K. G., Nehru, J. A., & Balasubramanian, S. (2021). *Artificial Intelligence and IoT*. Springer. doi:10.1007/978-981-33-6400-4

Manwell, J. F., McGowan, J. G., & Rogers, A. L. (2010). *Wind energy explained: theory, design and application*. John Wiley & Sons.

Mariano-Hernández, D., Hernández-Callejo, L., García, F. S., Duque-Perez, O., & Zorita-Lamadrid, A. L. (2020). A Review of Energy Consumption Forecasting in Smart Buildings: Methods, Input Variables, Forecasting Horizon and Metrics. *Applied Sciences (Basel, Switzerland)*, *10*(23), 8323. doi:10.3390/app10238323

Marrasso, E., Roselli, C., Sasso, M., & Tariello, F. (2018). Global and local environmental and energy advantages of a geothermal heat pump interacting with a low temperature thermal micro grid. *Energy Conversion and Management*, *172*, 540–553. doi:10.1016/j.enconman.2018.07.028

Martins, E., & Galegale, N. V. (2023). Sales forecasting using machine learning algorithms. *Revista de Gestão e Secretariado*, *14*(7), 11294–11308. doi:10.7769/gesec.v14i7.1670

Ma, Y.-J., & Zhai, M.-Y. (2011). A dual-step integrated machine learning model for 24h- ahead wind energy generation prediction based on actual measurement data and environmental factors. *Energy Procedia*, *12*, 733–740.

McCann, M. T., Jin, K. H., & Unser, M. (2017). Convolutional Neural Networks for Inverse problems in imaging. *IEEE Signal Processing Magazine*, *34*(6), 85–95. doi:10.1109/MSP.2017.2739299

Mekhilef, S., Saidur, R., & Safari, A. (2011). A review on solar energy use in industries. *Renewable & Sustainable Energy Reviews*, *15*(4), 1777–1790. doi:10.1016/j.rser.2010.12.018

Melakhsou, A. A., Batton-Hubert, M., & Casoetto, N. (2023). Welding fault detection and diagnosis using one-class SVM with distance substitution kernels and random convolutional kernel transform. *International Journal of Advanced Manufacturing Technology*, *128*(1-2), 459–477. doi:10.1007/s00170-023-11768-5

Melis, A., & Melnicki, M. R. (2006). Integrated biological hydrogen production. *International Journal of Hydrogen Energy*, *31*(11), 1563–1573. doi:10.1016/j.ijhydene.2006.06.038

Mellit, A., & Kalogirou, S. A. (2008). Artificial intelligence techniques for photovoltaic applications: A review. *Progress in Energy and Combustion Science*, *34*(5), 574–632. doi:10.1016/j.pecs.2008.01.001

Meng, A., Ge, J., Yin, H., & Chen, S. (2016). Wind speed forecasting based on wavelet packet decomposition and artificial neural networks trained by crisscross optimization algorithm. *Energy Conversion and Management*, *114*, 75–88. doi:10.1016/j.enconman.2016.02.013

Mert, İ. (2021). Agnostic deep neural network approach to the estimation of hydrogen production for solar-powered systems. *International Journal of Hydrogen Energy*, *46*(9), 6272–6285. doi:10.1016/j.ijhydene.2020.11.161

Mert, I., Karakuş, C., & Üneş, F. (2016). Estimating the energy production of the wind turbine using artificial neural network. *Neural Computing & Applications*, *27*(5), 1231–1244. doi:10.1007/s00521-015-1921-0

Microsoft. (2020). *The Next AI Guardians*. https://www.microsoft.com/en-cy/aibootcamps

Mikkelsen, L. R. (2015). Fire on the mountain: Action research for large scale change. Diss. Saint Mary's College of California.

Mills, D. (2004). Advances in solar thermal electricity technology. *Solar energy,* 76(1-3), 19-31.

Minerals, D. o., and Energy. (1998). White paper on the energy policy of the Republic of South Africa. In. Pretoria: Department of Minerals and Energy. doi:10.1007/s42979-021-00731-4

Mir, A. A., Alghassab, M., Ullah, K., Khan, Z. A., Lu, Y., & Imran, M. (2020). A Review of Electricity Demand Forecasting in Low and Middle Income Countries: The Demand Determinants and Horizons. *Sustainability (Basel)*, *12*(15), 5931. doi:10.3390/su12155931

Mira, K., Bugiotti, F., & Morosuk, T. (2023). Artificial intelligence and machine learning in energy conversion and management. *Energies*, *16*(23), 7773. doi:10.3390/en16237773

Miranda-Barbosa, E., Sigfússon, B., Carlsson, J., & Tzimas, E. (2017). Advantages from combining CCS with geothermal energy. *Energy Procedia*, *114*, 6666–6676. doi:10.1016/j.egypro.2017.03.1794

Mircetic, D., Rostami-Tabar, B., Nikolicic, S., & Maslaric, M. (2022). Forecasting hierarchical time series in supply chains: An empirical investigation. *International Journal of Production Research*, *60*(8), 2514–2533. doi:10.1080/00207543.2021.1896817

Mohammadian, H. D. (2022). Mapping the Future Global SMEs Growth via Hybrid SMEs/SME 5.0/Tomorrow's SMEs Concept through the 5th Wave, i-Sustainability Plus and DCT Theories. In *Management and Information Technology in the Digital Era* (Vol. 29, pp. 173–206). Emerald Publishing Limited. doi:10.1108/S1877-636120220000029011

Mohammadi-Ivatloo, B., Rabiee, A., Soroudi, A., & Ehsan, M. (2013). Iterative solution of multi-objective security-constrained unit commitment problem with high penetration of wind power using artificial bee colony algorithm. *International Journal of Electrical Power & Energy Systems*, *53*, 318–329.

Mohammadpourfard, M., Sami, A., & Weng, Y. (2018). Identification of False Data Injection Attacks with Considering the Impact of Wind Generation and Topology Reconfigurations. *IEEE Transactions on Sustainable Energy*, *9*(3), 1349–1364. doi:10.1109/TSTE.2017.2782090

Mohandes, M. A., Halawani, T. O., Rehman, S., & Hussain, A. A. (2004). Support vector machines for wind speed prediction. *Renewable Energy*, *29*(6), 939–947. doi:10.1016/j.renene.2003.11.009

Mohlakoana, N., Wolpe, P., & David, A. (2023). Unbundling Eskom: How would a new distribution system impact on energy poverty. *AFD Research Papers*, (269), 1–46.

Molokomme, D. N., Onumanyi, A. J., & Abu-Mahfouz, A. M. (2022). Edge intelligence in Smart Grids: A survey on architectures, offloading models, cyber security measures, and challenges. *Journal of Sensor and Actuator Networks*, *11*(3), 47. doi:10.3390/jsan11030047

Mona, S., Kumar, S. S., Kumar, V., Parveen, K., Saini, N., Deepak, B., & Pugazhendhi, A. (2020). Green Technology for Sustainable Biohydrogen Production (Waste to Energy): A Review. *The Science of the Total Environment*, *728*, 138481. doi:10.1016/j.scitotenv.2020.138481 PMID:32361358

Moosavian, S. F., Noorollahi, Y., & Shoaei, M. (2024). Renewable energy resources utilization planning for sustainable energy system development on a stand-alone island. *Journal of Cleaner Production*, *439*, 140892. doi:10.1016/j.jclepro.2024.140892

Moradzadeh, A., Mansour-Saatloo, A., Nazari-Heris, M., Mohammadi-Ivatloo, B., & Asadi, S. (2021). Introduction and literature review of the application of machine learning/deep learning to load forecasting in power system. In M. Nazari-Heris & ... (Eds.), *Application of Machine Learning and Deep Learning Methods to Power System Problems* (pp. 119–135). Springer., doi:10.1007/978-3-030-77696-1_6

Moroke, N. D., & Makatjane, K. (2022). Predictive Modelling for Financial Fraud Detection Using Data Analytics: A Gradient-Boosting Decision Tree. In Applications of Machine Learning and Deep Learning for Privacy and Cybersecurity (pp. 25-45). IGI Global. doi:10.4018/978-1-7998-9430-8.ch002

Morsy, F. M. (2015). CO_2-free biohydrogen production by mixed dark and photofermentation bacteria from sorghum starch using a modified simple purification and collection system. *Energy*, *87*, 594–604. doi:10.1016/j.energy.2015.05.044

Mostafaeipour, A., Jahangiri, M., Haghani, A., Dehshiri, S. J. H., Dehshiri, S. S. H., Issakhov, A., & Techato, K. (2020). Statistical evaluation of using the new generation of wind turbines in South Africa. *Energy Reports*, *6*, 2816–2827. doi:10.1016/j.egyr.2020.09.035

Muhammad, L., & Algehyne, E. A. (2021). Fuzzy-based expert system for diagnosis of coronary artery disease in Nigeria. Health and Technology, 11(2), 319-329. doi:10.1007/s12553-021-00531-z

Muhammad, L., Amshi, J. M., Usman, S. S., Badi, I. A., Mohammed, I., Dada, O., & Haruna, A. A. (2021). Deep learning models for classification and localization of COVID-19 abnormalities on chest radiographs. Paper presented at the 2021 4th International Conference on Computing and Information Sciences (ICCIS).

Muhammad, L., Badi, I., Haruna, A. A., Mohammed, I., & Dada, O. (2022). Deep learning models for classification of brain tumor with magnetic resonance imaging images dataset. In Computational Intelligence in Oncology: Applications in Diagnosis, Prognosis and Therapeutics of Cancers (pp. 159-176). Springer.

Muhammad, G., Alam, M. A., Mofijur, M., Jahirul, M. I., Lv, Y., Xiong, W., Ong, H. C., & Xu, J. (2021). Modern Developmental Aspects in the Field of Economical Harvesting and Biodiesel Production from Microalgae Biomass. *Renewable & Sustainable Energy Reviews*, *135*, 110209. doi:10.1016/j.rser.2020.110209

Muhammad, L. J., Algehyne, E. A., & Usman, S. S. (2020). Predictive Supervised Machine Learning Models for Diabetes Mellitus. *SN Computer Science*, *1*(5), 240.

Muhammad, L., Haruna, A. A., Sharif, U. S., & Mohammed, M. B. (2022). Cnn-lstm deep learning-based forecasting model for COVID-19 infection cases in Nigeria, South Africa, and Botswana. *Health and Technology*, *12*(6), 1259–1276.

Mujeeb, S., Alghamdi, T. A., Ullah, S., Fatima, A., Javaid, N., & Saba, T. (2019). Exploiting deep learning for wind power forecasting based on big data analytics. *Applied Sciences (Basel, Switzerland)*, *9*(20), 4417. doi:10.3390/app9204417

Munim, Z. H. (2022). State-space TBATS model for container freight rate forecasting with improved accuracy. Maritime Transport Research, 3, 100057. doi:10.1016/j.martra.2022.100057

Nadal, C., Legault, R., & Suen, C. Y. (1990). Complementary Algorithms for Recognition of totally Unconstrained Handwritten Numerals. in Proc. 10th Int. Conf. *Pattern Recognition*, *1*, 434–449.

Nagarajan, D., Dong, C.-D., Chen, C.-Y., Lee, D.-J., & Chang, J.-S. (2021). Biohydrogen production from microalgae-Major bottlenecks and future research perspectives. *Biotechnology Journal*, *16*(5), 2000124. doi:10.1002/biot.202000124 PMID:33249754

Nagarajan, D., Lee, D. J., Kondo, A., & Chang, J. S. (2017). Recent insights into biohydrogen production by microalgae – From biophotolysis to dark fermentation. *Bioresource Technology*, *227*, 373–387. doi:10.1016/j.biortech.2016.12.104 PMID:28089136

Nagy, Dicker, & Meyer-Wegener. (2011b). *Definition and Evaluation of the NEOCR Dataset for Natural-Image Text Recognition.* University of Erlangen, Dept. of Computer Science, Technical Reports, CS-2011-07.

Nagy, R., Dicker, A., & Meyer-Wegener, K. (2011a). NEOCR: A Configurable Dataset for Natural Image Text Recognition. *CBDAR Workshop 2011 at ICDAR 2011*, 53-58.

Naim, I., Mahara, T., & Idrisi, A. R. (2018). Effective short-term forecasting for daily time series with complex seasonal patterns. *Procedia Computer Science*, *132*, 1832–1841.

Nakata, T. (2004). Energy-economic models and the environment. *Progress in Energy and Combustion Science*, *30*(4), 417–475. doi:10.1016/j.pecs.2004.03.001

Naqvi, S. R., Jamshaid, S., Naqvi, M., Farooq, W., Niazi, M. B. K., Aman, Z., Zubair, M., Ali, M., Shahbaz, M., Inayat, A., & Afzal, W. (2018). Potential of biomass for bioenergy in Pakistan based on present case and future perspectives. *Renewable & Sustainable Energy Reviews*, *81*, 1247–1258. doi:10.1016/j.rser.2017.08.012

Nastasi, B., Markovska, N., Puksec, T., Duić, N., & Foley, A. (2022). Renewable and sustainable energy challenges to face for the achievement of Sustainable Development Goals. *Renewable & Sustainable Energy Reviews*, *157*, 112071. doi:10.1016/j.rser.2022.112071

National Center for Biotechnology. (n.d.). https://www.ncbi.nlm.nih.gov/pmc/articles/PMC4740163

National Resources Defense Council. (n.d.). *The Gulf Spill.* https://www.nrdc.org/sites/default/files/ gulfspill-impacts-summary-IP.pdf

Nawaz, S., Calefati, A., Ahmed, N., & Gallo, I. (2018). Hand Written Characters Recognition via Deep Metric Learning. *Proceedings of the 2018 13th IAPR International Workshop On Document Analysis Systems (DAS),* 417–422. 10.1109/DAS.2018.18

Netzer, Wang, Coates, Bissacco, Wu, & Ng. (n.d.). *Reading Digits in Natural Images with Unsupervised Feature Learning NIPS Workshop on Deep Learning and Unsupervised Feature Learning*. Academic Press.

Newell, A. J., & Griffin, L. D. (2011). Multiscale histogram of oriented gradient descriptors for robust character recognition. *Proceedings of the 2011 International Conference On Document Analysis And Recognition (ICDAR),* 1085–1089. 10.1109/ICDAR.2011.219

Nguyen, P. K. T., Das, G., Kim, J., & Yoon, H. H. (2020). Hydrogen production from macroalgae by simultaneous dark fermentation and microbial electrolysis cell. *Bioresource Technology*, *315*, 123795. doi:10.1016/j.biortech.2020.123795 PMID:32659424

Niall, O., Sean, C., Anderson, C., Suman, H., Gustavo, V. H., & Lenka, K. (2020). Deep learning vs. traditional computer vision. In A. Kohei & K. Supriya (Eds.), *Advances in computer vision* (pp. 128–144). Springer International Publishing.

Nie, M., & Wang, L. (2013). Review of condition monitoring and fault diagnosis technologies for wind turbine gearbox. *Procedia CIRP*, *11*, 287–290. doi:10.1016/j.procir.2013.07.018

Nobela, O., Bansal, R., & Justo, J. (2019). A review of power quality compatibility of wind energy conversion systems with the South African utility grid. *Renewable Energy Focus*, *31*, 63–72. doi:10.1016/j.ref.2019.10.001

Nti, I. K., Samuel, A.-A., & Michael, A. (2019). Predicting Monthly Electricity Demand Using Soft-Computing. *International Research Journal of Engineering and Technology*, *6*(6), 1967–1973.

O'Leary, D., & Kubby, J. (2017). Feature selection and ANN solar power prediction. *Journal of Renewable Energy*, *2017*, 1–7. doi:10.1155/2017/2437387

Ogunmodimu, O., & Okoroigwe, E. C. (2018). Concentrating solar power technologies for solar thermal grid electricity in Nigeria: A review. *Renewable & Sustainable Energy Reviews*, *90*, 104–119. doi:10.1016/j.rser.2018.03.029

Oncel, S. S. (2015). Biohydrogen from Microalgae, Uniting Energy, Life, and green Future. Handbook of Marine Microalgae. Elsevier. doi:10.1016/B978-0-12-800776-1.00011-X

Ozcanli, A. K., Yaprakdal, F., & Baysal, M. (2020). Deep learning methods and applications for electrical power systems: A comprehensive review. *International Journal of Energy Research*, *44*(9), 7136–7157. doi:10.1002/er.5331

Ozgoren, M., Bilgili, M., & Sahin, B. (2012). Estimation of global solar radiation using ANN over Turkey. *Expert Systems with Applications*, *39*(5), 5043–5051. doi:10.1016/j.eswa.2011.11.036

Ozturk, M., Saba, N., Altay, V., Iqbal, R., Hakeem, K. R., Jawaid, M., & Ibrahim, F. H. (2017). Biomass and bioenergy: An overview of the development potential in Turkey and Malaysia. *Renewable & Sustainable Energy Reviews*, *79*, 1285–1302. doi:10.1016/j.rser.2017.05.111

Paoli, C., Voyant, C., Muselli, M., & Nivet, M.-L. (2010). Forecasting of preprocessed daily solar radiation time series using neural networks. *Solar Energy*, *84*(12), 2146–2160. doi:10.1016/j.solener.2010.08.011

Park, J. H., Chandrasekhar, K., Jeon, B. H., Jang, M., Liu, Y. & Kim, S. H. (2021). State-of-the-art Technologies for Continuous High-Rate Biohydrogen Production. *Bioresource Technology*, 320(Part A), 124304. doi:10.1016/j.biortech.2020.124304

Parra, D., Valverde, L., Pino, F. J., & Patel, M. K. (2019). A review on the role, cost and value of hydrogen energy systems for deep decarbonisation. *Renewable & Sustainable Energy Reviews*, *101*, 279–294. doi:10.1016/j.rser.2018.11.010

Parthipan, V., & Dhanasekaran, D. (2019). Preventing and Monitoring of Framework for Forest Fire Detection and Data Analysis Using Internet of Things. *International Journal of Engineering and Advanced Technology*, *8*(3).

Pascaris, A. S. (2021). Examining existing policy to inform a comprehensive legal framework for agrivoltaics in the US. *Energy Policy*, *159*, 112620. doi:10.1016/j.enpol.2021.112620

Patel, R., Raval, S., & Shah, M. (2019). Anomaly Detection in Renewable Energy Systems Using One-Class SVM. *International Journal of Engineering and Advanced Technology*, *9*(2), 6358–6363.

Patil, S., Nemade, V., & Soni, P. K. J. P. s. (2018). Predictive modelling for credit card fraud detection using data analytics. *Procedia Computer Science*, *132*, 385–395.

Pedro, H. T. C., & Coimbra, C. F. M. (2012). Assessment of forecasting techniques for solar power production with no exogenous inputs. *Solar Energy*, *86*(7), 2017–2028. doi:10.1016/j.solener.2012.04.004

Pedro, H. T. C., Coimbra, C. F. M., David, M., & Lauret, P. (2018). Assessment of machine learning techniques for deterministic and probabilistic intra-hour solar forecasts. *Renewable Energy*, *123*, 191–203. doi:10.1016/j.renene.2018.02.006

Peng, L., Fu, D., Chu, H., Wang, Z., & Qi, H. (2019). Biofuel Production from Microalgae: A Review. *Environmental Chemistry Letters*, *18*(2), 285–297. doi:10.1007/s10311-019-00939-0

Pérez, J. M. P., Márquez, F. P. G., Tobias, A., & Papaelias, M. (2013). Wind turbine reliability analysis. *Renewable & Sustainable Energy Reviews*, *23*, 463–472. doi:10.1016/j.rser.2013.03.018

Peters, H. D. (2011). Statement by the Minister of Energy, Hon. Dipuo Peters, MP, on the Occasion of the Announcement of Preferred Bidders in Respect of the First Window of 3 725 MW under the Renewable Energy IPP Bidding Programme. 2011. Retrieved from https://www.energy.gov.za/files/media/pr/2011/MinisterRemarks_IPP_BidsAnnouncement_7Dec2011.pdf

Phaahla, E. (2023). The State of Electricity in South Africa-Part I: The problems in Eskom. Retrieved from https://hsf.org.za/publications/hsf-briefs/the-state-of-electricity-in-south-africa-part-i-part-i-the-problems-in-eskom

Pimitha, K., Anbarasi, S., & Balasubramanian, T. (2023). Machine Learning Techniques for Renewable Energy Resources. In IoT, Machine Learning and Blockchain Technologies for Renewable Energy and Modern Hybrid Power Systems (pp. 121-148). River Publishers.

Platzer, W. (2014). PV–enhanced solar thermal power. *Energy Procedia*, *57*, 477–486. doi:10.1016/j.egypro.2014.10.201

Pliuhin, V., Korobka, V., Karyuk, A., Pan, M., & Sukhonos, M. (2019a). Using Azure Machine Learning Studio with python scripts for induction motors optimization web-deploy project. In *2019 IEEE International Scientific-Practical Conference Problems of Infocommunications, Science and Technology (PIC S&T)* (pp. 631–634). IEEE. 10.1109/PICST47496.2019.9061447

Pliuhin, V., Pan, M., Yesina, V., & Sukhonos, M. (2018). Using Azure Machine Learning cloud technology for electric machines optimization. In *2018 International Scientific-Practical Conference Problems of Infocommunications. Science and Technology (PIC S&T)* (pp. 55–58). IEEE. 10.1109/INFOCOMMST.2018.8632093

Pliuhin, V., Sukhonos, M., Pan, M., Petrenko, O., & Petrenko, M. (2019b). Implementing of Microsoft Azure machine learning technology for electric machines optimization. *Electrical Engineering & Electromechanics*, *1*(1), 23–28. doi:10.20998/2074-272X.2019.1.04

Pliuhin, V., Teterev, V., & Lapko, A. (2021). Smart Grid technologies as a concept of innovative energy development: Initial proposals for the development of Ukraine. *Lighting Engineering & Power Engineering*, *60*(2), 47–65. doi:10.33042/2079-424X.2021.60.2.02

Prochnow, A., Heiermann, M., Plöchl, M., Linke, B., Idler, C., Amon, T., & Hobbs, P. J. (2009). Bioenergy from permanent grassland–A review: 1. Biogas. *Bioresource Technology*, *100*(21), 4931–4944. doi:10.1016/j.biortech.2009.05.070 PMID:19546001

Qing, X., & Niu, Y. (2018). Hourly day-ahead solar irradiance prediction using weather forecasts by LSTM. *Energy*, *148*, 461–468. doi:10.1016/j.energy.2018.01.177

Quentin, P., Anthony, H., Guillaume, A., Philippe, B., & Joan, L. (2022). SPIN: Simplifying polar invariance for neural networks application to vision-based irradiance forecasting. Proceedings of the IEEE/CVF conference on computer vision and pattern recognition workshops, 5182–91.

Radovanovic, M., Popov, S., & Dodic, S. (2012). *Sustainable Energy Management*. Academic Press.

Rahimi, I., Chen, F., & Gandomi, A. H. (2023). A review on COVID-19 forecasting models. *Neural Computing & Applications*, *35*(33), 23671–23681. doi:10.1007/s00521-020-05626-8 PMID:33564213

Rahman, M. M., Shakeri, M., Tiong, S. K., Khatun, F., Amin, N., Pasupuleti, J., & Hasan, M. K. (2021). Prospective methodologies in hybrid renewable energy systems for energy prediction using artificial neural networks. *Sustainability (Basel)*, *13*(4), 2393. doi:10.3390/su13042393

Raihan, A., Begum, R. A., Said, M. N. M., & Pereira, J. J. (2022). Relationship between economic growth, renewable energy use, technological innovation, and carbon emission toward achieving Malaysia's Paris agreement. *Environment Systems & Decisions*, *42*(4), 586–607. doi:10.1007/s10669-022-09848-0

Rai, P. K., Singh, S. P., Asthana, R. K., & Singh, S. (2014). Biohydrogen production from sugarcane bagasse by integrating dark- and photo-fermentation. *Bioresource Technology*, *152*, 140–146. doi:10.1016/j.biortech.2013.10.117 PMID:24291314

Rana, M., Koprinska, I., & Agelidis, V. G. (2016). Univariate and multivariate methods for very short-term solar photovoltaic power forecasting. *Energy Conversion and Management*, *121*, 380–390. doi:10.1016/j.enconman.2016.05.025

Rana, M., & Rahman, A. (2020). Multiple steps ahead solar photovoltaic power forecasting based on univariate machine learning models and data re-sampling. *Sustain. Energy Grids Netw.*, *21*, 100286. doi:10.1016/j.segan.2019.100286

Rangapuram, S. S., Kapoor, S., Nirwan, R. S., Mercado, P., Januschowski, T., Wang, Y., & Bohlke-Schneider, M. (2023). Coherent probabilistic forecasting of temporal hierarchies. Paper presented at the International Conference on Artificial Intelligence and Statistics.

Rangel-Martinez, D., Nigam, K. D. P., & Ricardez-Sandoval, L. A. (2021). Machine learning on sustainable energy: A review and outlook on renewable energy systems, catalysis, smart grid and energy storage. *Chemical Engineering Research & Design*, *174*, 414–441. doi:10.1016/j.cherd.2021.08.013

Rao, S., Katoch, S., Narayanaswamy, V., Muniraju, G., Tepedelenlioglu, C., & Spanias, A. (2022). *Machine learning for solar array monitoring, optimization, and control.* Springer Nature.

Raut, H. K., Ganesh, V. A., Nair, A. S., & Ramakrishna, S. (2011). Anti-reflective coatings: A critical, in-depth review. *Energy & Environmental Science*, *4*(10), 3779–3804. doi:10.1039/c1ee01297e

Raza, M. Q., & Khosravi, A. (2015). A review on artificial intelligence-based load demand forecasting techniques for smart grid and buildings. [doi:https://doi.org/10.1016/j.rser.2015.04.065]. *Renewable & Sustainable Energy Reviews*, *50*, 1352–1372.

Razavi, R., Gharipour, A., Fleury, M., & Akpan, I. J. (2019). Occupancy detection of residential buildings using smart meter data: A large-scale study. *Energy and Building*, *183*, 195–208. doi:10.1016/j.enbuild.2018.11.025

Razu, M. H., Hossain, F., & Khan, M. (2019). Advancement of Bio-Hydrogen Production from Microalgae, Microalgae Biotechnology for Development of Biofuel and Wastewater Treatment. Springer, 423–462. doi:10.1007/978-981-13-2264-8_17

Rego, Pina e Cunha, & Clegg. (2012). The virtues of leadership: Contemporary challenges for global managers. OUP Oxford.

Ren, B., Chi, Y., Zhou, N., Wang, Q., Wang, T., Luo, Y., Ye, J., & Zhu, X. (2024). Machine learning applications in health monitoring of renewable energy systems. *Renewable & Sustainable Energy Reviews*, *189*, 114039. doi:10.1016/j.rser.2023.114039

Rodriguez, F., Fleetwood, A., Galarza, A., & Fontán, L. (2018). Predicting solar energy generation through artificial neural networks using weather forecasts for microgrid control. *Renewable Energy*, *126*, 855–864. doi:10.1016/j.renene.2018.03.070

Rohaan, D., Topan, E., & Groothuis-Oudshoorn, C. G. (2022). Using supervised machine learning for B2B sales forecasting: A case study of spare parts sales forecasting at an after-sales service provider. *Expert Systems with Applications*, *188*, 115925. doi:10.1016/j.eswa.2021.115925

Rosas, M. A. T., Pérez, M. R., & Pérez, E. R. M. (2022). Itineraries for charging and discharging a BESS using energy predictions based on a CNN-LSTM neural network model in BCS, Mexico. *Renewable Energy*, *188*, 1141–1165.

Rosiello, A., & Maleki, A. (2021). A dynamic multi-sector analysis of technological catch-up: The impact of technology cycle times, knowledge base complexity and variety. *Research Policy*, *50*(3), 104194. doi:10.1016/j.respol.2020.104194

Ryan, G. (2019). *Detecting Fraud and Other Anomalies Using Isolation Forests*. SAS Institute Inc.

Sagel, V. N., Rouwenhorst, K. H., & Faria, J. A. (2023). Leveraging Green Ammonia for Resilient and Cost-Competitive Islanded Electricity Generation from Hybrid Solar Photovoltaic–Wind Farms: A Case Study in South Africa. *Energy & Fuels*, *37*(18), 14383–14392. doi:10.1021/acs.energyfuels.3c01950 PMID:37753452

Saini, G., Kannan, R., Benini, E., & Kumar, K. (Eds.). (2023). *Enabling Methodologies for Renewable and Sustainable Energy*. CRC Press. doi:10.1201/9781003272717

Sakr, G. E., Elhajj, I. H., Mitri, G., & Wejinya, U. C. (2010). Artificial intelligence for forest fire prediction. *IEEE/ASME International Conference on Advanced Intelligent Mechatronics*, 1311-1316. 10.1109/AIM.2010.5695809

Salakkam, A., Sittijunda, S., Mamimin, C., Phanduang, O., & Reungsang, A. (2021). Valorization of Microalgal Biomass for Biohydrogen Generation: A Review. *Bioresource Technology*, *322*, 124533. doi:10.1016/j.biortech.2020.124533 PMID:33348113

Salcedo-Sanz, S., Pastor-Sanchez, A., Prieto, L., Blanco-Aguilera, A., & Garcia-Herrera, R. (2014). Feature selection in wind speed prediction systems based on a hybrid coral reefs optimization-extreme learning machine approach. *Energy Conversion and Management*, *87*, 10–18. doi:10.1016/j.enconman.2014.06.041

Salem, A. H., Brunstermann, R., Mietzel, T., & Widmann, R. (2018). Effect of pre-treatment and hydraulic retention time on biohydrogen production from organic wastes. *International Journal of Hydrogen Energy*, *43*(10), 4856–4865. doi:10.1016/j.ijhydene.2018.01.114

Salleh, S. F., Mohd Roslan, M. E., Abd Rahman, A., Shamsuddin, A. H., Tuan Abdullah, T. A. R., & Sovacool, B. K. (2020). Transitioning to a sustainable development framework for bioenergy in Malaysia: Policy suggestions to catalyse the utilisation of palm oil mill residues. *Energy, Sustainability and Society*, *10*(1), 1–20. doi:10.1186/s13705-020-00269-y

Salman, S., Kuciauskas, D., & Dippo, P. (2020). Machine learning for extracting Cu doping profiles in CdTe solar cells. *Solar Energy Materials and Solar Cells*, *209*, 110527.

Saloux, E., & Candanedo, J. A. (2018). Forecasting district heating demand using machine learning algorithms. *Energy Procedia*, *149*, 59–68. doi:10.1016/j.egypro.2018.08.169

Salvarli, M. S., & Salvarli, H. (2020). For sustainable development: future trends in renewable energy and enabling technologies. In Renewable Energy-Resources, Challenges and Applications. London: IntechOpen. doi:10.5772/intechopen.91842

Sapitang, M., Ridwan, W. M., Kushiar, K. F., Ahmed, A. N., & El-Shafie, A. (2020). Machine learning application in reservoir water level forecasting for sustainable hydropower generation strategy. *Sustainability*, *12*, 6121. doi:10.3390/su12156121

Sarangi, P. K., & Nanda, S. (2020). Biohydrogen production through dark fermentation. *Chemical Engineering & Technology*, *43*(4), 601–612. doi:10.1002/ceat.201900452

Scarlat, N., Dallemand, J.-F., Monforti-Ferrario, F., & Nita, V. (2015). The role of biomass and bioenergy in a future bioeconomy: Policies and facts. *Environmental Development*, *15*, 3–34. doi:10.1016/j.envdev.2015.03.006

Scellato, S., Fortuna, L., Frasca, M., Gómez-Gardeñes, J., & Latora, V. (2010). Traffic optimization in transport networks based on local routing. *The European Physical Journal B*, *73*, 303–308.

Schlamadinger & Marland. (1996). The role of forest and bioenergy strategies in the global carbon cycle. *Biomass and Bioenergy,* 10(5-6), 275-300.

Schölkopf, B., Platt, J. C., Shawe-Taylor, J., Smola, A. J., & Williamson, R. C. (2001). Estimating the Support of a High-Dimensional Distribution. *Neural Computation*, *13*(7), 1443–1471. doi:10.1162/089976601750264965 PMID:11440593

Schölkopf, B., & Smola, A. J. (2018). *Learning with kernels: support vector machines, regularization, optimization, and beyond*. MIT Press. doi:10.7551/mitpress/4175.001.0001

Seetha, J., & Raja, S. S. (2018). Brain tumor classification using convolutional neural networks. *Biomedical & Pharmacology Journal*, *11*(3), 1457–1461.

Sengmee, D., Cheirsilp, B., Suksaroge, T. T., & Prasertsan, P. (2017). Biophotolysis-based Hydrogen and Lipid Production by Oleaginous Microalgae Using Crude Glycerol as Exogenous Carbon Source. *International Journal of Hydrogen Energy*, *42*(4), 1970–1976. doi:10.1016/j.ijhydene.2016.10.089

Senthil, K. P. (2019). Improved prediction of wind speed using machine learning. *EAI Endorsed Trans. Energy Web*, *6*(23), 1–7.

Shahzad, S., Abbasi, M. A., Chaudhry, M. A., & Hussain, M. M. (2022). Model Predictive Control Strategies in Microgrids: A Concise Revisit. *IEEE Access : Practical Innovations, Open Solutions*, *10*, 122211–122225. doi:10.1109/ACCESS.2022.3223298

Shams, M. H., Niaz, H., Na, J., Anvari-Moghaddam, A., & Liu, J. J. (2021). Machine Learning-Based Utilization of Renewable Power Curtailments under Un-certainty by Planning of Hydrogen Systems and Battery Storages. *Journal of Energy Storage*, *41*, 1–13. doi:10.1016/j.est.2021.103010

Sharma, A., & Singh, B. (2022). Measuring Impact of E-commerce on Small Scale Business: A Systematic Review. *Journal of Corporate Governance and International Business Law*, *5*(1).

Sharma, N. K., Tiwari, P. K., & Sood, Y. R. (2012). Solar energy in India: Strategies, policies, perspectives and future potential. *Renewable & Sustainable Energy Reviews*, *16*(1), 933–941. doi:10.1016/j.rser.2011.09.014

Sharma, N., Jain, V., & Mishra, A. (2018). An analysis of convolutional neural networks for image classification. *Procedia Computer Science*, *132*, 377–384. doi:10.1016/j.procs.2018.05.198

Sheng, Xiao, Cheng, Ni, & Wang. (2018). Short-term solar power forecasting based on weighted Gaussian process regression. *IEEE Trans. Ind. Electron.*

Shi, K., Qiao, Y., Zhao, W., Wang, Q., Liu, M., & Lu, Z. (2018). An improved random forest model of short-term wind-power forecasting to enhance accuracy, efficiency, and robustness. *Wind Energy (Chichester, England)*, *21*(12), 1383–1394. doi:10.1002/we.2261

Shi, M., Fujisawa, Y., Wakabayashi, T., & Kimura, F. (2002). Handwritten numeral recognition using gradient and curvature of gray scale image. *Pattern Recognition*, *35*(10), 2051–2059. doi:10.1016/S0031-3203(01)00203-5

Shimray, B. A., Singh, K., Khelchandra, T., & Mehta, R. (2017). Ranking of sites for installation of hydropower plant using MLP neural network trained with GA: A MADM approach. *Computational Intelligence and Neuroscience*, *2017*, 2017. doi:10.1155/2017/4152140 PMID:28331490

Shiri, Perumal, Mustapha, & Mohamed. (n.d.). A Comprehensive Overview and Comparative Analysis on Deep Learning Models: CNN, RNN, LSTM, GRU. https://arxiv.org/ftp/arxiv/papers/2305/2305.17473.pdf

Shivam, S., Kumar, A., & Singh, S. (2020). Artificial neural network based prediction of optimum thickness of anti-reflection coating for solar cell. *Materials Today: Proceedings*, *21*, 1628–1631.

Shoham, J. L. (1980). *Energy and pollution: Economic and technical aspects of fossil fuel power generation in the UK*. University of Surrey.

Show, K.-Y., Yan, Y., Ling, M., Ye, G., Li, T., & Lee, D.-J. (2018). Hydrogen Production from Algal Biomass - Advances, Challenges and Prospects. *Bioresource Technology*, *257*, 290–300. doi:10.1016/j.biortech.2018.02.105 PMID:29506887

Siano, P. (2014). Demand response and smart grids—A survey. *Renewable & Sustainable Energy Reviews*, *30*, 461–478. doi:10.1016/j.rser.2013.10.022

Sierra-García, J. E., & Santos, M. (2020). Exploring Reward Strategies for Wind Turbine Pitch Control by Reinforcement Learning. *Applied Sciences (Basel, Switzerland)*, *10*(21), 7462. doi:10.3390/app10217462

Sijm, J., Koutstaal, P., Özdemir, Ö., & van Hout, M. (2019). Energy Transition Implications for Demand and Supply of Power System Flexibility: A Case Study of the Netherlands Within an EU Electricity Market and Trading Context. In E. Gawel, S. Strunz, P. Lehmann, & A. Purkus (Eds.), *The European Dimension of Germany's Energy Transition: Opportunities and Conflicts* (pp. 363–394). Springer International Publishing. doi:10.1007/978-3-030-03374-3_21

Singaravel, S., Suykens, J., & Geyer, P. (2018). Deep-learning neural-network architectures and methods: Using component-based models in building-design energy prediction. *Advanced Engineering Informatics*, *38*, 81–90. doi:10.1016/j.aei.2018.06.004

Singh, B. (2023). Blockchain Technology in Renovating Healthcare: Legal and Future Perspectives. In Revolutionizing Healthcare Through Artificial Intelligence and Internet of Things Applications (pp. 177-186). IGI Global.

Singh, H., & Das, D. (2020). Biohydrogen from Microalgae, Handbook of Microalgae-Based Processes and Products. Elsevier.

Singh, O., Yadav, M., Yadav, P., & Rishiwal, V. (2021). AI-Based Renewable Energy with Emerging Applications: Issues and Challenges. In Introduction to AI Techniques for Renewable Energy System (pp. 29-42). CRC Press.

Singh, A., Singh, S., & Kumar, A. (2020). Artificial intelligence based optimal control of renewable energy integrated power system. *International Journal of Electrical Power & Energy Systems, 118*, 105794.

Singh, B. (2022). Relevance of Agriculture-Nutrition Linkage for Human Healthcare: A Conceptual Legal Framework of Implication and Pathways. *Justice and Law Bulletin*, *1*(1), 44–49.

Singh, B. (2023). Federated Learning for Envision Future Trajectory Smart Transport System for Climate Preservation and Smart Green Planet: Insights into Global Governance and SDG-9 (Industry, Innovation and Infrastructure). *National Journal of Environmental Law*, *6*(2), 6–17.

Singh, B. (2023). Tele-Health Monitoring Lensing Deep Neural Learning Structure: Ambient Patient Wellness via Wearable Devices for Real-Time Alerts and Interventions. *Indian Journal of Health and Medical Law*, *6*(2), 12–16.

Singh, B. (2024). Legal Dynamics Lensing Metaverse Crafted for Videogame Industry and E-Sports: Phenomenological Exploration Catalyst Complexity and Future. *Journal of Intellectual Property Rights Law*, *7*(1), 8–14.

Singh, K., & Bean, J. D. (2019). *Optimizing India's electricity grid for renewables using AI and machine learning applications*. Center for Strategic and International Studies.

Sinha, P., & Brophy, B. (2021). Life cycle assessment of renewable hydrogen for fuel cell passenger vehicles in California. *Sustainable Energy Technologies and Assessments*, *45*(21), 101188. doi:10.1016/j.seta.2021.101188

Sipahutar, R., Bernas, S. M., & Imanuddin, M. S. (2013). Renewable energy and hydropower utilization tendency worldwide. *Renewable & Sustainable Energy Reviews*, *17*, 213–215. doi:10.1016/j.rser.2012.09.010

Slater, L. J., Arnal, L., Boucher, M., Chang, A., Moulds, S., Murphy, C., Nearing, G., Shalev, G., Shen, C., Speight, L., Villarini, G., Wilby, R. L., Wood, A., & Zappa, M. (2022). Hybrid forecasting: Using statistics and machine learning to integrate predictions from dynamical models. *Hydrology and Earth System Sciences*. Advance online publication. doi:10.5194/hess-2022-334

Smith, A., & Johnson, B. (2020). Effectiveness of One-Class Support Vector Machine Model for Anomaly Detection in Solar Energy Systems. *Solar Energy*, *215*, 456–465.

Solyali, D. (2020). A Comparative Analysis of Machine Learning Approaches for Short-/Long-Term Electricity Load Forecasting in Cyprus. *Sustainability*, *12*(9), 3612.

Somu, N., Raman, M. R. G., & Ramamritham, K. (2021). A deep learning framework for building energy consumption forecast. *Renewable & Sustainable Energy Reviews*, *137*, 110591. doi:10.1016/j.rser.2020.110591

Song, W., Ding, L., Liu, M., Cheng, J., Zhou, J., & Li, Y.-Y. (2020). Improving Biohydrogen Production through Dark Fermentation of Steam-Heated Acid Pretreated *Alternanthera Philoxeroides* by Mutant *Enterobacter Aerogenes* ZJU1. *The Science of the Total Environment*, *716*, 134695. doi:10.1016/j.scitotenv.2019.134695 PMID:31837880

Srinivasan, S., Kumarasamy, S., Andreadakis, Z. E., & Lind, P. G. (2023). Artificial intelligence and mathematical models of power grids driven by renewable energy sources: A survey. *Energies*, *16*(14), 5383. doi:10.3390/en16145383

Staudemeyer, R. C., & Morris, E. R. (2019). Understanding LSTM—a tutorial into long short-term memory recurrent neural networks. arXiv preprint arXiv:1909.09586.

Steen, M., & Weaver, T. (2017). Incumbents' diversification and cross-sectorial energy industry dynamics. *Research Policy*, *46*(6), 1071–1086. doi:10.1016/j.respol.2017.04.001

Stetco, A., Dinmohammadi, F., Zhao, X., Robu, V., Flynn, D., Barnes, M., Keane, J., & Nenadic, G. (2019). Machine learning methods for wind turbine condition monitoring: A review. *Renewable Energy*, *133*, 620–635. doi:10.1016/j.renene.2018.10.047

Subramani, N., Veerappampalayam Easwaramoorthy, S., Mohan, P., Subramanian, M., & Sambath, V. J. B. D. (2023). A Gradient Boosted Decision Tree-Based Influencer Prediction in Social Network Analysis. *Big Data and Cognitive Computing*, *7*(1), 6. doi:10.3390/bdcc7010006

Sukhonos, M., Babaiev, V., Pliuhin, V., Teterev, V., & Khudiakov, I. (2023). Load forecasting and electricity consumption by regression model. In *Smart Technologies in Urban Engineering* (pp. 302–314). Springer. doi:10.1007/978-3-031-20141-7_28

Sun, S., Wang, S., Zhang, G., & Zheng, J. (2018). A decomposition-clustering-ensemble learning approach for solar radiation forecasting. *Solar Energy*, *163*, 189–199. doi:10.1016/j.solener.2018.02.006

Syed, S. A., Al-Turjman, F., & Alghamdi, S. A. (2018). IoT-Based Energy Management System for Wind-Powered Microgrids. *Energies*, *11*, 2625.

Szeliski, R. (2022). *Computer vision: algorithms and applications*. Springer Nature. doi:10.1007/978-3-030-34372-9

Tabor, D. P., Roch, L. M., Saikin, S. K., Kreisbeck, C., Sheberla, D., Montoya, J. H., Dwaraknath, S., Aykol, M., Ortiz, C., Tribukait, H., Amador-Bedolla, C., Brabec, C. J., Maruyama, B., Persson, K. A., & Aspuru-Guzik, A. (2018). Accelerating the discovery of materials for clean energy in the era of smart automation. *Nature Reviews. Materials*, *3*(5), 5–20. doi:10.1038/s41578-018-0005-z

Taluja, C., & Guha, P. (2018). Detection of Forest Fires at Early Stages using Wireless Sensor Network and Graphical User interfacing. *4th International Conference on Computers and Management (ICCM),* 1-6.

Tan, J., Sizov, M., & Pasolli, E. (2019). One-class support vector machines for detecting anomalies in renewable energy generation systems. *Renewable Energy*, *134*, 1180–1188.

Tarallo, E., Akabane, G. K., Shimabukuro, C. I., Mello, J., & Amancio, D. (2019). Machine learning in predicting demand for fast-moving consumer goods: An exploratory research. *IFAC-PapersOnLine*, *52*(13), 737–742. doi:10.1016/j.ifacol.2019.11.203

Tasnim, S., Rahman, A., Oo, A. M. T., & Haque, M. E. (2018). Wind power prediction in new stations based on knowledge of existing Stations: A cluster based multi source domain adaptation approach. *Knowledge-Based Systems*, *145*, 15–24. doi:10.1016/j.knosys.2017.12.036

Tchakoua, P., Wamkeue, R., Ouhrouche, M., Slaoui-Hasnaoui, F., Tameghe, T. A., & Ekemb, G. (2014). Wind turbine condition monitoring: State-of-the-art review, new trends, and future challenges. *Energies*, *7*(4), 2595–2630. doi:10.3390/en7042595

Teow, L. N., & Loe, K. F. (2002). Robust vision-based features and classification schemes for off-line handwritten digit recognition. *Pattern Recognition*, *35*(11), 2355–2364. doi:10.1016/S0031-3203(01)00228-X

Thelwall, M., Kousha, K., Wilson, P., Makita, M., Abdoli, M., Stuart, E., Levitt, J., Knoth, P., & Cancellieri, M. (2023). Predicting article quality scores with machine learning: The UK Research Excellence Framework. *Quantitative Science Studies*, *4*(2), 547–573. doi:10.1162/qss_a_00258

Theoharidis, A. F., Guillén, D. A., & Lopes, H. (2023). Deep learning models for inflation forecasting. *Applied Stochastic Models in Business and Industry*, *39*(3), 447–470. doi:10.1002/asmb.2757

Thorburn, K. (2006). *Electric energy conversion systems: wave energy and hydropower*. Diss. Acta Universitatis Upsaliensis.

Thornton, H. E., Scaife, A. A., Hoskins, B. J., & Brayshaw, D. J. (2017). The relationship between wind power, electricity demand and winter weather patterns in Great Britain. *Environmental Research Letters*, *12*(6), 064017. doi:10.1088/1748-9326/aa69c6

Tien, B. D., Tuan, T. A., Klempe, H., Pradhan, B., & Revhaug, I. (2016). Spatial prediction models for shallow landslide hazards: A comparative assessment of the efficacy of support vector machines, artificial neural networks, kernel logistic regression, and logistic model tree. *Landslides*, *13*(2), 361–378. doi:10.1007/s10346-015-0557-6

Tomar, A., Malik, H., Kumr, P., & Iqbal, A. (2022, September). Machine Learning, Advances in Computing, Renewable Energy and Communication (MARC). In *Proceedings of 3rd International Conference on Machine Learning, Advances in Computing, Renewable Energy and Communication: MARC 2021* (pp. 1-19). Singapore: Springer Nature Singapore. 10.1007/978-981-19-2828-4_1

Tong, C., & Tong, C. (2019). Advanced materials and devices for hydropower and ocean energy. Introduction to materials for advanced energy systems, 445-501. doi:10.1007/978-3-319-98002-7_7

Tong, W. (2010). *Fundamentals of wind energy* (Vol. 44). WIT Press.

Torres-Barran, A., Alonso, Á., & Dorronsoro, J. R. (2019). Regression tree ensembles for wind energy and solar radiation prediction. *Neurocomputing*, *326*, 151–160. doi:10.1016/j.neucom.2017.05.104

Torres, J. F., Martínez-Álvarez, F., & Troncoso, A. (2022). A deep LSTM network for the Spanish electricity consumption forecasting. *Neural Computing & Applications*, *34*(13), 10533–10545. doi:10.1007/s00521-021-06773-2

Torres, J. F., Troncoso, A., Koprinska, I., Wang, Z., & Martínez-Álvarez, F. (2019). Big- data solar power forecasting based on deep learning and multiple data sources. *Applied Energy*, *238*, 1312–1326.

Touati, F., Chowdhury, N. A., Benhmed, K., Gonzales, A. J. S. P., Al-Hitmi, M. A., Benammar, M., Gastli, A., & Ben-Brahim, L. (2017). Long-term performance analysis and power prediction of PV technology in the State of Qatar. *Renewable Energy*, *113*, 952–965. doi:10.1016/j.renene.2017.06.078

Toyota. (2020). New Mirai press information 2020. Author.

Tripathi, S. L., Dubey, M. K., Rishiwal, V., & Padmanaban, S. (Eds.). (2021). *Introduction to AI Techniques for Renewable Energy System*. CRC Press. doi:10.1201/9781003104445

Tsang, I. J., Tsang, I. R., & Dyck, D. V. (1998). Handwritten character recognition based on moment features derived from image partition. *Int. Conf. Image Processing, 2*, 939–942.

Tsegelnyk, Y., Kombarov, V., Plankovskyy, S., Aksonov, Y., Pliuhin, V., & Aksonov, O. (2022). Investigation of the portal-type machine tool gear-belt gearbox. *International Journal of Mechatronics and Applied Mechanics*, *2022*(11), 295–302. doi:10.17683/ijomam/issue11.41

TWI-Global. (n.d.). https://www.twi-global.com/technical-knowledge/faqs/what-is-green-energy

Tyagi, V. V., Kaushik, S. C., & Tyagi, S. K. (2012). Advancement in solar photovoltaic/thermal (PV/T) hybrid collector technology. *Renewable & Sustainable Energy Reviews*, *16*(3), 1383–1398. doi:10.1016/j.rser.2011.12.013

Uhunamure, J., & Shale, N. (2021). Significance of Anomaly Detection in Renewable Energy Systems: A Case Study in South Africa. *Energy Policy*, *158*, 112345.

Uz Zaman, M., Islam, A., & Sultana, N. (2018). Short-term load forecasting based on Internet of things (IoT). BRAC University. Retrieved from http://hdl.handle.net/10361/10170

Varanasi, J., & Tripathi, M. M. (2019). K-means clustering based photo voltaic power forecasting using artificial neural network, particle swarm optimization and support vector regression. *J. Inf. Optim. Sci.*, *40*(2), 309–328. doi:10.1080/02522667.2019.1578091

Vassallo, D., Krishnamurthy, R., Sherman, T., & Fernando, H. J. S. (2020). Analysis of Random Forest Modeling Strategies for Multi-Step Wind Speed Forecasting. *Energies*, *13*(20), 5488. doi:10.3390/en13205488

Venkatesan, Farooq, Imraan, Kumar, & Kumar. (2013). Animals and Vehicle Collision Avoidance Using Wireless Sensor Actuator Network. *International Journal of Scientific & Engineering Research, 4*(5). https://www.ijser.org/researchpaper/Animals-and-Vehicle-Collision-Avoidance-Using-Wireless-Sensor-Actuator-Network.pdf

Vignesh, V., Yuchi, S., & Brandt, R. A. (2019). Short-term solar PV forecasting using computer vision: The search for optimal CNN architectures for incorporating sky images and PV generation history. *Journal of Renewable and Sustainable Energy*, *11*(6), 066102. doi:10.1063/1.5122796

Vincent, L. G., & Nicolas, T. (2020). A deep physical model for solar irradiance forecasting with fisheye images. *Proceedings of the IEEE/CVF conference on computer vision and pattern recognition workshops*, 630–1.

Vinichenko, V. (2018). Mechanisms of energy transitions: National cases and the worldwide uptake of wind and solar power. *PhD diss., Central European University*.

Vishnu, S., Anirudh, J., Burak, U., & Stefano, E. (2020). Cloud removal from satellite images using spatiotemporal generator networks. *Proceedings of the IEEE/CVF winter conference on applications of computer vision*, 1796–805.

Walker & Jenkins.(1997). Wind energy technology. Academic Press.

Wang, Z., & Oates, T. (2015). Imaging time-series to improve classification and imputation. *arXiv preprint arXiv.00327*.

Wang, D., Gan, J., Mao, J., Chen, F., & Yu, L. (2023). Forecasting power demand in China with a CNN-LSTM model including multimodal information. *Energy*, *263*, 126012.

Wang, D., Gryshova, I., Balian, A., Kyzym, M., Salashenko, T., Khaustova, V., & Davidyuk, O. (2022a). Assessment of power system sustainability and compromises between the development goals. *Sustainability (Basel)*, *14*(4), 2236. doi:10.3390/su14042236

Wang, D., Gryshova, I., Kyzym, M., Salashenko, T., Khaustova, V., & Shcherbata, M. (2022b). Electricity price instability over time: Time series analysis and forecasting. *Sustainability (Basel)*, *14*(15), 9081. doi:10.3390/su14159081

Wang, F., Zhen, Z., Mi, Z., Sun, H., Su, S., & Yang, G. (2015). Solar irradiance feature extraction and support vector machines based weather status pattern recognition model for short-term photovoltaic power forecasting. *Energy and Building*, *86*, 427–438. doi:10.1016/j.enbuild.2014.10.002

Wang, F., Zhen, Z., Wang, B., & Mi, Z. (2017). Comparative study on KNN and SVM based weather classification models for day ahead short term solar PV power forecasting. *Applied Sciences (Basel, Switzerland)*, *8*(1), 28. doi:10.3390/app8010028

Wang, K., Khoo, K. S., Chew, K. W., Selvarajoo, A., Chen, W.-H., Chang, J.-S., & Show, P. L. (2021). Microalgae: The Future Supply House of Biohydrogen and Biogas. *Frontiers in Energy Research*, *9*, 1–9. doi:10.3389/fenrg.2021.660399

Wang, L., Zhang, J., Zhou, Y., & Li, G. (2019). A hybrid intelligent optimization algorithm for optimal power flow with renewable energy integration. *Applied Energy*, *242*, 1634–1647.

Wang, S., Tang, L., & Yu, L. (2011). SD-LSSVR-based decomposition-and-ensemble methodology with application to hydropower consumption forecasting. *Proceedings of the 4th International Joint Conference on Computational Sciences and Optimization (CSO)*, 603–607. 10.1109/CSO.2011.303

Wang, T., He, F., Yang, R., Ye, Z., & Huang, X. (2023). Consensus-Based Distributed Kernel One-class Support Vector Machine for Anomaly Detection. *International Joint Conference on Neural Networks (IJCNN)*, Gold Coast, Australia. 10.1109/IJCNN54540.2023.10191788

Wang, W., Zhang, M., & Li, P. (2014). Exploring temporal and spatial evolution of global energy production and consumption. *Renewable & Sustainable Energy Reviews*, *30*, 943–949. doi:10.1016/j.rser.2013.11.027

Wark, T. (2007). The Design and Evaluation of a Mobile Sensor/Actuator Network for Autonomous Animal Control. *6th International Symposium on Information Processing in Sensor Networks*, 206-215. 10.1109/IPSN.2007.4379680

Wasmer, S., Rajsrima, N., Geisemeyer, I., Fertig, F., Greulich, J. M., & Rein, S. (2018). Analytical modeling of the temporal evolution of hot spot temperatures in silicon solar cells. *arXiv preprint arXiv:1803.02463*.

Watanabe, T., Ito, S., & Yokoi, K. (2009). Co-occurrence histograms of oriented gradients for pedestrian detection. *Proceedings of the Pacific-Rim Symposium on Image and Video Technology,* 37–47. 10.1007/978-3-540-92957-4_4

Watson, R. (2022). The virtual economy of the metaverse: Computer vision and deep learning algorithms, customer engagement tools, and behavioral predictive analytics. *Linguistic and Philosophical Investigations*, (21), 41–56.

Wei, L., Jiang, J., & Yu, Y. (2018). A Deep Learning-Based Method for Anomaly Detection in Renewable Energy Systems. *Energy Procedia*, *154*, 289–295.

Welch, R. L., Ruffing, S. M., & Venayagamoorthy, G. K. (2009). Comparison of feedforward and feedback neural network architectures for short term wind speed prediction. *Proceedings of International Joint Conference on Neural Networks*, 3335–3340. doi:10.1109/IJCNN.2009.5179034

Wen, L. L., Zhou, K. L., Yang, S. L., & Lu, X. H. (2019). Optimal load dispatch of community micro grid with deep learning based solar power and load forecasting. *Energy*, *171*, 1053–1065. doi:10.1016/j.energy.2019.01.075

Wen, M., Zhou, C., & Konstantin, M. (2023). Deep neural network for predicting changing market demands in the energy sector for a sustainable economy. *Energies*, *16*(5), 2407. doi:10.3390/en16052407

Westman, L. K., Broto, V. C., & Huang, P. (2019). Revisiting multi-level governance theory: Politics and innovation in the urban climate transition in Rizhao, China. *Political Geography*, *70*, 14–23. doi:10.1016/j.polgeo.2019.01.002

White, D. E. (1966). Geothermal energy. *Bulletin of Volcanology*, *29*(1), 481–483. doi:10.1007/BF02597170

Will, A., Bustos, J., Bocco, M., Gotaya, J., & Lamelas, C. (2011). On the use of niching genetic algorithms for variable selection in solar radiation estimation. *Renewable Energy*, *50*, 168–176. doi:10.1016/j.renene.2012.06.039

Williams, R., & Zipser, D. (1989). A learning algorithm for continually running fully recurrent neural networks. *Neural Computation*, *1*(2), 270–280. doi:10.1162/neco.1989.1.2.270

Winfield, M., & Dolter, B. (2014). Energy, economic and environmental discourses and their policy impact: The case of Ontario's Green Energy and Green Economy Act. *Energy Policy*, *68*, 423–435. doi:10.1016/j.enpol.2014.01.039

Wongrin, W., Chaisee, K., & Suphawan, K. (2023). Comparison of Statistical and Deep Learning Methods for Forecasting PM2.5 Concentration in Northern Thailand. *Polish Journal of Environmental Studies*, *32*(2), 1419–1431. doi:10.15244/pjoes/157072

World Energy Council. (2016). World Energy Council. World Energy Resources Solar.

Wu, T. Y., Hay, J. X. W., Kong, L. B., Juan, J. C., & Jahim, J. M. (2012). Recent advances in reuse of waste material as substrate to produce biohydrogen by purple non-sulfur (PNS) bacteria. *Renewable & Sustainable Energy Reviews*, *16*(5), 3117–3122. doi:10.1016/j.rser.2012.02.002

Wu, Z., Liu, X., Chen, Y., Zhang, L., & Wang, Q. (2018). Anomaly Detection in Wind Power Generation Systems using One-Class Support Vector Machine Model. *Energy Conversion and Management*, *173*, 456–465.

Xiao, J., Bi, C., Bénard, P., Chahine, R., Zong, Y., Luo, M., & Yang, T. (2021). Neural network based optimization for cascade filling process of on-board hydrogen tank. *International Journal of Hydrogen Energy*, *46*(3), 2936–2951. doi:10.1016/j.ijhydene.2020.10.245

Xiao, J., Li, C., Fang, L., Böwer, P., Wark, M., Bénard, P., & Chahine, R. (2020). Machine learning–based optimization for hydrogen purification performance of layered bed pressure swing adsorption. *International Journal of Energy Research*, *44*(6), 4475–4492. doi:10.1002/er.5225

Xiao, M., Simon, S., & Pregger, T. (2019). Scenario analysis of energy system transition-A case study of two coastal metropolitan regions, eastern China. *Energy Strategy Reviews*, *26*, 100423. doi:10.1016/j.esr.2019.100423

Xiao, W. B., Li, X. F., Zhang, C., Yang, G. Q., & Wang, X. D. (2020). Power conversion efficiency prediction of silicon solar cells using artificial neural network. *Journal of Semiconductors*, *41*(2), 022101.

Xie, G. J., Liu, B. F., Wang, Q., Ding, J., & Ren, N. Q. (2016). Ultrasonic waste activated sludge disintegration for recovering multiple nutrients for biofuel production. *Water Research*, *93*, 56–64. doi:10.1016/j.watres.2016.02.012 PMID:26896823

Xie, Y., Wu, X., Hou, Z., Li, Z., Luo, J., Lüddeke, C. T., & Liao, J. (2023). Gleaning insights from German energy transition and large-scale underground energy storage for China's carbon neutrality. *International Journal of Mining Science and Technology*, *33*(5), 529–553. doi:10.1016/j.ijmst.2023.04.001

Xu, C., Liao, Z., Li, C., Zhou, X., & Xie, R. (2022). Review on interpretable machine learning in smart grid. *Energies*, *15*(12), 4427. doi:10.3390/en15124427

Yang, W. (2022). *Machine learning based wind power forecasting for operational decision support* [Doctoral thesis, University of Vaasa]. OSUVA. https://urn.fi/URN:NBN:fi-fe2022091559103

Yang, D., Liu, D., Huang, A., Lin, J., & Xu, L. (2021). Critical transformation pathways and socio-environmental benefits of energy substitution using a LEAP scenario modeling. *Renewable & Sustainable Energy Reviews*, *135*, 110116. doi:10.1016/j.rser.2020.110116

Yang, L., Gao, J., Zhao, Y., & Shao, L. (2021). Decentralized One-Class Support Vector Machine for anomaly detection. *Pattern Recognition*, 119.

Yao, C., Bai, X., Liu, W., Ma, Y., & Tu, Z. (2012). *Detecting Texts of Arbitrary Orientations in Natural Images*. CVPR.

Yao, G., Lei, T., & Zhong, J. (2019). A review of Convolutional-Neural-Network-based action recognition. *Pattern Recognition Letters*, *118*, 14–22. doi:10.1016/j.patrec.2018.05.018

Ye, C. (2024). Using Time Series Analysis to Study the Changing Trend of Business English Teaching Effectiveness in Colleges and Universities. *Applied Mathematics and Nonlinear Sciences*, *9*(9), 1–15. doi:10.2478/amns.2023.2.01108

Yi, J., Dai, S., Li, L., & Cheng, J. (2024). How does digital economy development affect renewable energy innovation? *Renewable & Sustainable Energy Reviews*, *192*, 114221. doi:10.1016/j.rser.2023.114221

Yousif, J. H., Kazem, H. A., Alattar, N. N., & Elhassan, I. I. (2019). A comparison study based on artificial neural network for assessing PV/T solar energy production. *Case Studies in Thermal Engineering*, *13*, 1–13. doi:10.1016/j.csite.2019.100407

Yu, X., & Zhou, Y. (2024). Machine learning and artificial intelligence-distributed renewable energy sources: technologies, perspectives, and challenges. Advances in Digitalization and Machine Learning for Integrated Building-Transportation Energy Systems, 17-30.

Yue, M., Lambert, H., Pahon, E., Roche, R., Jemei, S., & Hissel, D. (2021). Hydrogen energy systems: A critical review of technologies, applications, trends and challenges. *Renewable & Sustainable Energy Reviews*, *146*, 111180. doi:10.1016/j.rser.2021.111180

Yu, R., Gao, J., Yu, M., Lu, W., Xu, T., Zhao, M., Zhang, J., Zhang, R., & Zhang, Z. (2019). LSTM-EFG for wind power forecasting based on sequential correlation features. *Future Generation Computer Systems*, *93*, 33–42. doi:10.1016/j.future.2018.09.054

Zadeh, P. A., & Jadid, S. (2016). A systematic approach for assessment of renewable energy using analytic hierarchy process: A case study of Iran. *Energy, Sustainability and Society*, *6*(1), 1–14.

Zambrano, A. F., & Giraldo, L. F. (2020). Solar-irradiance forecasting models without onsite training measurements. *Renewable Energy*, *152*, 557–566. doi:10.1016/j.renene.2020.01.092

Zhang, C., Wei, H., Zhao, J., Liu, T., Zhu, T., & Zhang, K. (2016). Short-term wind speed forecasting using empirical mode decomposition and feature selection. *Renewable Energy*, *96*, 727–737. doi:10.1016/j.renene.2016.05.023

Zhang, T., Jiang, D., Zhang, H., Jing, Y., Tahir, N., Zhang, Y., & Zhang, Q. (2020). Comparative Study on Bio-Hydrogen Production from Corn stover: Photo-Fermentation, Dark-Fermentation and Dark-Photo Co-fermentation. *International Journal of Hydrogen Energy*, *45*(6), 3807–3814. doi:10.1016/j.ijhydene.2019.04.170

Zhang, W., Maleki, A., Rosen, M. A., & Liu, J. (2018). Optimization with a simulated annealing algorithm of a hybrid system for renewable energy including battery and hydrogen storage. *Energy*, *163*, 191–207. doi:10.1016/j.energy.2018.08.112

Zhang, X. Y., Zhang, L. W., Wu, J., Fu, H., & Dian, L. Y. (2022). Tunnel stability analysis of coral reef limestone stratum in ocean engineering. *Ocean Engineering*, *265*, 112636. doi:10.1016/j.oceaneng.2022.112636

Zhang, X., Peng, Y., Xu, W., & Wang, B. (2017). An optimal operation model for hydropower stations considering inflow forecasts with different Lead-times. *Water Resources Management*, *33*(1), 173–188. doi:10.1007/s11269-018-2095-1

Zhang, Y., Satapathy, S. C., Zhu, L.-Y., Górriz, J. M., & Wang, S. (2020). A seven-layer convolutional neural network for chest CT-based COVID-19 diagnosis using stochastic pooling. *IEEE Sensors Journal*, *22*(18), 17573–17582.

Zhang, Y., Wang, J., Wang, J., & Zhao, H. (2019). A deep neural network-based approach for short-term wind power forecasting using spatial-temporal data. *Applied Energy*, *236*, 220–231.

Zhou, L., Pan, S., Wang, J., & Vasilakos, A. V. (2017). Machine learning on big data: Opportunities and challenges. *Neurocomputing*, *237*, 350–361. doi:10.1016/j.neucom.2017.01.026

Zhuang, F., Liu, Y., Liu, S., Xu, C., & Zhu, Q. (2019). Anomaly Detection Method for Industrial Big Data Based on Improved One-Class SVM Algorithm. *IEEE Access : Practical Innovations, Open Solutions*, *7*, 57088–57097.

Zhu, G., Yang, Y., Liu, J., Liu, F., Lu, A., & He, W. (2017). Enhanced Photocurrent Production by the Synergy of Hematite Nanowire-Arrayed Photoanode and Bioengineered *Shewanella oneidensis* MR-1. *Biosensors & Bioelectronics*, *94*, 227–234. doi:10.1016/j.bios.2017.03.006 PMID:28285200

Zhu, Y., Jing, H., & Wang, F. (2017). Anomaly Detection in Wind Turbine Based on One-Class Support Vector Machine. *Procedia Computer Science*, *107*, 156–161.

Zhu, Y., Zhang, J., Wang, X., & Zhang, Y. (2020). Machine learning for key variables affecting the performance of CIGS solar cells. *Journal of Alloys and Compounds*, *827*, 154226.

Zwane, N., Tazvinga, H., Botai, C., Murambadoro, M., Botai, J., De Wit, J., & Mabhaudhi, T. (2022). A bibliometric analysis of solar energy forecasting studies in Africa. *Energies*, *15*(15), 5520. doi:10.3390/en15155520

About the Contributors

Pinaki Pratim Acharjya earned his Doctorate in Philosophy in 2016 from National Institute of Technology, Agartala, Tripura, India. He is currently employed as a Professor in the department of Computer Science and Engineering, Haldia Institute of Technology, Haldia, West Bengal, India. In several AICTE-approved engineering institutions across India, he has more than 15 years of teaching experience and more than fourteen years of research experience. In numerous national and international journals, conferences, books (written/edited), and book chapters, he has published more than hundred research publications. He is currently conducting research in the fields of Deep Learning, Machine Learning, Data Analytics, Blockchain Technology, Digital Image Processing, Computer Vision, and Artificial Intelligence.

Santanu Koley earned his doctorate in philosophy (PhD) from CSJM University in Kanpur, Uttar Pradesh, India in 2013, and he is currently employed as a professor in the department of computer science and engineering at Haldia Institute of Technology in Haldia, West Bengal, India. In addition to sixteen years of teaching experience, he has more than fourteen years of research experience from several AICTE-approved engineering colleges across India. Dr. Koley has published more than 30 research papers in journals and conferences from throughout the nation and the world. The areas of cloud computing, digital image processing, artificial intelligence, and machine learning are where he is currently concentrating his research efforts.

Subhabrata Barman is an Assistant Professor with the Department of Computer Science & Engineering, Haldia Institute of Technology, West Bengal, India. His research interests are in the field of Wireless Networks, Computational Intelligence, Remote Sensing and Geo-Informatics, Parallel and Grid Computing. He has published research papers at various International and National Journals and Conferences. He is a Professional Member of IEEE, IACSIT, IAENG and a reviewer of International Journal of Wireless Networks (Springer).

* * *

Ihor Biletskyi is a First Vice-Rector at the O.M. Beketov National University of Urban Economy in Kharkiv (Ukraine). He received his Ph.D. degree in Project Management from O.M. Beketov National University of Urban Economy in Kharkiv (Ukraine) in 2016 and a D.Sc. degree in Economics from Odesa

National University of Technology (Ukraine) in 2023. He is an expert of the Ministry of Education and Science of Ukraine for examining research and development projects of the thematic direct "Energy and energy efficiency", and a Member of the Expert Council of the Sikorsky Challenge Startup Competition. His research activities mainly focus on Energy Efficiency, BIM Technologies, Smart Services, and Project Management.

AnganaChakraborty is M-Tech (IT) from Indian Institute of Engineering, Science and Technology (IIEST), Shibpur and B-Tech (CSE) from West Bengal University of Technology (WBUT). Her research interests include Mobile WiMAX, Wireless Networks, Mobility Management. She published 3 papers in international conference proceedings of repute.

Shamba Chatterjee is currently serving as Associate Professor & HoD at Department of Biotechnology, HIT Haldia (An Autonomous Institution under MAKAUT), India. He has successfully completed his Doctoral degree at Indian Institute of Technology, Roorkee (IITR) with prestigious DAAD Fellowship (Germany), CSIR Fellowship (CSIR) and National Doctoral Fellowship (AICTE, India). His Post-Doctoral research was from Seoul National University, Seoul, Republic of Korea. He has published several research article and book chapters in reputed peer reviewed national and international journals. He has also received NPTEL Discipline Star Award in Biotechnology from IIT Madras. He is the life member of Society of Biological Chemists and IICHE, India.

Himadri Sekhar Das is presently working as Assistant professor in Haldia Institute of Technology. He did his post doc from Haldia Institute of Technology. He got his Ph. D degree from Indian Institute of Engineering Science and Technology (IIEST), Shibpur in 2021, M. Tech from Maulana Abul Kalam Azad University of Technology, West Bengal in 2010. M.Sc and B.Sc from Vidyasagar University in 2006 and 2008 respectively. His research area was Doped ZnO based thin Film for different optoelectronics applications. Synthesis and characterization of organic-inorganic Perovskite material for Solar cell. ZnO based transparent conducting oxide thin film for mesoporous TiO 2 based dye sensitized solar cell. Solid state electrolyte based dye sensitized solar cell: stability and device performance. Solvo thermal synthesis of ZnO nanoparticle. Polymer based Organic Solar Cell, Photonic Materials and devices (LED & Thin Film Display). He has published more than 30 research articles and 4 book chapter. He is also DST Fellow and life member and Fellow of Indian Chemical Society (FICS) and Indian Photobiology society (IPS) also the Member of Royal Society of Chemistry-UK.

Sucheta Das is currently working as Associate Professor at Department of Biotechnology, HIT Haldia (An Autonomous Institution under MAKAUT), India. She has successfully completed her Doctoral degree from Visva-Bharati University and M.Tech degree from Jadavpur University. She has worked on the metal reducing property of some mesophilic and thermophilic bacteria and also on Lactic acid bacteria, bacteriocin and probiotics. She has published several research articles and book chapters in reputed peer reviewed national and international journals. She is the life member of Association of Microbiologists of India (AMI).

Prasad G. is currently working as faculty in the department of Aerospace Engineering Chandigarh University, India. He has 10 years of experience in Teaching and research. He received his Bachelor of Engineering and Master of Engineering degrees with first class and distinction and Doctor of Philosophy in the field of Aeronautical Engineering from the Anna University, Chennai, India. Prof. Prasad G is the author of over 30 technical publications, 1 Patent Published and 1 Book published. His research interests include Unmanned Aerial Vehicles, Aircraft Structures, Computational Fluid Dynamics and Interdisciplinary research. He has completed three funded project sponsored by The Institution of Engineers (India) and Tamilnadu State Council for Science and Technology. Awarded Indian National Science Academy (INSA) Visiting Scientist Programme 2019 and Awarded Science Academies' Summer Research Fellowship Programme (SRPF) 2019. Further he is SPOC of NPTEL- IIT Madras, SPOC of ISRO IIRS outreach coordinator and SPOC for NASA International SpaceApps Challenge. Further key positions such as coordinator of Board of Studies, Outcome Based Education, Project Coordinator, Research Coordinator NBA, NAAC, ABET. He is a Professional Membership of American Institute of Aeronautics and Astronautics, Institution of Engineers, Life Member in Indian Cryogenic Council and Life Member in Shock Wave Society. Reviewer in Aircraft Engineering and Aerospace Technology, International Journal of Engine Research and Journal of Institution of Engineers (India): Series C.

Christian Kaunert is professor of international security at Dublin City University and Professor of policing and Security at the University of South Wales, where he holds a Jean Monnet Chair of EU counter-terrorism and directs a Jean Monnet Centre of Excellence. He is also the co-ordinator of a research-focused Jean Monnet Network of EU counter-terrorism, which brings together fourteen universities and research centres across Europe and its neighbourhood. He has published extensively in the field of European security, and also collaborated with major international news outlets as an expert in security, migration and counter-terrorism.

Illia Khudiakov is the Head of the Project Group in Marketing and Analytics of MEGAPOLIS Technology Transfer Center and the Assistant Professor of the Oil and Gas Engineering and Technologies Department at O.M. Beketov National University of Urban Economy in Kharkiv (Ukraine). He received his Ph.D. degree in Computer Science from O.M. Beketov National University of Urban Economy in Kharkiv (Ukraine) in 2024. His research activities are mainly focused on Program Management, Machine Learning, Decision-making Support, and Infrastructure Projects.

Jayeeta Majumdar earned her Doctorate in Philosophy in 2022 in Computer Science and Engineering. She is currently employed as an Associate Professor in the department of Computer Science and Engineering, Haldia Institute of Technology, Haldia, West Bengal, India. She has more than 15 years of teaching. In numerous national and international journals, conferences, and book chapters, she has published research articles/papers. She is currently conducting research in the fields of Cryptography & Network Security, Machine Learning, Computer Vision, and Artificial Intelligence.

Katleho Makatjane is a trained quantitative business and risk analyst with more than four years of experience. His research interest is in data mining with applications to business and finance. In particular, he specializes in Statistical methods used for quantitative business and risk analysis, financial engineering and extreme value theory. He also specializes in the area of causal inference in business and finance.

Ntebo Moroke earned her Ph.D in Statistics from North West University in 2014. She joined the faculty in 2002 as a junior lecturer. Her areas of research are on using Multivariate and Time Series Data Analysis methods to analyse real-world high-dimensional data. She has published a number of papers in accredited peer-reviewed journals as well as book chapters. Her ability to execute research has garnered her numerous accolades in the field of statistic.

Anshit Mukherjee is the student pursuing BTech in computer sciences and Engineering from Abacus Institute of Engineering and Management.

Neeta Nathani is a Professor and Head of the Department in Electronics and Communication Engineering in Gyan Ganga Institute of Technology and Sciences, Jabalpur, Madhya Pradesh. She received the Graduate degree in Electronics and Telecommunication and M.Tech. in Digital Communication from Rajiv Gandhi Proudyogiki Vishwavidyalaya, Bhopal, Madhya Pradesh. She has completed her Ph.D. in Wireless Communication under faculty of Electronics Engineering from Rashtrasant Tukadoji Maharaj Nagpur University, Nagpur, Maharashtra. She is a member of CSI, IETE and IEEE. She has organised many national and international conferences and seminar. She received Outstanding Paper Award on "Assessment of Quasi-Permanently Vacant Channels in Mobile Communication Bands for Cognitive Radio" in the 15th IEEE International Conference on Advanced Communications Technology (ICACT-2013) in Phoenix Park, South Korea. With a teaching experience of over 17 years, she has international and national paper publications to her credit.

Sergiy Plankovskyy is originally from Volodymyr-Volynsky, Ukraine. He has 35 years of international experience in the research and development of processing with intense energy fluxes (mainly technologies using plasma, laser, and detonation sources). After being Head of the Faculty of Aircraft Engineering at the National Aerospace University "Kharkiv Aviation Institute", he is now Head of the School of Energy, Information and Transport Infrastructure at the O.M. Beketov National University of Urban Economy in Kharkiv, Ukraine. He has over 200 refereed publications and has supervised 11 Ph.D. students to successful completion. His research interests cover innovation technologies, including the methods of manufacturing for the digital industry. He is a member of the Expert Council on Mechanical Engineering of the National Agency for Higher Education Quality Assurance (NAQA), Ukraine.

Vladyslav Pliuhin is the Head of the Department of Urban Electrical Energy Supply and Consumption Systems at the O.M. Beketov National University of Urban Economy in Kharkiv (Ukraine). He received his Ph.D. degree in Electrical Machines and Apparatus from Donetsk National University (Ukraine) in 2004 and a D.Sc. degree in Electrical Machines and Apparatus from National Technical University "Kharkiv Polytechnic Institute" (Ukraine) in 2016. He had an international traineeship as a Post-Doctorate at the Czech Technical University in Prague (Czech Republic) in 2012. He is an expert of the Ministry of Education and Science of Ukraine for examining research and development projects of the thematic direct "Energy and Energy Efficiency", and a member of the Ministry of Education and Science of Ukraine Council for examining Ph.D. and D.Sc. thesis. His research activities are mainly focused on Electrical Machines Design and Optimization, CAD Software Development, Energy Efficiency, Machine Learning, and Digital Twins.

Pampa Saha is working as assistant professor in Brainware University. She is working in research also.

Modisane Bennett Seitshiro graduated PhD in Business Mathematics and Informatics under the centre for Business Mathematics and Informatics at the North-West University, South Africa in 2020. He was permanently employed by Standard Bank Group, Corporate and Investment Banking - Market Risk for 4 years as a Quantitative Analyst and Market Risk Manager for 2 years until 2012. In 2013, Modisane started a permanent lecturing position in the Department of Mathematical Statistics, North-West University. Dr Seitshiro is currently a Senior Lecturer at the Centre for Business Mathematics and Informatics at the North-West University. His research interests are Quantitative Risk Management, Statistical Learning and Financial Time Series Analysis. He is involved with teaching and learning, industry related postgraduate supervision and projects. He is passionate about enhancing the discipline of Mathematical Sciences in South Africa, especially in breaching the gap between industry and academic research.

Claris Shoko is a Mathematical Statistics lecturer at the University of Botswana. Her areas of specialization include Stochastic modelling (Multistate modelling, Markov Modelling), Time Series Analysis, Machine Learning, Epidemiology, and Biostatistics.

Maria Sukhonos is a Vice-Rector for Research at the O.M. Beketov National University of Urban Economy in Kharkiv (Ukraine). She received her Ph.D. degree in Project Management from Prydniprovska State Academy of Civil Engineering and Architecture (Ukraine) in 2007 and a D.Sc. degree in Project Management from the National Aerospace University "Kharkiv Aviation Institute" (Ukraine) in 2013. She is the Curator of the "Energy Innovation Hub" implemented by Deutsche Geselschaft für Internationale Zusammenarbeit (GIZ) GmbH and funded by the Federal Ministry of Economic Cooperation and Development of Germany (BMZ) in the O.M. Beketov National University of Urban Economy in Kharkiv (Ukraine). Her research activities are mainly focused on Energy Efficiency, Sustainable Electrical Energy Systems, Smart Services for Smart Cities, Digital Project Management.

Yevgen Tsegelnyk is a Deputy Head (Research) at the School of Energy, Information and Transport Infrastructure of the O.M. Beketov National University of Urban Economy in Kharkiv (Ukraine). He received his Ph.D. degree in Aerospace Engineering from the National Aerospace University "Kharkiv Aviation Institute" (Ukraine) in 2010. He has more than 15 years of international experience in research and development. He had an international traineeship as a Post-Doctorate at the Czech Technical University in Prague (Czech Republic) in 2014. His research interests cover innovation technologies, including the methods of manufacturing for the digital industry, processing with intense energy fluxes (technologies using plasma, laser, and detonation sources), additive and subtractive manufacturing technologies, and applied computing techniques.

Diteboho Xaba is an emerging researcher and lecturer at the University of South Africa (Unisa) in Pretoria, South Africa. He pursued a Ph.D. in Operations Research and Statistics because he was eager to learn more about the relationship between data and decision-making. His doctoral research explored innovative approaches to modelling BRICS exchange rates using MRS-GARCH models with various estimation algorithms, a critical area in operations research. He is a researcher who has written and published academic journal papers as well as presented research papers at national and international conferences. He has published a book chapter from his Ph.D. work, which is being published by Springer Nature. Dr. Xaba is interested in research fields such as operations research, statistics, and applied econometrics. Dr. Xaba made contributions to statistics and applied econometrics by developing models that captured the nonlinear dynamics of economic variables. His works, published in reputable journals, demonstrated an understanding of economic systems, incorporating behavioural factors, and addressing challenges posed by volatile markets.

Index

K

L

M

N

O

P

R

S

T

U

W

Ensure Quality Research is Introduced to the Academic Community

Become a Reviewer for IGI Global Authored Book Projects

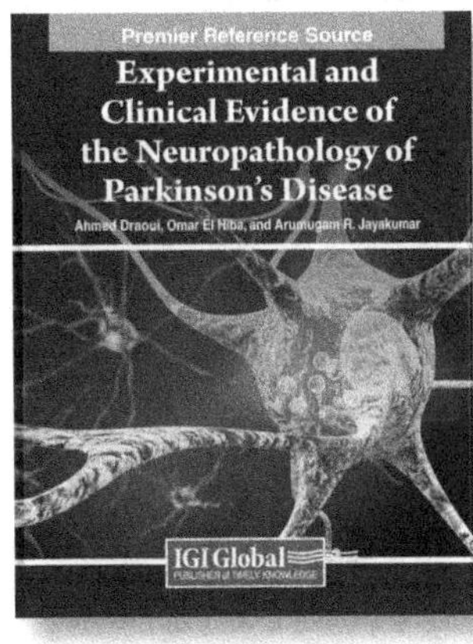

The overall success of an authored book project is dependent on quality and timely manuscript evaluations.

Applications and Inquiries may be sent to:

development@igi-global.com

Applicants must have a doctorate (or equivalent degree) as well as publishing, research, and reviewing experience. Authored Book Evaluators are appointed for one-year terms and are expected to complete at least three evaluations per term. Upon successful completion of this term, evaluators can be considered for an additional term.

If you have a colleague that may be interested in this opportunity, we encourage you to share this information with them.

www.ingramcontent.com/pod-product-compliance
Lightning Source LLC
LaVergne TN
LVHW080637011026
849160LV00014B/1248